High Performance Liquid Chromatography: Advances in Chromatography

High Performance Liquid Chromatography: Advances in Chromatography

Edited by
Courtney Wingfield

STATES
ACADEMIC PRESS
www.statesacademicpress.com

Published by States Academic Press,
109 South 5th Street,
Brooklyn, NY 11249, USA

ISBN: 978-1-63989-269-3

Cataloging-in-Publication Data

High performance liquid chromatography : advances in chromatography / edited by Courtney Wingfield.
p. cm.
Includes bibliographical references and index.
ISBN 978-1-63989-269-3
1. High performance liquid chromatography. 2. Liquid chromatography. 3. High pressure (Science).
4. Chromatographic analysis. 5. Chemistry, Analytic. I. Wingfield, Courtney.
QD79.C454 H54 2021
543.84--dc23

For information on all States Academic Press publications
visit our website at www.statesacademicpress.com

STATES
ACADEMIC PRESS

Contents

Preface

This book was inspired by the evolution of our times; to answer the curiosity of inquisitive minds. Many developments have occurred across the globe in the recent past which has transformed the progress in the field.

The technique used in analytical chemistry to identify, separate and quantify each element in a mixture is known as high performance liquid chromatography. It was initially known as high pressure liquid chromatography. For the separation of the component, it relies on the process of pumping to pass the liquid through the adsorbent. The various parameters in this field are particle size, pump pressure, internal diameter, pore size, autosamplers and detectors. Partition chromatography, displacement chromatography, normal-phase chromatography, size exclusion chromatography, bioaffinity chromatography and aqueous normal phase chromatography are some of its types. The field has applications in manufacturing, detecting illicit drugs, research and medicine. This book is a valuable compilation of topics, ranging from basic to the most complex advancements in high performance liquid chromatography. The various studies that are constantly contributing towards the advancing technologies and evolution of this field are examined in detail. For all those readers who are interested in this area of study, the case studies included in this book will serve as an excellent guide to develop a comprehensive understanding.

This book was developed from a mere concept to drafts to chapters and finally compiled together as a complete text to benefit the readers across all nations. To ensure the quality of the content we instilled two significant steps in our procedure. The first was to appoint an editorial team that would verify the data and statistics provided in the book and also select the most appropriate and valuable contributions from the plentiful contributions we received from authors worldwide. The next step was to appoint an expert of the topic as the Editor-in-Chief, who would head the project and finally make the necessary amendments and modifications to make the text reader-friendly. I was then commissioned to examine all the material to present the topics in the most comprehensible and productive format.

I would like to take this opportunity to thank all the contributing authors who were supportive enough to contribute their time and knowledge to this project. I also wish to convey my regards to my family who have been extremely supportive during the entire project.

Editor

1

Quantification of Temozolomide in Nonhuman Primate Fluids by Isocratic Ultra-High Performance Liquid Chromatography-Tandem Mass Spectrometry to Study Brain Tissue Penetration Following Intranasal or Intravenous Delivery

Cody J. Peer [1], Lukas Ronner [1,2], Louis Rodgers [1,2], Cynthia M. Lester McCully [3], Katherine E. Warren [2,3] and William D. Figg [1,*]

1 Clinical Pharmacology Program, Office of the Clinical Director, National Cancer Institute, 9000 Rockville Pike, Building 10, Room 5A01, Bethesda, MD 20892, USA; cody.peer@nih.gov (C.J.P.); lukasronner@gmail.com (L.R.); louis.rodgers@nih.gov (L.R.)

2 Neuro-Oncology Branch, National Cancer Institute, Bethesda, MD 20892, USA; warrenk@mail.nih.gov

3 Pediatric-Oncology Branch, National Cancer Institute, Bethesda, MD 20892, USA; mccullyc@mail.nih.gov

* Correspondence: figgw@helix.nih.gov

Academic Editor: Frank L. Dorman

Abstract: A sensitive and selective ultra-high performance liquid chromatography-tandem mass spectrometric method was developed for the quantification of temozolomide (TMZ) in nonhuman primate (NHP) plasma, cerebrospinal fluid (CSF), and brain extracellular fluid (ECF) following microdialysis. Ethyl acetate was used to extract the plasma and CSF samples, using theophylline as the internal standard (IS). ECF samples were diluted with acetonitrile prior to analysis. TMZ was separated on a Waters UPLC® BEH C18 column with an isocratic mobile phase of ammonium acetate (10 mM)-0.1% formic acid/acetonitrile (30:70, v/v) in a positive-ion multiple reaction monitoring mode (m/z 195.5→137.6 for TMZ; m/z 181.5→124.2 for IS). The retention time of TMZ and theophylline was 0.45 min with a total run time of 2.5 min. The method was validated over the range from 5–2000 ng/mL in NHP plasma, CSF, and ECF with respect to linearity, accuracy, precision, selectivity, and stability. This method was successfully applied toward the measurement of pharmacokinetic samples following various routes of drug administration.

Keywords: ultra-high performance liquid chromatography; tandem mass spectrometry; temozolomide; nonhuman primates

---

1. Introduction

Temozolomide (TMZ) is a prodrug that is ultimately metabolized to an alkylating agent. Upon exposure to physiological pH, TMZ is spontaneously hydrolyzed to the intermediate 5-(3-methyl-triazen-1-yl) imidazole-4-carboxamide (MTIC). MTIC further fragments to 5-amino-4-imidazolecarboxamide and the extremely reactive methyldiazonium cation, which methylates DNA at several nucleophilic sites.

N^7-guanine, N^3-adenine, and O^6-guanine [1]. Although the formation of O^6-methylguanine accounts for a minority of DNA adducts formed, it is particularly cytotoxic and is correlated with TMZ-mediated cytotoxicity, resulting in mispairing during replication and subsequent strand breakage and tumor cell death. TMZ is FDA approved as a first-line treatment for patients with newly diagnosed glioblastoma multiforme alongside radiotherapy and as a second-line treatment for refractory anaplastic astrocytoma [2].

Although some drugs have shown modest efficacy, there is continued interest in studying the brain penetration of certain anti-glioma agents to improve therapeutic efficacy and clinical outcome. Intranasal delivery is one route being explored as a means to bypass the blood-brain barrier, as agents can undergo direct transport along the olfactory and trigeminal nerves. Studies are ongoing at the National Cancer Institute (NCI) using rhesus macaques as a nonhuman primate (NHP) model administering several agents in a variety of routes (intranasal, intravenous, oral) to understand the extent of drug penetration into different areas of the brain, and if intranasal administration improves brain penetration *vs.* systemic delivery. TMZ is being studied as one such agent, since existing research demonstrated relatively high brain penetration (20%–40%) [3,4].

Nonhuman primates are an advantageous model for studying the pharmacokinetics (PK) and brain penetration of neuro-oncology agents, as the pharmacology has been predictive of that in humans [5,6]. As more researchers wade into this field, there is a lack of robust published LC-MS/MS assays for measuring TMZ in NHP fluids (plasma, CSF) as well as extracellular fluid (ECF) resulting from microdialysis (MD). MD is a technique that allows the opportunity to measure TMZ concentrations directly in the ECF of the tissue or tumor site sampled, as previously described [7]. It has previously been used to study if TMZ brain penetration is enhanced with bevacizumab [8] and cediranib in rats [9], but there is no published study for measuring TMZ from microdialysis in NHP. Several bioanalytical assays measuring TMZ have been published using LC-MS/MS instrumentation; however, these were all based on human [10] or rodent [8,11,12] biological matrices. Several additional studies have used less selective and sensitive instrumentation as well to study TMZ pharmacokinetics in rodents [13] and humans [3,14–18]. Only one study on TMZ pharmacokinetics used the NHP model, and plasma and CSF levels were measured using a less sensitive HPLC-UV assay (lower limit of quantification (LLOQ):19 ng/mL) [6]. The purpose of this report is to describe a sensitive LC-MS/MS assay that can be used to quantify TMZ concentrations in NHP plasma, CSF, and brain tissue or tumor ECF fluid *in situ* following MD.

2. Experimental Section

2.1. Materials

Temozolomide (>98% purity), theophylline (>99% purity), hydrochloric acid, formic acid, and ammonium acetate were purchased from Sigma (St. Louis, MO, USA). Optima (LC-MS) grade methanol and acetonitrile were purchased from Fisher Scientific (Pittsburgh, PA, USA). NHP plasma and CSF was purchased from Bioreclamation IVT (Baltimore, MD, USA). CNS perfusion fluid was purchased from MDialysis (North Chelmsford, MA, USA). Ultra-pure deionized water was produced from a Hydro-Reverse Osmosis system (Durham, NC, USA) connected to a Mili-Q UV Plus purifying system (Billerica, MA, USA).

2.2. Preparation of Stock Solutions

To prepare 1 mg/mL stocks for TMZ, aqueous ammonium acetate solution (pH 3.5; 10 mM)-methanol (20:80, *v*/*v*) (henceforth referred to as "acidic methanol") was prepared and used to dissolve TMZ. Subsequent serial dilutions of this stock were prepared in acidic methanol as working stocks for final concentrations in each matrix of 5, 10, 25, 100, 250, 500, 1000 and 2000 ng/mL. These same working stocks were used to prepare quality control (QC) standards for both plasma and CSF at low (15 ng/mL), mid (800 ng/mL), high (1600 ng/mL), and a 50-fold dilution (50,000 ng/mL diluted to 1000 ng/mL) level for plasma. ECF fluid (from MD) had the same low, mid, and high QCs, but used a 4-fold dilution QC up to 5000 ng/mL. The internal standard theophylline was prepared in a 1 mg/mL stock in methanol. This stock was diluted into a 1000 ng/mL solution in ethyl acetate for extractions.

2.3. Sample Preparation

2.3.1. Plasma and CSF

Calibration standards (in duplicate), QC standards (in quintuplet) and study samples were extracted from 100 μL of plasma or CSF. Because plasma and CSF samples acidified with 10% of 1 normal (N) HCl solution to prevent degradation of TMZ [15], all standards were acidified with 10% 1 N HCl to match study samples. A liquid-liquid extraction using ethyl acetate (containing 1000 ng/mL theophylline) was performed, with samples vortexed, centrifuged, and the supernatent (extract) dried before reconstitution in 100 μL of water-acetonitrile-formic acid (50:50:0.1, *v*/*v*/*v*).

2.3.2. MD Extracellular Fluid

Extracellular brain fluid was drawn from various locations using MD technique, as previously described [7], that typically only produces ~10–20 μL of ECF per draw (depending on collection time [over several hours] and pump flow rate). Further, ECF itself can vary in its salinity based on sampling site. To account for such low volumes of high salinity (that suppresses ion signals in the mass spectrometer) and acidity (that can non-uniformly affect compound ionization), while maintaining a lower limit of quantification (LLOQ) of 5 ng/mL, a new sample preparation procedure was developed. Drug free perfusion fluid was acidified with 1 N HCl before use in preparation of calibration and QC standards. A volume of 20 μL was needed to perform this procedure, thus if study samples were below this mark, acidified perfusion fluid (same as that used to make standards) was used to dilute study samples up to the 20 μL mark. Acetonitrile (1 mL) was added to each sample to aid in aqueous evaporation via the formation of an azeotropic mixture. The leftover dried residue contains drug and salts that is then reconstituted in 100 μL of ACN, vortexed, and centrifuged to form a pellet of precipitated salts, thereby avoiding injection from the UHPLC system. Injection volumes were 20 μL. No IS in the initial preparation of MD samples due to the non-uniform ionization that occurred from varying levels of sample acidity, which was difficult to control for, and for the sake of simplicity the IS was left out.

2.4. Instrument Conditions

A Waters ACQUITY UPLC® (Waters Corp, Milford, MA, USA) was used to chromatographically separate TMZ and the IS from unwanted biological remnants post extraction on a Waters UPLC® BEH C18 (2.1 × 50 mm, 1.7 μm) column (Waters Corp, Milford, MA, USA) using an isocratic composition of 10 mM ammonium acetate (10 mM)-acetonitrile (ACN)-formic acid (30:70:0.1, *v*/*v*/*v*) delivered at a flow rate of 0.3 mL/min. Retention times were 0.45 min for both TMZ and the IS. The column eluent was directed into a Micromass (Waters) Quattro Micro® triple quadrupole mass spectrometer (Waters Corp, Milford, MA, USA) with an electrospray ionization source operated in the positive ion mode. Mass spectrometer conditions were: capillary voltage 3000 V, cone voltage 35 V, RF lens 5, extractor 5, source temperature 120 °C, desolvation temperature 350 °C, cone gas flow 30 L/h, desolvation gas flow 900 L/h. TMZ and theophylline (IS) were both selectively identified based on their multiple reaction monitoring (MRM) transitions: m/z 195.5→137.6 (TMZ); m/z 181.5→124.2 (IS) (Figure 1). Both MRM transitions had a dwell time of 200 ms and collision energy of 15 eV.

2.5. Validation

2.5.1. Linearity

This assay was validated according to FDA recommended guidelines for bioanalytical validation [19] for linearity, accuracy, precision, selectivity, and various stability tests. The calibration range in all three matrices (NHP plasma, CSF, MD perfusion fluid) was 5–2000 ng/mL, with dilution capabilities up to 50,000 ng/mL in plasma and CSF, and up to 5000 ng/mL in MD perfusion fluid. A least squares linear regression model was used to assess the calibration linearity, with a weighting factor of $1/x^2$, with x being the ratio of peak areas for analyte/IS. Fresh calibration standards were prepared daily in drug-free matrices in duplicate. A linear regression correlation of at least $r^2 > 0.99$ was deemed acceptable.

2.5.2. Accuracy and Precision

Accuracy and precision were evaluated by determining TMZ at four different concentrations of QC samples (low-range, mid-range, high-range, 50-fold dilution for plasma; 4-fold dilution for MD perfusion fluid), in five replicates analyzed over 3–4 different days, depending on scarcity of matrix. Each run consisted of blank matrix extracts, internal standard only extracts, calibration standards in duplicate, and QC samples in replicates of five. Accuracy (% DEV) was defined as the percent mean observed concentration relative to the nominal concentration. Assay repeatability, or precision, was determined by both the within-run precision (WRP) and between-run (BRP), as calculated below.

$$\text{WRP} = 100 \times [(\text{MS}_{\text{WIT}})^{0.5}/\text{GM}] \quad (1)$$

$$\text{BRP} = 100 \times [(\{\text{MS}_{\text{BET}} - \text{MS}_{\text{WIT}}\}/n)^{0.5}/\text{GM}] \quad (2)$$

GM represents the grand mean, MS_{WIT} represents within-group mean squared, MS_{BET} represents between-group mean squared, and n represents the number of repetitions. FDA guidelines for bioanalytical validation were followed, with $\pm 15\%$ variability in accuracy and precision allowed [19].

2.5.3. Stability

Stability tests were performed to examine the potential for degradation of TMZ in plasma and CSF during freeze/thaw cycles. Samples were assayed at three concentrations (15, 800, 1600 ng/mL) in quintuplet in each matrix. The samples were subjected to three freeze/thaw cycles at −80 °C (two cycles for CSF due to scarcity of matrix), with each freeze cycle lasting at least 12 h. The analyte concentration after each storage period was compared to the concentration of freshly prepared samples in the same analytical run. Freeze/thaw stability was not assessed in MD perfusion fluid, as these study samples were immediately processed and analyzed the same day they were drawn, and due to low collection volumes, there was no opportunity for freezing and re-running.

2.5.4. Extraction Recovery and Matrix Effects

For the calculation of the extraction recovery and matrix effects, three sets were prepared. Set 1 was composed of stock solutions of TMZ spiked into the reconstitution solution ([50/50/0.1, *v*/*v*/*v*] water/ACN/formic acid) at low (25 ng/mL) and high (1000 ng/mL) concentrations in triplicate; set 2 was composed of stock solutions of TMZ spiked into NHP plasma at low (25 ng/mL) and high (1000 ng/mL) concentrations after extraction in triplicate; set 3 was composed of stock solutions of TMZ spiked into NHP plasma at low (25 ng/mL) and high (1000 ng/mL) concentrations before extraction in triplicate. Only the peak area of the analyte was taken into consideration for the calculation of each evaluation. The extraction recovery and matrix effects of TMZ in NHP plasma were calculated as shown below.

$$\%\,\text{Extraction Recovery} = A/B \times 100 \quad (3)$$

$$\begin{aligned} A &= \text{mean area count of Set 3 for one concentration level} \\ B &= \text{mean area count of Set 2 for one concentration level} \end{aligned} \quad (4)$$

$$\begin{aligned} \%\,\text{Matrix Effect} = {} & (\text{slope of calibration curve from standards prepared via Set 2} - \\ & \text{standards prepared via Set 1})/\text{standards prepared via Set 1} \times 100 \end{aligned} \quad (5)$$

2.6. Application

This validated analytical method was tested for applicability by analyzing plasma and CSF from a male rhesus macaque given a 20 mg dose of TMZ via intranasal injection (in a 1 mL solution; ~1.75 mg/kg). Blood (eventually processed into 1 mL of plasma acidified with 100 μL HCl) and CSF (300 μL acidified with 30 μL HCl) were collected at the following times in order to capture the full *in vivo* disposition: pre-dose, 5 min, 10 min, 15 min, 30 min, 1 h, 2 h, 3 h, 4 h, 6 h, 8 h, and 24 h post-dose. Acidification of study samples and standards was based on previous data regarding the instability of TMZ in human plasma [17]. Additionally, the same animal was later dosed with a second dose (10 mg/kg) of TMZ administered via 1-h intravenous infusion, with plasma and CSF sampling occurring at pre-dose, 30 min, 1 h (end of infusion), 1.25, 1.5, 2, 3, and 4 h post infusion-start. Microdialysis probes were placed in the animal's cortex and pons, and brain perfusion dialysate was collected in HCl-coated tubes from each area for approximately 3.5 h and 4 h, respectively. The data presented here were simply for assay applicability and were not meant to represent a comprehensive pharmacokinetic analysis.

3. Results and Discussion

3.1. Selectivity

Temozolomide and theophylline (Figure 1) eluted at the same retention time of 0.45 min, however due to their differing precursor and product ion masses, this LC-MS/MS assay was able to identify each compound based on its fragmentation. Figure 2 depicts the chromatograms for drug-free plasma extract (A), with internal standard (B) showing no cross-talk to the analyte transition, the LLOQ of 5 ng/mL (C) and a study sample (D). There was a minor plasma contaminant that eluted near (0.6 min) the analyte retention time, but did not interfere with peak area estimation or quantification.

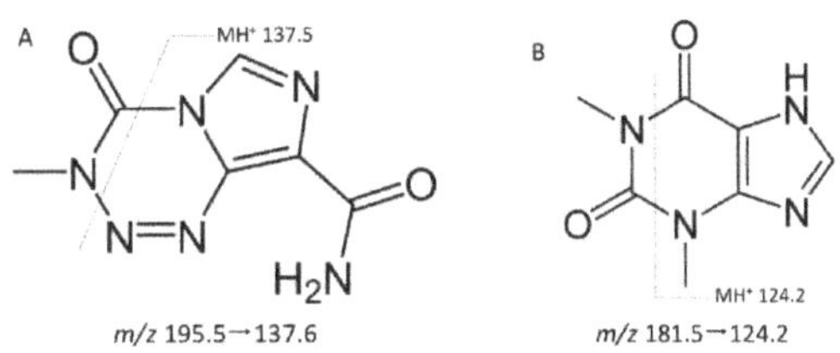

Figure 1. Structures of temozolomide (TMZ) (**A**) and the internal standard theophylline (**B**), with proposed mass spectral fragmentation patterns.

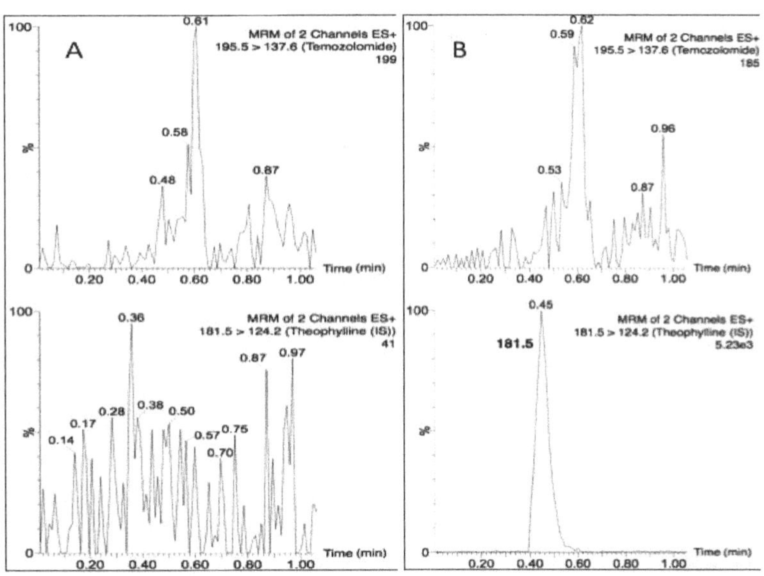

Figure 2. *Cont.*

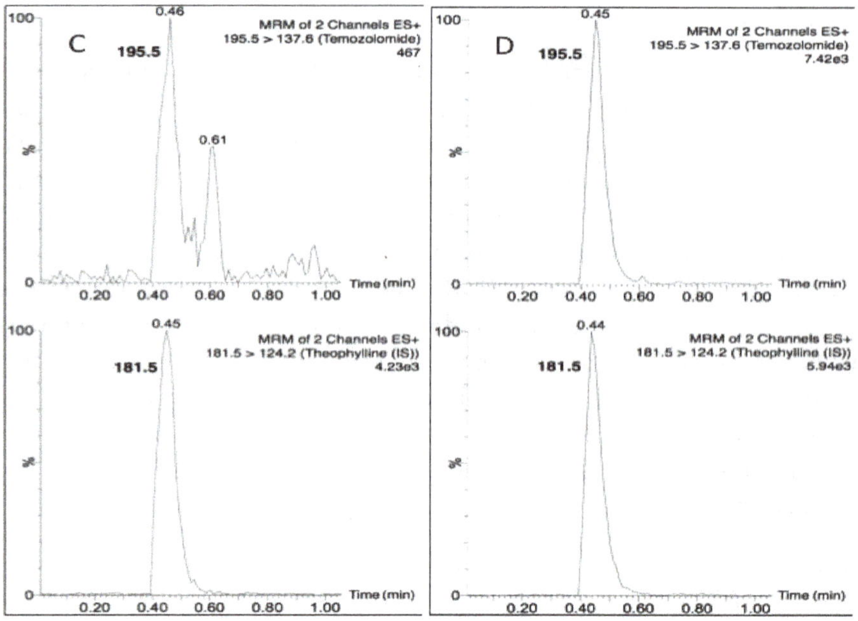

Figure 2. Chromatograms of (**A**) a blank plasma extract; (**B**) internal standard only extract; (**C**) the lower limit of quantification (LLOQ); and (**D**) a single plasma sample from a nonhuman primate (NHP) measured 15 min following a 20 mg dose.

3.2. Validation

The calibration standards, ranging from 5–2000 ng/mL, were prepared in duplicate on each of four days (n = 8), except in CSF (three days, n = 6). The eight calibration standards proved accurate and precise in each matrix with a mean ($\pm$SD) calibration curve linear correlation (r^2) of 0.9984 $\pm$ 0.0006 (n = 4) in plasma, 0.9975 $\pm$ 0.0007 (n = 3) in CSF, and 0.9949 $\pm$ 0.0019 (n = 4) in ECF (Figure 3).

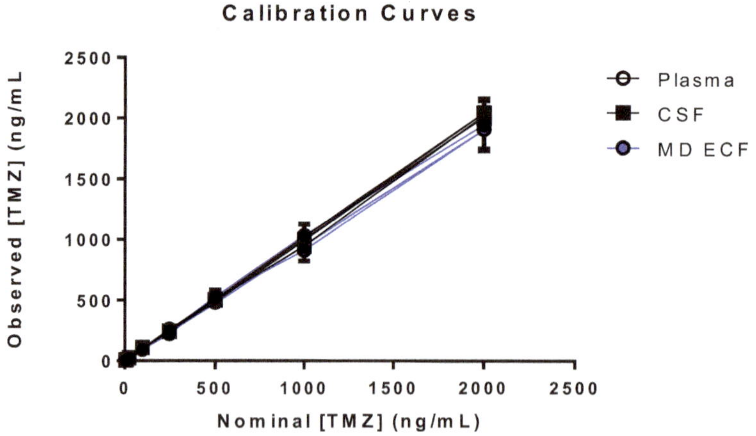

Figure 3. Calibration curves in NHP plasma, cerebrospinal fluid (CSF), and microdialysis extracellular fluid over the range of 5–2000 ng/mL.

Four quality control (QC) samples, at low (15 ng/mL), mid (800 ng/mL), high (1600 ng/mL), and 50-fold diluted (50,000 ng/mL diluted to 1000 ng/mL) concentrations, were run in quintuplet daily over four days (n = 20) in plasma and ECF, three days (n = 15) in CSF. All QC levels proved accurate and precise, within the required $\leqslant$15% (Tables 1–3).

Table 1. Accuracy and Precision in Nonhuman Primate Plasma.

Nominal (ng/mL)	GM (ng/mL)	SD (ng/mL)	Accuracy (%)	WRP (%)	BRP (%)	*n*
15	15.5	2.9	103	13.7	14.7	20
800	795	45.2	99.4	5.2	2.6	20
1600	1593	76.3	99.5	4.5	1.9	20
50,000	50,325	5,150	101.	7.0	8.4	20

Table 2. Accuracy and Precision in Nonhuman Primate cerebrospinal fluid (CSF).

Nominal (ng/mL)	GM (ng/mL)	SD (ng/mL)	Accuracy (%)	WRP (%)	BRP (%)	*n*
15	13.3	1.4	88.6	10.0	3.1	15
800	809	54.4	101	4.2	6.2	15
1600	1583	102	99	6.2	1.8	15
50,000	51,580	5373	96.8	9.6	4.8	15

Table 3. Accuracy and Precision in Nonhuman Primate microdialysis (MD) Perfusion Fluid.

Nominal (ng/mL)	GM (ng/mL)	SD (ng/mL)	Accuracy (%)	WRP (%)	BRP (%)	*n*
15	15.1	1.8	101	8.0	8.6	20
800	8109	59.4	101	6.8	2.3	20
1600	1646	149	103	8.7	3.9	20
5000	5289	331	106	4.9	4.2	20

Note: GM, grand mean; SD, standard deviation; DEV (%) relative deviation from nominal value; WRP, within-run precision; *n*, number of replicate observations within each validation run.

3.3. Stability

Temozolomide was stable through at least two freeze/thaw cycles in NHP CSF and three cycles in plasma, where changes in measured values at each of the three concentrations after each freeze/thaw cycle were below 5% in plasma and 11% in CSF, indicating no significant degradation. The 24-h post-preparative stability measurements were consistent with the initial run (<10% change), allowing samples to be reanalyzed on the following day when necessary without significant degradation.

3.4. Extraction Recovery and Matrix Effects

The extraction recoveries for the low (25 ng/mL) and high (1000 ng/mL) concentrations were 88.6% and 89.4%, respectively. Matrix effects induced a 47% decrease in signal due to ion suppression. These data suggest the ethyl acetate liquid extraction is adequate for sample cleanup for sensitive, quantitative measurements.

3.5. Clinical Application

Following a single intranasal bolus delivery of 20 mg in a 1 mL solution (~1.75 mg/kg), both plasma and CSF concentrations of TMZ decreased in a mono-exponential manner over a 24-h sampling period, with CSF penetrance of approximately 40% in this individual specimen (Figure 4A). Additionally, this assay was capable of measuring TMZ from *in situ* microdialysate in the pons and cortex ECF following a 1-h intravenous infusion (10 mg/kg) (Figure 4B).

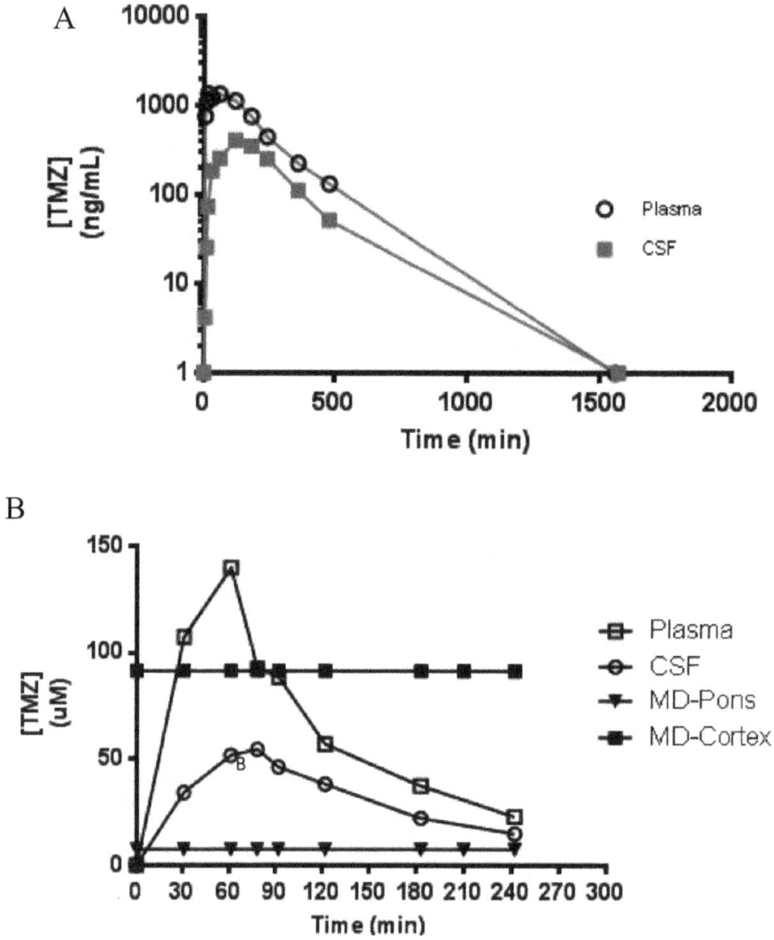

Figure 4. TMZ plasma and cerebrospinal fluid (CSF) concentration vs time curve from a single NHP dosed with 20 mg intranasally with 24 h sampling (**A**); and TMZ plasma, CSF, and microdialysis (MD) sampling of extracellular fluid (ECF) from the cortex (over a 3.5 h sampling period) and pons (4-h sampling period) following a 1-h continuous intravenous infusion of 10 mg/kg (**B**).

4. Conclusions

This paper describes an ultra HPLC-MS/MS method designed for the quantitative measurement of TMZ in nonhuman primate plasma, CSF, and ECF taken directly from designated areas in the brain inserted with MD probes. This method proved sensitive (LLOQ 5 ng/mL), selective, accurate, precise, and efficient at recovering drug from NHP plasma (~90% recovery). TMZ was found to be stable through multiple freeze/thaw cycles in both NHP plasma and CSF and post-preparation after 24-h. Ultimately, this concentration data can be used to describe the PK in each matrix, as well as brain penetration analysis and direct drug brain concentration measurement *in situ*.

Acknowledgments: Disclaimer: The content of this publication does not necessarily reflect the views or policies of the Department of Health and Human Services, nor does mention of trade names, commercial products, or organization imply endorsement by the US Government. The views in this manuscript are those of the authors and may not necessarily reflect NIH policy. No official endorsement is intended nor should be inferred.

Acknowledgments: Acknowledgments: The authors would like to acknowledge the contributions of John Bacher and Marvin T. Thomas (Division of Research Services, NCI) for completing the MD surgical procedures, and Rafael Cruz (Pediatric Oncology Branch) for collecting and processing all animal samples. Funding for this work was supported through the Intramural Research Program within the NCI.

Author Contributions: Author Contributions: Cody Peer and Lukas Ronner designed the assay; Cynthia McCully and Katherine Warren designed the animal experiments; Lukas Ronner, Louis Rodgers, and Cynthia

McCully performed the experiments; Cody Peer, Lukas Ronner, and Louis Rodgers analyzed the data; Cody Peer, Louis Rodgers, and William D. Figg wrote the paper.

Conflicts of Interest: Conflicts of Interest: The authors declare no conflicts of interest.

Abbreviations

The following abbreviations are used in this manuscript:

UHPLC	ultra-high performance liquid chromatography
MS	mass spectrometry
MS/MS	tandem mass spectrometry
QC	Quality control
LLOQ	Lower limit of quantification
NHP	nonhuman primates
CSF	cerebrospinal fluid
MD	microdialysis
PK	pharmacokinetics
ECF	extracellular fluid

References

1. Denny, B.J.; Wheelhouse, R.T.; Stevens, M.F.; Tsang, L.L.; Slack, J.A. NMR and molecular modeling investigation of the mechanism of activation of the antitumor drug temozolomide and its interaction with DNA. *Biochemistry* **1994**, *33*, 9045–9051. [PubMed]
2. Stupp, R.; Mason, W.P.; van den Bent, M.J.; Weller, M.; Fisher, B.; Taphoorn, M.J.; Belanger, K.; Brandes, A.A.; Marosi, C.; Bogdahn, U.; *et al.* Radiotherapy plus concomitant and adjuvant temozolomide for glioblastoma. *New Engl. J. Med.* **2005**, *352*, 987–996. [PubMed]
3. Ostermann, S.; Csajka, C.; Buclin, T.; Leyvraz, S.; Lejeune, F.; Decosterd, L.A.; Stupp, R. Plasma and cerebrospinal fluid population pharmacokinetics of temozolomide in malignant glioma patients. *Clin. Cancer Res.* **2004**, *10*, 3728–3736. [PubMed]
4. Newlands, E.S.; Stevens, M.F.; Wedge, S.R.; Wheelhouse, R.T.; Brock, C. Temozolomide: A review of its discovery, chemical properties, pre-clinical development and clinical trials. *Cancer Treat. Rev.* **1997**, *23*, 35–61. [PubMed]
5. McCully, C.L.; Balis, F.M.; Bacher, J.; Phillips, J.; Poplack, D.G. A rhesus monkey model for continuous infusion of drugs into cerebrospinal fluid. *Lab. Anim. Sci.* **1990**, *40*, 520–525. [PubMed]
6. Patel, M.; McCully, C.; Godwin, K.; Balis, F.M. Plasma and cerebrospinal fluid pharmacokinetics of intravenous temozolomide in non-human primates. *J. Neurooncol.* **2003**, *61*, 203–207. [PubMed]
7. McCully, C.M.; Pastakia, D.; Bacher, J.; Thomas, M.L., 3rd; Steffen-Smith, E.A.; Saleem, K.; Murphy, R.F.; Walbridge, S.; Brinster, L.; Widemann, B.C.; *et al.* Model for concomitant microdialysis sampling of the pons and cerebral cortex in rhesus macaques (*Macaca mulatta*). *Comp. Med.* **2013**, *63*, 355–360. [PubMed]
8. Grossman, R.; Rudek, M.A.; Brastianos, H.; Zadnik, P.; Brem, H.; Tyler, B.; Blakeley, J.O. The impact of bevacizumab on temozolomide concentrations in intracranial U87 gliomas. *Cancer Chemother. Pharmacol.* **2012**, *70*, 129–139. [PubMed]
9. Grossman, R.; Tyler, B.; Rudek, M.A.; Kim, E.; Zadnik, P.; Khan, U.; Blakeley, J.O.; Pathak, A.P.; Brem, H. Microdialysis measurement of intratumoral temozolomide concentration after cediranib, a pan-VEGF receptor tyrosine kinase inhibitor, in a U87 glioma model. *Cancer Chemother. Pharmacol.* **2013**, *72*, 93–100. [CrossRef] [PubMed]
10. Meany, H.J.; Warren, K.E.; Fox, E.; Cole, D.E.; Aikin, A.A.; Balis, F.M. Pharmacokinetics of temozolomide administered in combination with O^6-benzylguanine in children and adolescents with refractory solid tumors. *Cancer Chemother. Pharmacol.* **2009**, *65*, 137–142. [CrossRef] [PubMed]
11. Goldwirt, L.; Zahr, N.; Farinotti, R.; Fernandez, C. Development of a new UPLC-MSMS method for the determination of temozolomide in mice: Application to plasma pharmacokinetics and brain distribution study. *Biomed. Chromatogr.* **2013**, *27*, 889–893. [PubMed]

12. Chowdhury, S.K.; Laudicina, D.; Blumenkrantz, N.; Wirth, M.; Alton, K.B. An LC/MS/MS method for the quantitation of MTIC (5-(3-*N*-methyltriazen-1-yl)-imidazole-4-carboxamide), a bioconversion product of temozolomide, in rat and dog plasma. *J. Pharm. Biomed. Anal.* **1999**, *19*, 659–668. [PubMed]
13. Reyderman, L.; Statkevich, P.; Thonoor, C.M.; Patrick, J.; Batra, V.K.; Wirth, M. Disposition and pharmacokinetics of temozolomide in rat. *Xenobiotica* **2004**, *34*, 487–500. [CrossRef] [PubMed]
14. Jain, D.; Athawale, R.; Bajaj, A.; Shrikhande, S. Double-salting out assisted liquid-liquid extraction (SALLE) HPLC method for estimation of temozolomide from biological samples. *J. Chromatogr. B* **2014**, *970*, 86–94.
15. Gilant, E.; Kaza, M.; Szlagowska, A.; Serafin-Byczak, K.; Rudzki, P.J. Validated HPLC method for determination of temozolomide in human plasma. *Acta Pol. Pharm.* **2012**, *69*, 1347–1355. [PubMed]
16. Panetta, J.C.; Kirstein, M.N.; Gajjar, A.; Nair, G.; Fouladi, M.; Heideman, R.L.; Wilkinson, M.; Stewart, C.F. Population pharmacokinetics of temozolomide and metabolites in infants and children with primary central nervous system tumors. *Cancer Chemother. Pharmacol.* **2003**, *52*, 435–441. [PubMed]
17. Kim, H.; Likhari, P.; Parker, D.; Statkevich, P.; Marco, A.; Lin, C.C.; Nomeir, A.A. High-performance liquid chromatographic analysis and stability of anti-tumor agent temozolomide in human plasma. *J. Pharm. Biomed. Anal.* **2001**, *24*, 461–468. [PubMed]
18. Shen, F.; Decosterd, L.A.; Gander, M.; Leyvraz, S.; Biollaz, J.; Lejeuneb, F. Determination of temozolomide in human plasma and urine by high-performance liquid chromatography after solid-phase extraction. *J. Chromatogr. B* **1995**, *667*, 291–300.
19. U.S. Food and Drug Administration. *Guidance for Industry: Bioanalytical Method Validation*; U.S. Food and Drug Administration: Silver Spring, MD, USA, 2015.

2

HPLC-UV Polyphenolic Profiles in the Classification of Olive Oils and Other Vegetable Oils via Principal Component Analysis

Mireia Farrés-Cebrián [1], Raquel Seró [1], Javier Saurina [1,2] and Oscar Núñez [1,2,3,*]

1 Department of Chemical Engineering and Analytical Chemistry, University of Barcelona, Martí i Franquès 1-11, E08028 Barcelona, Spain; mirefarres28@gmail.com (M.F.-C.); rsero@ub.edu (R.S.); xavi.saurina@ub.edu (J.S.)

2 Research Institute in Food Nutrition and Food Safety, University of Barcelona, Recinte Torribera, Av. Prat de la Riba 171, Edifici de Recerca (Gaudí), E08921 Santa Coloma de Gramenet, Barcelona, Spain

3 Serra Húnter Program, Generalitat de Catalunya, Rambla de Catalunya 19-21, E08007 Barcelona, Spain

* Correspondence: oscar.nunez@ub.edu

Academic Editor: Didier Thiébaut

Abstract: High performance liquid chromatography-ultraviolet (HPLC-UV) was applied to the analysis and characterization of olive oils and other vegetable oils. A chromatographic separation on a Zorbax Eclipse XDB-C8 reversed-phase column was proposed under gradient elution, employing 0.1% formic acid aqueous solution and methanol as mobile phase, for the determination of 14 polyphenols and phenolic acids, allowing us to obtain compositional profiles in less than 20 min. Acceptable sensitivity (limit of detection (LOD) values down to 80 μg/L in the best of cases), linearity (r^2 higher than 0.986), good run-to-run and day-to-day precisions (relative standard deviation (RSD) values lower than 11.5%), and method trueness (relative errors lower than 6.8%) were obtained. The proposed HPLC-UV method was then applied to the analysis of 72 oil samples (47 olive oils and 27 vegetable oils including sunflower, soy, corn, and mixtures of them). Analytes were recovered using a liquid–liquid extraction method employing ethanol:water 70:30 (v/v) solution and hexane as extracting and defatting solvents, respectively. HPLC-UV polyphenolic profiles using peak areas were then analysed by principal component analysis (PCA) to extract information from the most significant data contributing to the characterization and classification of olive oils against other vegetable oils, as well as among Arbequina and Picual olive oil varieties. PCA results showed a noticeable difference between olive oils and the other classes. In addition, a reasonable discrimination of olive oils as a function of fruit varieties was also encountered.

Keywords: high performance liquid chromatography; UV-detection; polyphenols; principal component analysis; food characterization; olive oils; vegetable oils

---

1. Introduction

The health of the individual and the population in general is the result of interactions between genetics and a number of environmental factors, nutrition being one of the factors of major importance [1,2]. An appropriate diet can prevent non-communicable disease linked to chronic inflammation such as cancer, diabetes, metabolic syndrome, obesity, pulmonary diseases, and cognition disorders. The importance of incorporating functional foods in the diet or following beneficial dietary patterns to improve health is fairly well-demonstrated [3]. Dietary patterns comprise all foods and beverages consumed in various characteristic combinations by different populations across the world. Among the dietary patterns that are known to have beneficial health effects, the Mediterranean diet is the best-studied and most evidence-based diet for preventing not only cardiovascular diseases but also other chronic diseases, and today has become a standard for healthy eating and a dietary template of particular value [4]. Moreover, due to its widely proven beneficial health effects, the Mediterranean diet has been on UNESCO's Representative List of the Intangible Cultural Heritages of Humanity since 2013 [5].

Olive oil has been produced for about 6000 years, but in the last 30 years there has been a growing interest in the use of olive oil in cooking due to the increased knowledge about Mediterranean food and public awareness of the health benefits of the Mediterranean diet and, particularly, of olive oil [6]. In fact, virgin olive oil (VOO) is the primary fat source and a fundamental component of the Mediterranean diet. VOO is obtained through physical procedures either by mechanical or by direct pressing of the pulp of the olive fruit (*Olea europaea* L.). The olives, after being crushed to form a pomace, are homogenized and pressed. This oil is not subjected to any other treatment than washing, decantation, centrifugation or filtration [7]. The oil produced from this first press is known as extra-virgin olive oil (EVOO), and it is the highest qualitative olive oil containing the highest levels of beneficial constituents [8]. Apart from the natural unrefined state, olive oil can also be consumed as a refined product which is made either from virgin olive oil and called refined olive oil (ROO) or refined olive–pomace oil [7], or from solvent-extracted oil [9] which is called refined husk oil (RHO).

Regarding EVOO constituents, triglycerides are the major components, representing more than 98% of the total weight. The fatty acid composition of triglycerides is 11% saturated fatty acids (e.g., palmitic acid (C16:0)), around 80% monounsaturated fatty acids (being oleic acid (C18:1), and from 55% to 83% total fatty acids, the most important feature of olive oils in comparison to other vegetable oils), and 9% polyunsaturated fatty acids (i.e., linoleic acid (C18:2 ω6)) [10,11]. In addition to the fatty acids, EVOO contains antioxidant and anti-inflammatory compounds, such as squalene, phytosterols, and mainly polyphenols and phenolic acids, which provide important beneficial health effects [12,13]. The total amount of polyphenols and phenolic acids in olive oils depends on several parameters, such as the growing area, cultivation techniques, water resources, soil management, degree of maturation, and climate, but the oil production method (malaxation and extraction) also affects the final amount of these compounds [14–16]. Apart from the sensory and functional properties attributed to these compounds, their impact on the characterization, classification, and authentication studies cannot be underestimated [17]. Polyphenol and phenolic acid contents seem to be related to several food features such as the cultivar geographical areas, variety, and extraction conditions, and as a result, polyphenol content can be exploited as a source of analytical data to establish vegetable oil product classifications and even to achieve detection of product adulterations.

Different analytical methodologies have been developed for the determination of polyphenols and phenolic acids in vegetable oils [18,19]. In this kind of matrix, one of the most critical points is the extraction procedure employed because it will influence the results obtained after analysis. Liquid–liquid extraction (LLE) and solid phase extraction (SPE) are among the most widely used extraction methods in the determination of polyphenols in olive oils [18–28]. As regards the analytical technique, chromatographic and electrophoretic methods are employed to achieve separation and quantification of these compounds after sample extraction. However, the fact that a derivatization step is commonly necessary when gas chromatography (GC) is employed [21], and the low sensitivity attainable with capillary electrophoresis (CE) methods [29,30] in the determination of polyphenols make liquid chromatography (LC) the most widely employed technique for that purpose either with UV detection [22,31,32] or coupled to mass spectrometry (LC-MS) systems [20,33–35]. Liquid chromatography in combination with high-resolution mass spectrometry (HRMS) has also been described for the identification and determination of phenolic compounds in vegetable oils [23,24,28,36]. Despite the number of applications described in the literature dealing with the determination of polyphenols in vegetable oil samples by LC, few works are devoted to classification and authentication purposes [25].

The aim of this work is to develop an economic, simple, and robust high-performance liquid chromatography (HPLC) method with UV detection for the determination of several polyphenols and phenolic acids in a variety of vegetable oils, including olive, sunflower, soy, and corn oils as well as vegetable oils prepared by mixtures of different seeds and fruits. HPLC-UV polyphenolic profiles using peak area data will be exploited to carry out exploratory principal component analysis (PCA) to achieve vegetable oil sample characterization and classification.

2. Materials and Methods

2.1. Chemicals and Standards

Unless otherwise specified, analytical grade reagents were used. The polyphenols and phenolic acids studied, which are shown in Table 1, were all obtained from Sigma-Aldrich (Steinheim, Germany) except for rosmanol, which was purchased from Cymit Quimica S.L. (Barcelona, Spain). Stock standard solutions of all polyphenols and phenolic acids at ca. 1000 mg/L were prepared in methanol and store in amber glass vials. The intermediate working solutions were prepared weekly by dilution of these stock standard solutions using Milli-Q water. All stock solutions were stored at 4 °C for no more than 1 month.

Methanol (Chromasolv® for HPLC, ≥99.9%), hexane, and formic acid (≥98%) was also obtained from Sigma-Aldrich, and ethanol (absolute) was purchased from VWR International Eurolab S.L. (Barcelona, Spain).

The water used in this work was purified by employing an Elix 3 coupled to a Milli-Q purification system (Millipore, Bedford, MA, USA) and filtered through a 0.22 μm nylon membrane integrated into the Milli-Q system.

Table 1. Chemical structures of studied polyphenols and phenolic acids.

Peak	Compound	Family	Structure	CAS Number
1	Homogentisic acid	Phenolic acid	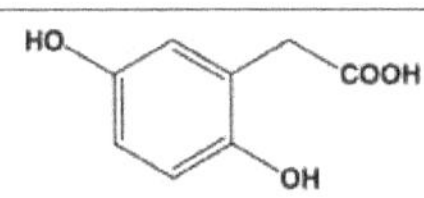	451-13-8
2	(+)-Catechin hydrate	Flavanol	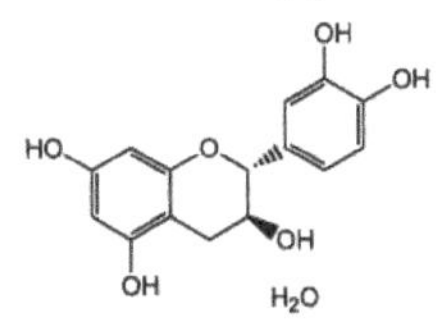	225937-10-0
3	Tyrosol	Phenolic alcohol	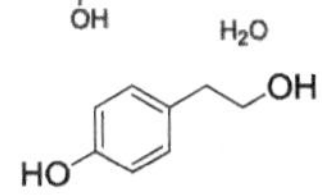	501-91-0
4	4-Hydroxybenzoic acid	Phenolic acid	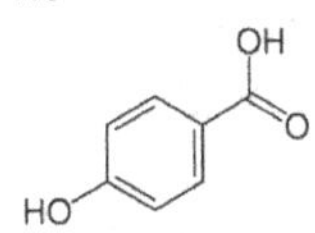	99-96-7
5	Caffeic acid	Phenolic acid	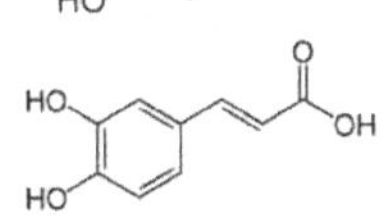	331-39-5
6	Syringic acid	Phenolic acid	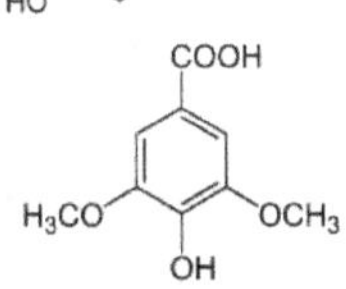	530-57-4
7	Vanillin	Phenolic aldehyde	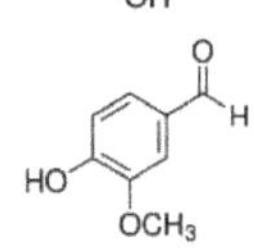	121-33-5

Table 1. *Cont.*

Peak	Compound	Family	Structure	CAS Number
8	Syringaldehyde	Phenolic aldehyde	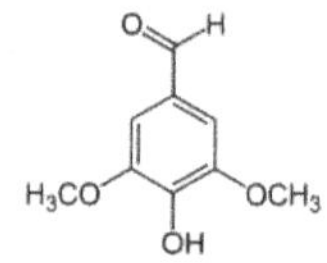	134-96-3
9	*p*-Coumaric acid	Phenolic acid	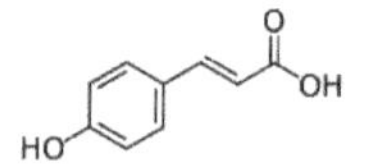	501-98-4
10	Sinapic acid	Phenolic acid	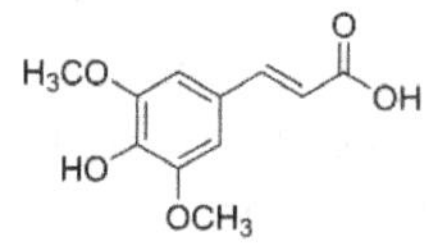	530-59-6
11	Ferulic acid	Phenolic acid	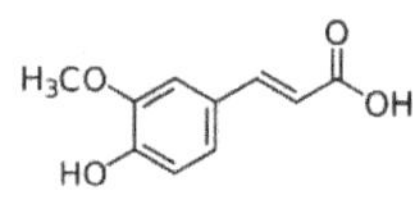	537-98-4
12	Veratric acid	Benzoic acid	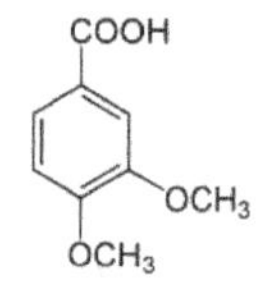	93-07-2
13	Kaempferol	Flavonol	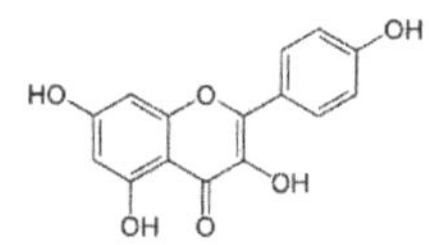	520-18-3
14	Rosmanol	Phenolic diterpene	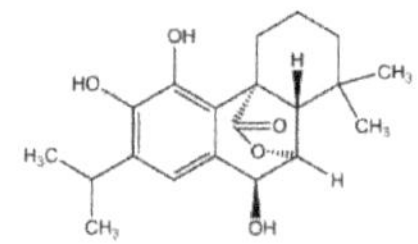	80225-53-2

2.2. Instrumentation and Methods

Chromatographic separation was performed on a HPLC instrument from Dionex (Dionex Softron GmbH, Gemering, Germany) equipped with a P680 pump, an ASI-100 automatic sample injector, a UVD170U UV-Vis detector, and a TCC-100 thermostatted column compartment. A Zorbax Eclipse XDB-C8 reversed-phase column (150 × 4.6 mm i.d., 5 μm particle size) provided by Agilent Technologies (Santa Clara, CA, USA) was used for the proposed method. A 0.1% (v/v) formic acid aqueous solution (solvent A) and methanol (solvent B) were used to establish the gradient elution as follows: 0–2 min at 10% B (initial conditions); 2–4.5 min linear gradient from 10% B to 25% B; 4.5–7 min at 25% B; 7–22 min linear gradient from 25% B to 90% B; 22–24 min at 90% B; 24–25 min back to the initial conditions at 10% B; and 25–30 min at 10% B for column equilibration. The mobile phase flow rate was 1 mL·min^{-1} and the injection volume was 10 μL. Direct UV absorption detection was carried out at 257 nm (for 4-hydroxybenzoic and veratric acids, and kaempferol), 280 nm (for catechin, tyrosol, vanillin, rosmanol, and homogentisic and syringic acids) and 316 nm (for syringaldehyde and caffeic, *p*-coumaric, sinapic, and ferulic acids). Chromeleon software version 6.7 (Dionex, Softron GmbH, Gemering, Germany) was used for HPLC system control and data acquisition.

2.3. Samples and Sample Treatment

The set of samples to be analysed consisted of the following vegetable oils from different trademarks purchased from Barcelona markets: 47 olive oils, 16 sunflower oils, 2 corn oils, 2 soy oils, and 5 vegetable oils produced from mixtures of seeds (3 sunflower/corn oils and 2 sunflower/soy oils). Among the 47 analysed olive oils, 10 were obtained from Arbequina olives and 5 from Picual olives. No information regarding the olive variety was provided by producers regarding the other 32 olive oils.

Sample treatment was carried out following a previously described method with some modifications [20]. Briefly, 2.00 g of vegetable oil were weighed into a 15 mL polytetrafluoroethylene (PTFE) centrifuge tube (Serviquimia, Barcelona, Spain), and 2 mL of an ethanol:water 70:30 v/v solution were added. The mixture was shaken vigorously for 2 min using a vortex mixer (Stuart, Stone, UK). The extract was then centrifuged at 3500 rpm for 5 min using a Rotanta 460 RS centrifuge (Hettich, Tuttlingen, Germany). Then the mixture was frozen for 24 h at −18 °C to facilitate the quantitative recovery of the extract solution. The corresponding extract was transferred into another 15 mL PTFE centrifuge tube for further clean-up. Then, 2 mL of hexane were added as the defatting agent and the mixture was shaken vigorously for 2 min in a vortex mixer and centrifuged at 3500 rpm for 5 min. Then, the aqueous ethanolic extracts were directly analysed with the proposed HPLC-UV method.

Also, a quality control (QC) consisting of a mixture of 50 μL of each ethanolic aqueous sample extract was prepared to evaluate the repeatability of the method and the robustness of the chemometric results.

2.4. Data Analysis

SOLO chemometric software from Eigenvector Research was used for calculations with principal component analysis (PCA) [37]. A detailed description of the theoretical background of this method is given elsewhere [38].

Data matrices to be treated consisted on the peak area values of the 14 studied polyphenols and phenolic acids detected (at the registered wavelengths) in the different samples under study. Thus, the dimension of the matrix was 72 samples × 14 analytes. Normalization pretreatment with respect to the overall polyphenolic concentration was applied to provide similar weighs to all the samples. For additional details see Reference [39].

Peak identification in the analysed vegetable oil samples was performed by comparison of retention times and UV spectra with those of a polyphenolic standard solution. Peak purity was confirmed by comparison of UV spectra through the entire peak signal.

3. Results and Discussion

3.1. HPLC Conditions

The present work aimed to develop a HPLC-UV method for the separation and determination of fourteen polyphenols and phenolic acids. We first employed a gradient elution program with an increasing methanol component from 10% to 90% over 19 min (after an isocratic step of 1 min at initial conditions), followed by an isocratic step of 3 min at 90%. The HPLC-UV chromatogram obtained under these conditions for a standard mixture of ~25 mg/L of studied compounds in water as well as the gradient profile are shown in Figure 1a. The gradient elution program continued up to 30 min to get back to mobile phase initial conditions and to guarantee column re-equilibration. As can be seen, baseline chromatographic separation was achieved for almost all compounds in less than 17 min, although coelutions between vanillin and syringaldehyde (peaks 7 and 8, respectively) and between *p*-coumaric, sinapic and ferulic acids (peaks 9, 10 and 11, respectively) occurred.

The gradient elution program was modified in order to improve the chromatographic separation of the fourteen studied compounds. For that purpose, methanol content was increased more slowly from 10% to 25% (from min 1 to min 7), and then up to 90% methanol (from min 7 to min 20). The chromatographic separation obtained is shown in Figure 1b. From this modification, a full resolution of vanillin and syringaldehyde was attained and a considerable improvement on the separation between *p*-coumaric and sinapic acids (peaks 9 and 10, respectively) was achieved, although partial coelution of these two last phenolic acids with ferulic acid still persisted. Regarding analysis time, a small increase up to 19 min was obtained.

Up to this point, several combinations of linear gradients and isocratic steps were evaluated to try to get a full resolution of all compounds. It should be mentioned that none of the gradient programs evaluated led to the baseline separation of *p*-coumaric, sinapic, and ferulic acids, even when increasing analysis time. In all the conditions studied, sinapic acid coeluted either with *p*-coumaric acid or with ferulic acid. For this reason, as a compromise between analysis time and peak chromatographic resolution, the gradient elution program finally selected consisted of an isocratic step at initial conditions (10% methanol) for 2 min, a linear gradient up to 25% methanol in 2.5 min, an isocratic step at 25% methanol for 2.5 min, and then up to 90% methanol till 22 min.

Figure 1c shows the HPLC-UV chromatogram obtained under these conditions. Although partial coelutions are still observed between *p*-coumaric, sinapic and ferulic acids, the chromatographic separation was considered acceptable for the intended purposes.

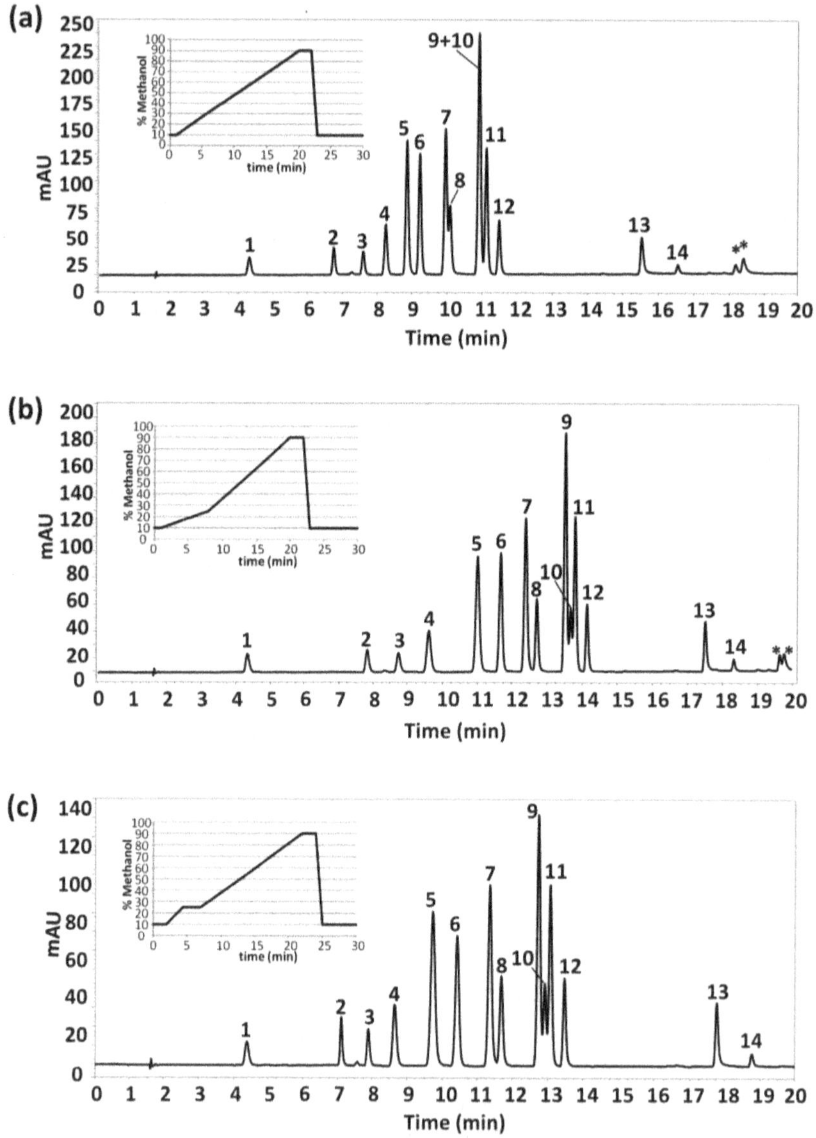

Figure 1. High performance liquid chromatography-ultraviolet (HPLC-UV) chromatograms of a mixture of studied polyphenols (~25 mg/L in water) under different gradient elution conditions. UV detection: λ 280 nm. Peak identification: **1**, homogentisic acid; **2**, (+)-catechin hydrate; **3**, tyrosol; **4**, 4-hydroxybenzoic acid; **5**, caffeic acid; **6**, syringic acid; **7**, vanillin; **8**, syringaldehyde; **9**, *p*-coumaric acid; **10**, sinapic acid; **11**, ferulic acid; **12**, veratric acid; **13**, kaempferol; and **14**, rosmanol. (*) system peaks.

3.2. Instrumental Quality Parameters

The proposed HPLC-UV method was validated by determining instrumental quality parameters under optimal conditions. Figures of merit are summarized in Table 2. The LODs, based on a signal-to-noise ratio of 3:1, were obtained by analyzing polyphenol standard solutions at decreasing concentration levels. Most of the analysed polyphenols showed LOD values below 0.9 mg/L (in the range between 80 μg/L and 900 μg/L), and only five polyphenols showed lower sensitivity, being rosmanol the one with the poorest LOD (11.5 mg/L). The limits of quantification (LOQs) determined, based on a signal-to-noise ratio of 10:1, were in the range of 0.25 to 38.4 mg/L.

Run-to-run and day-to-day precisions for polyphenol quantification were calculated at a concentration level of ~30 mg/L. In order to estimate the run-to-run precision, five replicate determinations were carried out using the proposed method under optimal conditions. The relative standard deviations (% RSDs) were between 0.6% and 2.9%.

Table 2. HPLC-UV instrumental quality parameters.

Compound	LOD (mg/L)	LOQ (mg/L)	Linearity (r^2)	Run-to-Run Precision (RSD%, n = 5)		Day-to-Day Precision (RSD%, n = 5 × 3)		Trueness (n = 5)		
				t_r	Concentration	t_r	Concentration	Conc. value (mg/L)	Calc. value (mg/L)	Error (%)
1) Homogentisic acid	2.6	8.7	0.987	0.12	2.9	0.32	4.3	4.3	29.3	4.5
2) (+)-catechin hydrate	3.5	11.6	0.995	0.05	1.1	0.16	10.1	10.1	31.9	2.3
3) Tyrosol	3.3	11.0	0.986	0.05	2.0	0.15	7.3	7.3	35.1	6.8
4) 4-hydroxybenzoic acid	0.23	0.8	0.995	0.06	1.0	0.23	6.9	6.9	32.6	0.4
5) Caffeic acid	0.5	1.5	0.992	0.07	0.6	0.23	4.2	4.2	41.0	6.7
6) Syringic acid	0.5	1.7	0.996	0.05	0.7	0.20	6.6	6.6	33.2	3.9
7) Vanillin	0.6	1.9	0.995	0.04	0.6	0.14	7.7	7.7	31.1	2.8
8) Syringaldehyd	0.9	3.2	0.988	0.04	1.0	0.13	4.8	4.8	33.4	2.8
9) *p*-coumaric acid	0.08	0.25	0.997	0.04	0.8	0.12	3.0	3.0	33.4	0.9
10) Sinapic acid	0.15	0.5	0.995	0.04	0.6	0.11	5.7	5.7	33.1	2.4
11) Ferulic acid	0.08	0.27	0.999	0.04	0.6	0.11	11.5	11.5	32.6	3.2
12) Veratric acid	0.15	0.5	0.987	0.04	0.7	0.10	8.6	8.6	32.5	0.3
13) Kaempferol	1.4	4.8	0.991	0.02	0.9	0.06	6.6	6.6	33.9	2.1
14) Rosmanol	11.5	38.4	0.999	0.02	2.1	0.06	3.6	3.6	39.1	1.7

[a] Calculated at a concentration ~30 mg/L.

A total of 15 replicate determinations were carried out on three non-consecutive days to calculate the day-to-day precision. As expected, RSD values slightly higher than those observed for run-to-run precision were achieved, with percentages between 3.0% and 11.5%. Run-to-run and day-to-day precisions of retention times were also calculated obtaining RSD values lower than 0.12% and 0.32%, respectively, thus showing the good precision performance of the proposed HPLC-UV method.

External calibration curves based on peak areas at concentrations between LOQ and 80 mg/L were built. Good linearity was observed (r^2 > 0.986) for all the compounds. The trueness of the proposed HPLC-UV method was also evaluated via quintuplicate analyses, using external calibration, of standard solutions at concentrations of ~30 mg/L, achieving very good results, with relative absolute errors ranging from 0.3% to 8.3%.

3.3. Exploratory Analysis by PCA

PCA provided plots of scores and loadings, showing the distribution of the samples and variables employed on the principal components (PCs), respectively. The study of the plot of scores revealed patterns that may be correlated to sample characteristics, such as the type of seed employed on the vegetable oil, etc. The study of the distribution of variables from the plot of loadings provided information dealing with their correlations as well as dependencies of polyphenols and phenolic acids on vegetable oil.

Characterization of vegetable oils was attempted using the peak areas of the fourteen polyphenols and phenolic acids under study found in the analysed samples. For that purpose, the 72 vegetable oil samples were processed as indicated in the sample treatment section and the resulting extracts were randomly analysed with the proposed HPLC-UV method. Figure 2 shows the HPLC-UV chromatograms of (a) an extra virgin olive oil sample within the 47 olive oils analysed; (b) a sunflower oil sample; (c) a soybean oil sample; and (d) a corn oil sample, registered at 280 nm under optimal conditions.

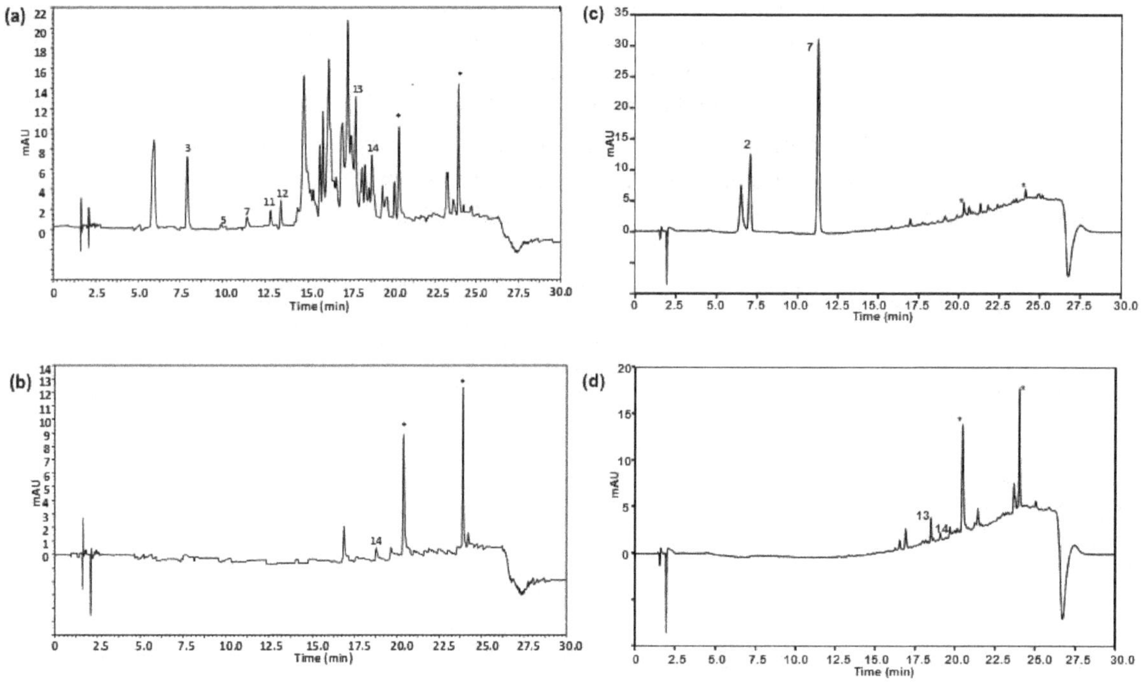

Figure 2. HPLC-UV chromatograms of (**a**) an extra virgin olive oil sample within the 47 olive oils analysed (olive type not specified); (**b**) a sunflower oil sample; (**c**) a soy oil sample, and (**d**) a corn oil sample, obtained under optimal conditions. UV detection: λ 280 nm. Peak identification as in Figure 1.

As can be seen, chromatographic profiles are very different, olive oils being richer in absorbing signals in comparison to the other vegetable oils. Additionally, only a few of the fourteen analysed polyphenols were detected being tyrosol, caffeic acid, vanillin, ferulic acid, veratric acid and kaempferol, and in a lower proportion rosmanol, the most frequently found in olive oils. This is the first time that veratric acid, kaempferol and rosmanol have been detected in extra virgin olive oils. This is only a tentative identification of these chemical compounds and further investigations will be carried out to confirm their presence. Regarding other oils, rosmanol, and in some cases vanillin among others, prevailed in sunflower and the other seed-based oils. With the information of the detected targeted polyphenols and phenolic acids, a peak area data matrix with a dimension of 72 samples × 14 compounds was built to be treated by PCA. Figure 3 shows the scatter plots of scores and loadings of PC1 vs. PC2 of the model so obtained.

As can be seen, QCs appeared in a compact group in the center of the graph, thus demonstrating the good repeatability and robustness of the proposed method. Regarding this preliminary exploratory classification model, vegetable oils other than olive oils appeared in a very compact group in the center-left area of the PC1 vs. PC2 plot, and from this point olive oils were widely distributed to the right of the plot, showing that PCA employing HPLC-UV polyphenolic profiles targeting only 14 simple polyphenols and phenolic acids allows discrimination between olive oils and other vegetable oils.

The attention was then focused only on olive oil samples to see if some discrimination occurred regarding the type variety of olives. Figure 4 shows, as an example, the HPLC-UV chromatograms registered at 280 nm under optimal conditions of (a) an EVOO obtained from Arbequina olives and (b) an EVOO obtained from Picual olives. As can be seen, the chromatographic profiles were quite different, with Picual olive oils characterized to present more intense peak signals between 14 and 20 min in comparison to Arbequina olive oils.

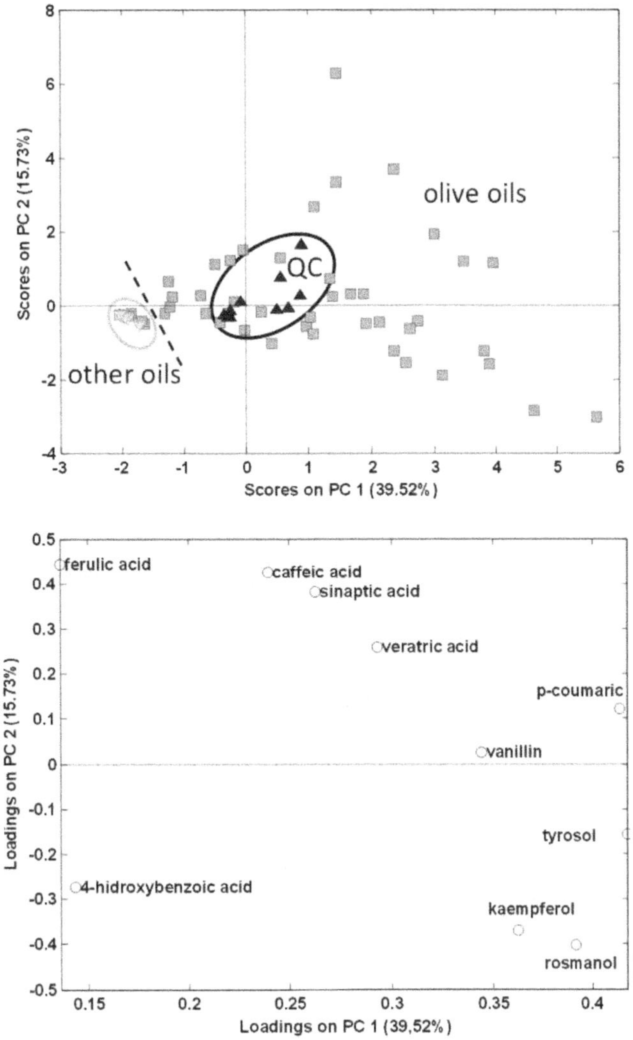

Figure 3. Principal component analysis (PCA) results (score and loading plots of PC1 vs. PC2) using as analytical data the normalized peak areas of the fourteen targeted polyphenols and phenolic acids found in the analysed vegetable oils. ■ olive oils, other vegetable oils, and ▲ quality controls.

A PCA model was then made only using EVOOs; to do this, peak areas of targeted polyphenols and phenolic acids in the olive oil samples were used to build a data matrix with a dimension of 47 samples × 14 compounds to be subjected to PCA. The scatter plots of scores and loadings of PC1 vs. PC2 are shown in Figure 5. PC1 is discriminating samples according to the content on analysed polyphenols, being samples distributed to the right area of the plot richer in the targeted compounds than samples located to the left. In contrast, it seems that PC2 is differentiating samples according to the variety of olive employed, with 100% Arbequina olive oils located to the top area of the plot, while 100% Picual olive oils tend to be located to the bottom area. Information regarding type of olive fruits employed in the production of the other olive oil samples analysed was not provided by the producers, and most of them will probably be obtained from mixtures of different varieties.

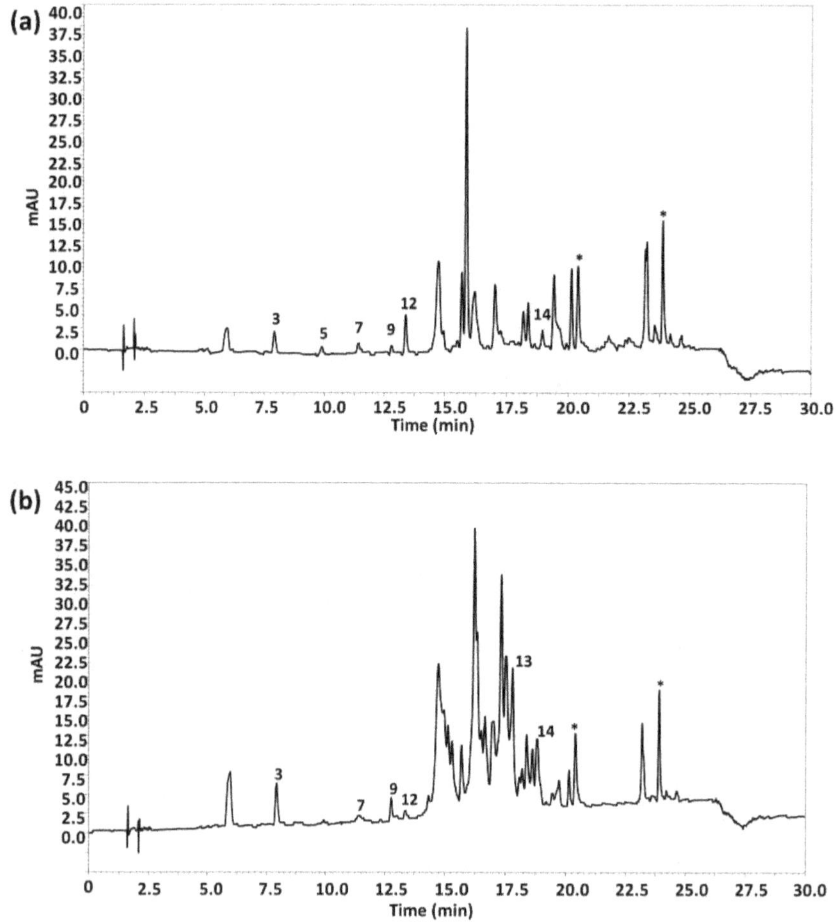

Figure 4. HPLC-UV chromatograms of (**a**) an extra-virgin olive oil (EVOO) obtained from 100% Arbequina olives and (**b**) an EVOO obtained from 100% Picual olives, obtained under optimal conditions. UV detection: λ 280 nm. Peak identification as in Figure 1.

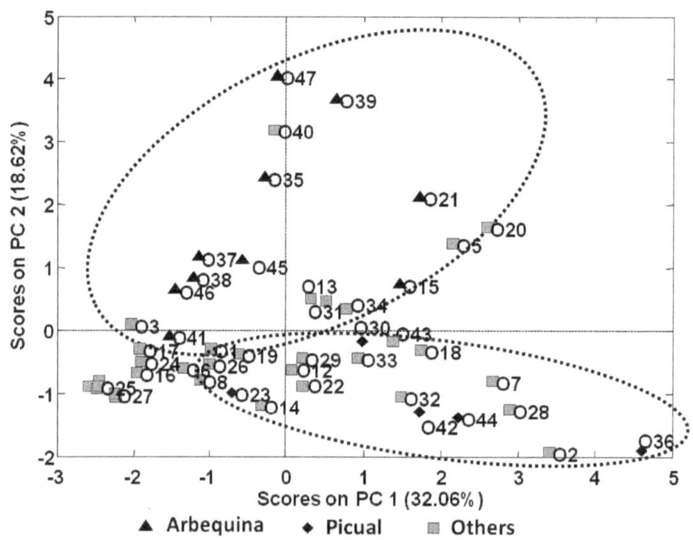

Figure 5. *Cont.*

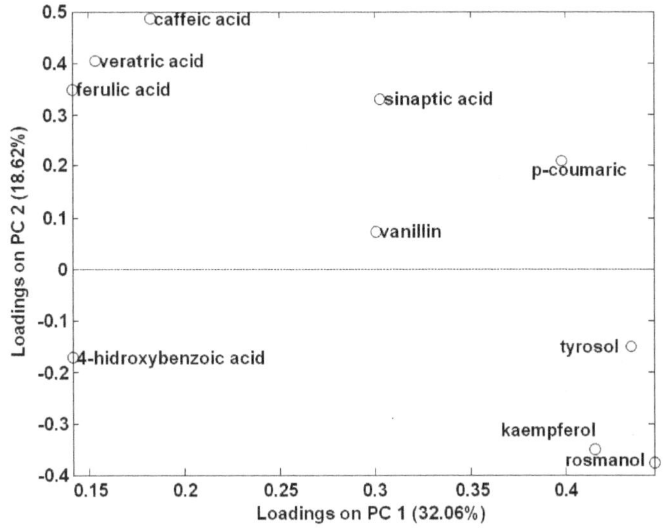

Figure 5. PCA results (score and loading plots of PC1 vs. PC2) using as analytical data the normalized peak areas of the fourteen targeted polyphenols and phenolic acids found in the analysed olive oils.

4. Conclusions

The results obtained in this work show that HPLC-UV polyphenolic profiles can be proposed as the analytical data for a simple and suitable characterization and classification of olive oils and other vegetable oils. HPLC-UV method performance was evaluated, producing acceptable LOD and LOQ values, with good linearity, run-to-run and day-to-day precisions, and method trueness.

PCA applied to phenolic peak area data and the resulting models provided an acceptable classification and discrimination between olive oils with respect to any other kind of vegetable oil sample, although no differentiation among these latter was achieved. In addition, an acceptable discrimination among olive oils regarding the olive variety was also observed, being possible also to slightly classify samples among Arbequina and Picual olives.

Exploratory PCA of the data obtained by HPLC-UV demonstrated that the method was useful for the discrimination of olive oil and other vegetable oils, and it could be employed in future studies to identify olive oil samples adulterated with other vegetable oils of poorer quality, as well as a starting point to guarantee the Protected Designation of Origin of these products. Future experiments will be carried out in this sense by employing partial least square regression to resolve authentication and adulteration issues.

Acknowledgments: The authors gratefully acknowledge the financial support received from the Spanish Ministry of Economy and Competitiveness under the projects CTQ2014-65324 and CTQ2015-63968-C2-1-P, and from the Agency for Administration of University and Research Grants (Generalitat de Catalunya, Spain) under the projects 2014 SGR-377 and 2014 SGR-539.

Author Contributions: Mireia Farrés-Cebrián performed the experiments and analysed the data regarding HPLC-UV optimization with the supervision of Raquel Seró. Mireia Farrés-Cebrián also peformed the application analysis and the chemometric study. Both Mireia Farrés-Cebrián and Raquel Seró worked under the close supervision of Javier Saurina and Oscar Núñez, who designed the experiments and wrote the paper.

Conflicts of Interest: The authors declare no conflict of interest.

References

1. Solomons, N.W. Evolutionary Aspects of Nutrition and Health: Diet, Exercise, Genetics and Chronic Disease, volume 84 of World Review of Nutrition and Dietetics edited by Artemis P. Simopoulos. *Am. J. Clin. Nutr.* **2000**, *71*, 854–855.
2. Simopoulos, A.P. The mediterranean diets: What is so special about the diet of Greece? The scientific evidence. *J. Nutr.* **2001**, *131*, 3065S–3073S. [PubMed]

3. Stradling, C.; Hamid, M.; Fisher, K.; Taheri, S.; Thomas, G.N. A review of dietary influences on cardiovascular health: Part 1: The role of dietary nutrients. *Cardiovasc. Hematol. Disord. Drug Targets* **2013**, *13*, 208–230. [CrossRef] [PubMed]
4. Widmer, R.J.; Flammer, A.J.; Lerman, L.O.; Lerman, A. The Mediterranean diet, its components, and cardiovascular disease. *Am. J. Med.* **2015**, *128*, 229–238. [CrossRef] [PubMed]
5. Mediterranean diet. United Nation Educational, Scientific and Cultural Organization (UNESCO). Available online: http://www.unesco.org/culture/ich/en/RL/mediterranean-diet-00884 (accessed on 15 October 2016).
6. Helsing, E. Trends in fat consumption in Europe and their influence on the Mediterranean diet. *Eur. J. Clin. Nutr.* **1993**, *47* (Suppl. 1), S4–S12. [PubMed]
7. European Union Commission. Council Regulation (EC) No. 1513/2001 of 23 July 2001 amending Regulations No. 136/66/EEC and (EC) No. 1638/98 as regards the extension of the period of validity of the aid scheme and the quality strategy for olive oil. *Off. J. Eur. Communities* **2001**, *L201*, 4–7.
8. Rafehi, H.; Ververis, K.; Karagiannis, T.C. Mechanisms of Action of Phenolic Compounds in Olive. *J. Diet. Suppl.* **2012**, *9*, 96–109. [CrossRef] [PubMed]
9. Fedeli, E. Lipids of olives. *Prog. Chem. Fats Other Lipids* **1977**, *15*, 57–74. [CrossRef]
10. Rossell, J.B. *Frying: Improving Quality*; Woodhead Publishing Limited, CRC Press: Boca Raton, FL, USA, 2001.
11. Visioli, F.; Galli, C. Olive Oil Phenols and Their Potential Effects on Human Health. *J. Agric. Food Chem.* **1998**, *46*, 4292–4296. [CrossRef]
12. Bester, D.; Esterhuyse, A.J.; Truter, E.J.; van Rooyen, J. Cardiovascular effects of edible oils: A comparison between four popular edible oils. *Nutr. Res. Rev.* **2010**, *23*, 334–348. [CrossRef] [PubMed]
13. Waterman, E.; Lockwood, B. Active components and clinical applications of olive oil. *Altern. Med. Rev.* **2007**, *12*, 331–342. [PubMed]
14. Caponio, F.; Alloggio, V.; Gomes, T. Phenolic compounds of virgin olive oil: Influence of paste preparation techniques. *Food Chem.* **1999**, *64*, 203–209. [CrossRef]
15. Vekiari, S.A.; Koutsaftakis, A. The effect of different processing stages of olive fruit on the extracted olive oil polyphenol content. *Grasas Aceites* **2002**, *53*, 304–308. [CrossRef]
16. Motilva, M.J.; Tovar, M.J.; Romero, M.P.; Alegre, S.; Girona, J. Evolution of oil accumulation and polyphenol content in fruits of olive tree (*Olea europaea* L.) related to different irrigation strategies. *Acta Hortic.* **2002**, *586*, 345–348. [CrossRef]
17. Saurina, J.; Sentellas, S. Determination of Phenolic Compounds in Food Matrices: Applications to Characterization and Authentication. In *Fast Liquid Chromatography-Mass Spectrometry Methods in Food and Environmental Analysis*; Núñez, O., Gallart-Ayala, H., Martins, C.P.B., Lucci, P., Eds.; Imperial College Press: London, UK, 2015; pp. 517–547.
18. Carrasco-Pancorbo, A.; Cerretani, L.; Bendini, A.; Segura-Carretero, A.; Gallina-Toschi, T.; Fernandez-Gutierrez, A. Analytical determination of polyphenols in olive oils. *J. Sep. Sci.* **2005**, *28*, 837–858. [CrossRef] [PubMed]
19. Segura-Carretero, A.; Carrasco-Pancorbo, A.; Bendini, A.; Cerretani, L.; Fernandez-Gutierrez, A. Analytical determination of polyphenols in olive oil. *Olives Olive Oil Health Dis. Prev.* **2010**, 509–523.
20. Gosetti, F.; Bolfi, B.; Manfredi, M.; Calabrese, G.; Marengo, E. Determination of eight polyphenols and pantothenic acid in extra-virgin olive oil samples by a simple, fast, high-throughput and sensitive ultrahigh performance liquid chromatography with tandem mass spectrometry method. *J. Sep. Sci.* **2015**, *38*, 3130–3136. [CrossRef] [PubMed]
21. Purcaro, G.; Codony, R.; Pizzale, L.; Mariani, C.; Conte, L. Evaluation of total hydroxytyrosol and tyrosol in extra virgin olive oils. *Eur. J. Lipid Sci. Technol.* **2014**, *116*, 805–811. [CrossRef]
22. Gouvinhas, I.; Machado, J.; Gomes, S.; Lopes, J.; Martins-Lopes, P.; Barros, A.I.R.N. Phenolic composition and antioxidant activity of monovarietal and commercial Portuguese olive oils. *J. Am. Oil Chem. Soc.* **2014**, *91*, 1197–1203. [CrossRef]
23. De la Torre-Robles, A.; Rivas, A.; Lorenzo-Tovar, M.L.; Monteagudo, C.; Mariscal-Arcas, M.; Olea-Serrano, F. Estimation of the intake of phenol compounds from virgin olive oil of a population from southern Spain. *Food Addit. Contam. A* **2014**, *31*, 1460–1469. [CrossRef] [PubMed]
24. Capriotti, A.L.; Cavaliere, C.; Crescenzi, C.; Foglia, P.; Nescatelli, R.; Samperi, R.; Lagana, A. Comparison of extraction methods for the identification and quantification of polyphenols in virgin olive oil by ultra-HPLC-QTOF mass spectrometry. *Food Chem.* **2014**, *158*, 392–400. [CrossRef] [PubMed]

25. Longobardi, F.; Ventrella, A.; Casiello, G.; Sacco, D.; Tasioula-Margari, M.; Kiritsakis, A.K.; Kontominas, M.G. Characterisation of the geographical origin of Western Greek virgin olive oils based on instrumental and multivariate statistical analysis. *Food Chem.* **2012**, *133*, 169–175. [CrossRef]
26. Bayram, B.; Esatbeyoglu, T.; Schulze, N.; Ozcelik, B.; Frank, J.; Rimbach, G. Comprehensive Analysis of Polyphenols in 55 Extra Virgin Olive Oils by HPLC-ECD and Their Correlation with Antioxidant Activities. *Plant Foods Hum. Nutr.* **2012**, *67*, 326–336. [CrossRef] [PubMed]
27. Dais, P.; Boskou, P. Detection and quantification of phenolic compounds in olive oil, olives, and biological fluids. *Olive Oil* **2009**, 55–107.
28. Garcia-Villalba, R.; Carrasco-Pancorbo, A.; Zurek, G.; Behrens, M.; Baessmann, C.; Segura-Carretero, A.; Fernandez-Gutierrez, A. Nano and rapid resolution liquid chromatography-electrospray ionization-time of flight mass spectrometry to identify and quantify phenolic compounds in olive oil. *J. Sep. Sci.* **2010**, *33*, 2069–2078. [CrossRef] [PubMed]
29. Bonoli, M.; Montanucci, M.; Toschi, T.G.; Lercker, G. Fast separation and determination of tyrosol, hydroxytyrosol and other phenolic compounds in extra-virgin olive oil by capillary zone electrophoresis with ultraviolet-diode array detection. *J. Chromatogr. A* **2003**, *1011*, 163–172. [CrossRef]
30. Bonoli, M.; Toschi, T.G.; Lercker, G. Analysis and quantification of phenolic compounds in vegetable extracts by high-performance capillary electrophoresis (HPCE). *Prog. Nutr.* **2005**, *7*, 31–35.
31. Beltran, G.; Jimenez, A.; Aguilera, M.P.; Uceda, M. HPLC analysis of the phenolic fraction of Arbequina virgin olive oil. Relations to the bitterness index K225 and oil stability. *Grasas Aceites* **2000**, *51*, 320–324. [CrossRef]
32. Marini, F.; D'Aloise, A.; Bucci, R.; Buiarelli, F.; Magri, A.L.; Magri, A.D. Fast analysis of 4 phenolic acids in olive oil by HPLC-DAD and chemometrics. *Chemom. Intell. Lab. Syst.* **2011**, *106*, 142–149. [CrossRef]
33. Lozano-Sanchez, J.; Segura-Carretero, A.; Menendez, J.A.; Oliveras-Ferraros, C.; Cerretani, L.; Gutierrez, A. Prediction of Extra Virgin Olive Oil Varieties through Their Phenolic Profile. Potential Cytotoxic Activity against Human Breast Cancer Cells. *J. Agric. Food Chem.* **2010**, *58*, 9942–9955. [CrossRef] [PubMed]
34. Alarcon Flores, M.I.; Romero-Gonzalez, R.; Garrido Frenich, A.; Martinez Vidal, J.L. Analysis of phenolic compounds in olive oil by solid-phase extraction and ultrahigh performance liquid chromatography-tandem mass spectrometry. *Food Chem.* **2012**, *134*, 2465–2472. [CrossRef] [PubMed]
35. Gutierrez-Rosales, F.; Rios, J.J.; Gomez-Rey, M.L. Main polyphenols in the bitter taste of virgin olive oil. Structural confirmation by on-line high-performance liquid chromatography electrospray ionization mass spectrometry. *J. Agric. Food Chem.* **2003**, *51*, 6021–6025. [CrossRef] [PubMed]
36. Garcia-Villalba, R.; Carrasco-Pancorbo, A.; Nevedomskaya, E.; Mayboroda, O.A.; Deelder, A.M.; Segura-Carretero, A.; Fernandez-Gutierrez, A. Exploratory analysis of human urine by LC-ESI-TOF MS after high intake of olive oil: Understanding the metabolism of polyphenols. *Anal. Bioanal. Chem.* **2010**, *398*, 463–475. [CrossRef] [PubMed]
37. Eigenvector Research Incorporated. Powerful Resources for Intelligent Data Analysis. Available online: http://www.eigenvector.com/software/solo.htm (accessed on 15 October 2016).
38. Massart, D.L.; Vandeginste, B.G.M.; Buydens, L.M.C.; de Jong, S.; Lewi, P.J.; Smeyers-Verbeke, J. *Handbook of Chemometrics and Qualimetrics*; Elsevier: Amsterdam, The Netherlands, 1997.
39. Pérez-Rafols, C.; Saurina, J. Liquid chromatographic fingerprints and profiles of polyphenolic compounds applied to the chemometric characterization and classification of beers. *Anal. Methods* **2015**, *7*, 8733–8739. [CrossRef]

3

Determination of Posaconazole in Plasma/Serum by High-Performance Liquid Chromatography with Fluorescence Detection

Peter H. Tang

Departments of Pediatrics, Pathology and Laboratory Medicine, Cincinnati Children's Hospital Medical Center, University of Cincinnati College of Medicine, 3333 Burnet Avenue, Cincinnati, OH 45229-3039, USA; peter.tang@cchmc.org

Academic Editor: Fabio Gosetti

Abstract: A sensitive high-performance liquid chromatographic (HPLC) method is described for the determination of posaconazole in human plasma/serum. The method is based on a single dilution step by treating the sample with methanol, and followed by the direct injection of the sample into the HPLC system. Posaconazole and internal standard ketoconazole in the methanol extract are subsequently analyzed by using a fluorescence (FL) detector at optimized wavelengths (excitation 245 nm and emission 380 nm). The method achieves a linear detector response for peak height measurements over the concentration range of 0.1–10 µg/mL which adequately covers the therapeutic range for appropriate patient monitoring. The chromatographic time is less than 8 min per injection, an improvement over most published HPLC/FL or HPLC/UV methods. The method's limit of quantitation, linearity, imprecision, and accuracy met all criteria required by the Guidance for Industry Bioanalytical Method Validation. In comparison to other published methods, the current method would be of interest to analytical and clinical laboratories because it employs simple, rapid, and cost-effective procedures.

Keywords: posaconazole; ketoconazole; HPLC; fluorescence

---

1. Introduction

Posaconazole is a triazole antifungal drug with enhanced activity against major microorganisms, including the *Candida*, *Aspergillus*, and *Zygomycetes* fungal groups [1–4]. It is approved for antifungal prophylaxis of invasive *Candida* and *Aspergillus* infections in patients who are immunocompromised, such as those with hematologic malignancies with prolonged neutropenia from chemotherapy, or hematopoietic stem cell transplantation recipients with graft versus host disease. Posaconazole is believed to inhibit lanosterol 14α-demethylase, an enzyme responsible for the conversion of lanosterol to ergosterol, which is an essential component of the fungal cell membrane [5]. This inhibition results in depleting ergosterol and accumulating toxic methylated sterol precursors within the fungal cell membrane, and thus weakening the structure and function of the cell membrane and leading to fungal cell death.

For treating patients in various conditions, posaconazole is formulated into three forms including an oral suspension, a delayed-release tablet and an intravenous formulation which can be readily dictated by the clinicians. The pharmacokinetics and pharmacodynamics of each formulation have been outlined extensively [6–11]. The oral formulations have exhibited significant inter-patient and intra-patient variations in bioavailability due to inconsistent absorption, metabolism, elimination, or drug-drug interactions. The intravenous formulation is suitable for patients who cannot tolerate oral formulation, however, its solubilizing excipient may cause additional renal toxicity for patients with renal impairment [12]. Posaconazole could accumulate in tissues due to its hydrophobic nature; as such, its disposition could perhaps augment its efficacy. To successfully manage invasive fungal infections, the clinician strives for antifungal therapeutic drug monitoring. Therapeutic drug monitoring of a patient's serum or plasma has been a useful tool to detect drug exposures outside of the therapeutic window and to help individualize a dosage regimen and treatment, increasing therapeutic efficacy, and decreasing side effects [13–35]. Reported methods for the analysis of posaconazole are mostly high-performance liquid chromatography (HPLC) with mass spectrometry [13–25], ultraviolet [26–31] and fluorescence [32–35] detections. Mass spectrometry (MS) and tandem mass spectrometry (MS/MS) are superior in sensitivity and specificity to ultraviolet (UV) and fluorescence (FL) detections. However, the purchase and operating costs of LC-MS and LC-MS/MS are high. Therefore, the need for the therapeutic drug monitoring of posaconazole in pediatric samples with a simple, rapid, and cost-effective procedure prompted the development of the current method.

2. Materials and Methods

2.1. Chemicals and Reagents

Posaconazole, ketoconazole, trifluoroacetic acid (TFA) and ammonium acetate were obtained from Sigma-Aldrich (St. Louis, MO, USA). HPLC-grade organic solvents were purchased from Fisher Scientific (Fair Lawn, NJ, USA) and used without further purification. Lyphochek® Therapeutic Drug Monitoring Control was from Bio-Rad (Hercules, CA, USA).

2.2. Instrumentation and Chromatographic Conditions

The Hitachi Chromaster™ HPLC system was equipped with a Model 5110 quaternary pump, Model 5210 autosampler, Model 5310 column oven, and Model 5440 FL detector. The output signal was monitored and processed using the EZChrom Elite® software. The analytical column was a 250-mm × 4.6-mm ODS HYPERSIL column (Thermo Scientific) with 5-µm spherical particles equipped with a Phenomenex Security Guard (C18 cartridge, 4 × 3-mm). Details of chromatographic and detection conditions are described in Table 1. The mobile phase was filtered through a 0.2-µm MAGNA nylon filter (GE Water & Process Technologies, Minnetonka, MN, USA) under reduced pressure.

Table 1. HPLC conditions.

Parameters	Conditions
Column	ODS HYPERSIL, 5-µm, 250 × 4.6 mm
Mobile phase	Ammonium acetate (0.1 M): water:acetonitrile:TFA (409/590/1, *v*/*v*/*v*)
Mode	Isocratic elution
Oven	45 °C
Autosampler	5 °C
Flow rate	1.1 mL/min
Injection	10 µL
Detector settings	245 nm (Ex) and 380 nm (Em)
Run time	7 min

2.3. Preparation of Calibrator and Quality Control Samples

Individual stock solutions of posaconazole (100 µg/mL) and ketoconazole (100 µg/mL) were prepared in methanol. The stock solution of posaconazole was diluted with analyte-free pooled human serum (blank sample) to prepare calibrator samples over the range of 0.1–10 µg/mL. Quality control (QC) samples were similarly prepared containing 0.3, 2, 4, and 8 µg/mL. Portions of calibrator and QC samples were stored in 1.5-mL polypropylene screw-top tubes. Internal standard (IS) ketoconazole solution (2 µg/mL) was prepared in methanol from the stock solution (100 µg/mL). All solutions and samples were stored at −20 °C until required.

2.4. Sample Preparation Procedure

A single dilution step was employed for sample preparation in order to avoid the time-consuming extraction procedure such as solid-phase extraction, filtration, evaporation and reconstitution. Because organic solvent could extract posaconazole from the sample matrix and precipitate proteins, several organic solvents including acetonitrile, methanol, and mixtures of acetonitrile-methanol (3/1, 2/2, and 1/3, *v*/*v*) were examined. After protein precipitation and centrifugation, a clear supernatant was obtained without further manipulation.

A patient's serum, calibrator, or QC sample (100 μL) was pipetted to a microcentrifuge tube, followed by internal standard solution (100 μL) and methanol or other organic solvents (300 μL). After vortex-mixing for 1 min, the mixture was centrifuged at 4 °C for 10 min at 10,000 rpm. After protein precipitation, the supernatant was transferred to an autosampler vial for analysis. A 10-μL aliquot of supernatant was injected onto the HPLC system.

2.5. Method Validation

2.5.1. Selectivity

To explore possible sources of interference, several lyophilized products from human blood containing endogenous components, metabolites, decomposition products, concomitant medication and other exogenous xenobiotics (included in the Lyphochek® Therapeutic Drug Monitoring Control) were examined. These substances included acetaminophen, amikacin, amiodarone, amitriptyline, caffeine, carbamazepine, chloramphenicol, cortisol, cyclosporine, desipramine, digoxin, disopyramide, estriol, ethosuximide, flecainide, gentamicin, haloperidol, imipramine, lidocaine, lithium, methotrexate, *N*-acetylprocainamide, netilmicin, nortriptyline, phenobarbital, phenytoin, primidone, procainamide, propranolol, quinidine, salicylate, T3, T4, theophylline, tobramycin, tricyclic antidepressants, thyroid-stimulating hormone (TSH), valproic acid, and vancomycin. These lyophilized products were treated as patient samples and processed in an identical fashion, as described in the sample preparation procedure section. Analyses of patients' samples (n = 10) were also carried out, which were previously tested for felbamate, lacosamide, lamotrigine, methsuximide, micafungin, milrinone, oxcarbazepine, rufinamide, voriconazole, and zonisamide.

2.5.2. Calibration Curve

Peak heights of posaconazole and ketoconazole obtained in the chromatographic run were used to generate a serum-based calibration curve for posaconazole by using a least squares linear regression of the posaconazole-ketoconazole peak height ratio versus the posaconazole concentration. The calibration curve was used for calculating the concentration of posaconazole in the unknown samples. A calibration curve range of 0.1–10 μg/mL was used to cover the entire range of concentrations in the unknown samples. Plasma is closely related to serum, but the main difference is that it contains clotting factors that help in the clotting of blood. To detect matrix effects on the determination of posaconazole concentration, the plasma calibration curves were compared with the corresponding serum calibration curves.

2.5.3. Recovery

Recoveries of posaconazole were evaluated with analyte-free pooled human serum spiked with concentrations varying from 0.3 to 8 μg/mL. Ketoconazole was added at the beginning of the extraction process. In one series, the posaconazole standards were spiked before extraction, and in the other series, they were spiked only after the extraction. Samples were analyzed in triplicate. The recovery estimates were based on the comparison of the posaconazole-ketoconazole peak height ratios obtained in both series of samples. Recovery of ketoconazole was evaluated with analyte-free pooled human serum spiked with a concentration of 2 μg/mL before and after the extraction. The recovery estimates were based on the comparison of ketoconazole peak heights obtained in both series of samples.

2.5.4. Precision, Reproducibility, and Stability

Within-run precision was evaluated by the analysis of the QC samples (n = 5 at each concentration). To evaluate between-run reproducibility, the QC samples were analyzed on different days (n = 5 at each concentration). Stabilities of posaconazole and ketoconazole in QC samples were evaluated under a variety of storage and handling conditions: freeze-thaw cycles at room temperature (under chemical/biological hood); bench-top stability (experiments was conducted at room temperature under normal laboratory light for up to 24 h); long-term stability (freeze samples for 5 days at –20 °C and then thaw for 30 min at room temperature); the stability of stock solutions of posaconazole and ketoconazole were evaluated; and finally the stability of processed samples sitting on the bench-top for up to 24 h were tested.

2.5.5. Cross-Validation

Cross-validation was performed on 20 patients' samples which were separated into paired sets and stored at −20 °C until required. One set of samples was submitted to analysis by LC-MS/MS at the reference laboratory. The second set of patients' samples were analyzed using the current method. To confirm the reliability of this method, paired results from the two different methods were compared.

3. Results and Discussion

3.1. Internal Standard Alternative

An internal standard is commonly used when many sample preparation steps are required before the sample solution can be injected onto the analytical column. The addition of an internal standard to the sample is known to compensate for any losses during sample preparation and injection onto an analytical instrument. Previously published HPLC/FL methods for the determination of serum posaconazole levels have used itraconazole [32,35] and R051012 [34] as internal standards. While R051012 was not available, itraconazole was first tested in our initial trials. However, the analysis of posaconazole required a lengthy chromatography due to the late elution of itraconazole (>14 min). After scrutinizing reports for ketoconazole, it was found to exhibit native fluorescent properties at 345–440 nm with excitation at 240–320 nm [36,37]. As such, ketoconazole was selected and proved to be an excellent alternative which allowed for a shorter chromatography time (<8 min).

3.2. Selectivity

Figures 1 and 2 are the representative chromatograms obtained after extraction of a blank sample and a patient's sample, respectively. The overall HPLC analysis took 7 min; the retention time of posaconazole was ~6.4 min and that of the internal standard was ~5.6 min. The selectivity of the FL detection allowed for a good resolution of internal standard and posaconazole without ambiguity in the identification in the analyzed samples. Optimal chromatography conditions had produced satisfactory separations of posaconazole, internal standard, and other unknown peaks. Furthermore, no interference was observed from the analyte-free samples and the Lyphochek® Therapeutic Drug Monitoring Control samples that contained endogenous matrix components, concomitant medications, drug metabolites, decomposition products, and other xenobiotics.

3.3. Solvent Extraction in a Single Dilution Step

A single dilution step using organic solvent for simultaneous extraction and protein precipitation was developed in our clinical laboratory in 2001 [38], and it has been used for therapeutic drug monitoring since then. The technique has been proved to be the most rapid and simplest procedure for sample preparation, and is widely accepted in many clinical and analytical laboratories. Quantitative extraction of posaconazole from human plasma or serum has been reported previously by using acetonitrile [22,26,29,31], methanol [23,31,33] or mixtures of both solvents [19–21,25] as the extracting reagent in sample preparation. The use of perchloric acid as a precipitating agent had also been previously examined by Zhang et al. [31], and the addition of perchloric acid to methanol extraction proved to be advantageous. Since perchloric acid is extremely hazardous, additional efforts must be made to avoid toxic exposure. For safety purposes, its use has been banned in many clinical laboratories. To find a better alternative, various extraction solvents including acetonitrile, methanol, and mixtures of acetonitrile/methanol (3/1, 2/2, and 1/3, v/v) were evaluated in our preliminary study. No significant differences on the extraction efficiency were observed among these solvents. Methanol alone was selected because its extracts produced the best chromatographic behaviors of both analytes. Methanol extraction produced mean absolute recoveries of 92.4% $\pm$ 1.3% and 91.6% $\pm$ 1.3% for posaconazole and ketoconazole, respectively.

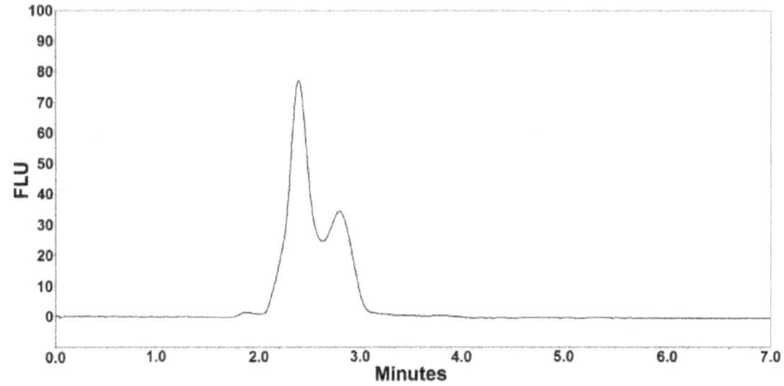

Figure 1. Blank chromatogram after extraction of analyte-free pooled human serum (blank sample).

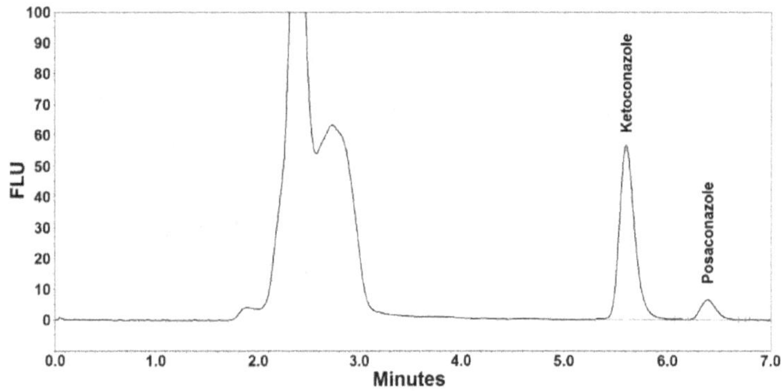

Figure 2. Representative chromatogram (ketoconazole at ~5.6 min, posaconazole at ~6.4 min) after extraction of a patient's serum. The posaconazole concentration was 0.98 μg/mL.

3.4. Calibration Curve

The peak height ratios of posaconazole to ketoconazole were plotted against posaconazole concentration using least squares linear regression. The serum calibration curve was linear over a concentration range of 0.1–10 μg/mL with a correlation coefficient of (r^2) >0.995. The slope was 7.876 $\pm$ 0.047 (mean $\pm$ SD) and the intercept was 0.022 $\pm$ 0.005 (mean $\pm$ SD). The plasma calibration curve had a least squares linear regression equation: $y = 7.885x + 0.034$, where $r^2 = 0.996$. Because the serum and plasma calibration curves were parallel, no matrix effects was detected. The linear regression of the serum calibration curve was selected and used to derive the posaconazole concentration in both types of samples. The limit of detection (LOD) was found to be ~0.04 μg/mL, which was determined at a signal-to-noise (S/N) ratio of 3:1, and the limit of quantification (LOQ) was determined to be 0.1 μg/mL. The validated linearity was set up to 10 μg/mL for monitoring patients under posaconazole therapy. Table 2 presents the coefficients of variation (CV), which averaged 19% and 3.9% for lower limit of quantification (LLOQ) and upper limit of quantification (ULOQ), respectively. The accuracy of

LLOQ and ULOQ samples ranged from 80% to 120% and from 96.5% to 104.4%, respectively. Based on the Guidance for Industry Bioanalytical Method Validation [39] published by the FDA, the following criteria were met in developing a calibration curve: accuracy within 80%–120% of the target value for the LLOQ and 15% deviation of standards other than LLOQ from the nominal concentration.

Table 2. Method validation.

Sample	Sample Concentration (μg/mL)	Observed Concentration (μg/mL)	Standard Deviation (μg/mL)	CV (%)
LLOQ (n = 10)	0.10	0.08–0.12	0.02	19.0
ULOQ (n = 10)	10.00	9.65–10.44	0.39	3.9
Patient #1 (n = 5)	1.23	1.15–1.30	0.06	4.9
Patient #2 (n = 5)	0.67	0.63–0.72	0.04	6.0
Patient #3 (n = 5)	0.96	0.92–1.01	0.05	5.2

3.5. Reproducibility

Reproducibility of the method was assessed by repeated measurements using the method, including three patient samples (Table 2) and three QCs (Table 3). The reproducibility data for three patient samples containing posaconazole concentrations ranged from 0.63 to 1.30 μg/mL are summarized in Table 2. All CVs for patient samples were <6%, indicating the reproducibility of the method. The experimental results of the within- and between-run precision tests for QC samples are summarized in Table 3. Within-run precision was between 2.9% and 5.2%, and between-run precision ranged between 4.0% and 5.8%. Overall accuracy was between 97.4% and 105.5%.

Table 3. Method precision and accuracy.

Run	Nominal Concentration (μg/mL)	Observed Concentration (μg/mL)	Recovery (%)	CV (%)
Within-run (n = 5)	0.300	0.303	101.1	4.5
	2.00	1.95	97.4	5.2
	4.00	4.16	104.1	3.6
	8.00	8.04	100.6	2.9
Between-run (n = 5)	0.300	0.293	97.7	5.3
	2.00	2.11	105.5	5.8
	4.00	4.09	102.3	4.1
	8.00	8.17	102.2	4.0

3.6. Stability

Based on the chemistry review on posaconazole conducted by the Center for Drug Evaluation and Research of the FDA, the drug is stable during a period of 24 months stored at 25 °C. Stability-indicating experiments have also been previously conducted by other authors [13–35]. Overall, posaconazole has good stability and degrades only under stress conditions. In the preliminary study, ketoconazole was found to be unstable in an acidic medium, but quite stable in neutral and alkaline media. Therefore, posaconazole and ketoconazole were prepared in a neutral medium and their stability was evaluated under the designed storage and handling conditions as described in the methods section. At −20 °C, no significant change in concentrations could be observed during 24 h for both compounds. Stored at room temperature for 24 h, the posaconazole concentration remained constant while the ketoconazole concentration decreased by 5%. However, ketoconazole was stable in prepared samples on the autosampler (at 4 °C) of the HPLC system for 24 h. Table 4 presents the results for stability over 5 days at −20 °C; no significant changes in the posaconazole and ketoconazole concentrations were observed during repeated freeze-thaw cycles. The experiment was carried out using QC samples that were stored in a freezer (at −20 °C) for 5 days. During this period, the freeze-and-thaw cycles were repeated three times. The method was proved relevant for the rugged analysis of posaconazole. Our data support the previous claims that posaconazole is stable in all storage and handling conditions between sample collection and analysis (Table 4).

Table 4. Stability data for posaconazole in QC samples.

Nominal (μg/mL)	Measured Concentration (μg/mL)		
Freeze-thaw stability (n = 3 at each concentration)			
	Cycle 1	Cycle 2	Cycle 3
0.300	0.296	0.302	0.304
2.00	2.16	1.98	2.02
4.00	4.05	4.12	3.88
8.00	7.85	8.07	7.96
Short-term stability at room temperature (n = 3 at each concentration)			
		24 h	
0.300		0.284	
2.00		2.15	
4.00		4.18	
8.00		8.29	
Short-term stability at refrigeration (2–8 °C, n = 3 at each concentration)			
	Day 2	Day 5	
0.300	0.288	0.292	
2.00	2.08	2.14	
4.00	4.26	4.09	
8.00	7.95	8.16	

3.7. Method Comparison

When compared with two validated HPLC/UV methods for posaconazole quantification in human serum or plasma samples [29,31], the current method provides a sensitive, cost-effective and time-saving procedure. Cendejas-Bueno et al. [29] used 150 μL of serum sample and 50 μL of injection volume to obtain an LOQ of 0.125 μg/mL (6.25 ng on column). Zhang et al. [31] reported a sensitive detection with an LLOQ of 0.05 μg/mL (1.5 ng on column) by using 100 μL of serum sample and 30 μL of injection volume. Based on their results, which showed that posaconazole's retention time was ~6.95 min, the current method could be modified for the determination of posaconazole only with a gradient elution of 8 min or less. However, the modified procedure still requires an extra 5 min to allow the initial eluent to equilibrate the column after completing a gradient elution, which is unnecessary for the current method using isocratic elution. The current method took the advantage of sensitive fluorescence detection to achieve an LOQ of 0.1 μg/mL (1 ng on column) with a smaller injection volume (10 μL) and 100 μL of serum sample. Most important, an injection volume of 10 μL methanol extract allowed the HPLC system to load ~1500 injections without replacing pre-filter columns. According to our experience, the replacement of pre-filter columns was frequently needed after injecting 300–500 samples with an injection volume of 30 μL or 50 μL.

Prior to our current procedure, four HPLC/FL methods have been reported for measuring posaconazole in biological samples. The first method was developed by Muller et al. [32], in which the excitation and emission wavelengths were set at 260 and 365 nm, respectively. Their method required a long chromatography run (>23 min) to elute the IS (itraconazole). Sample preparation was tedious and labor-intensive. Although the method was sensitive, it needed 250 μL of plasma sample and 50 μL of injection volume to achieve an LOQ of 0.05 μg/mL (2.5 ng on column). Neubauer et al. [33] used different detector settings at 261 and 357 nm for the excitation and emission, respectively. Although the method needed only 50 μL serum, a complicate column-switching technique was used to isolate posaconazole from the serum matrix. This method required a total of 16 min to complete the chromatographic run per injection. Another method was reported by Buckner et al. [34], in which the excitation and emission wavelengths were set at 260 and 350 nm, respectively. Again, the method employed tedious sample preparation procedures consisted of liquid-liquid extraction, solvent evaporation and reconstitution prior to analysis. Although a chromatographic run could be completed in 8 min, the R051012 (used as internal standard) was not readily available. The most recent assay was reported by Hens et al. [35], in which the detection was at 385 nm with excitation at 240 nm, and a gradient run of 19 min was performed for the analysis. Again, their sample preparation was tedious and labor-intensive; sample volume was 1000 μL. In comparison, the current method used a simpler sample preparation, a smaller sample volume (100 μL), and a shorter chromatographic run time (<8 min). The optimized fluorescence detection was at 380 nm with excitation at 245 nm. Figure 3 shows the fluorescence intensities of posaconazole and ketoconazole at different settings. The current method offers the most sensitive detection.

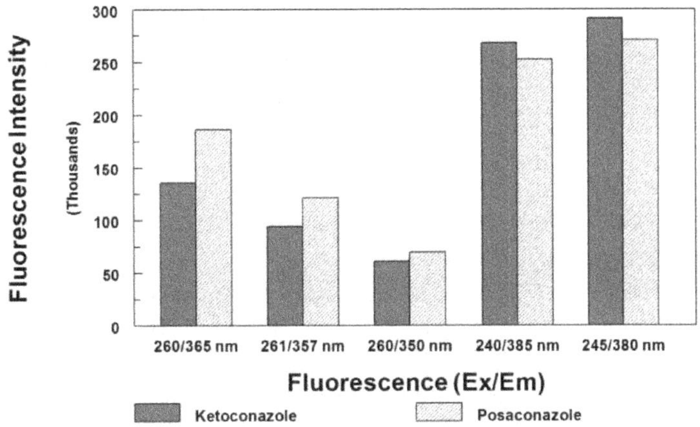

Figure 3. Comparison of fluorescence intensity for posaconazole (7.5 μg/mL) and ketoconazole (2 μg/mL) at different settings used in previously published methods and the current method (mean measured peak height, $n = 3$).

3.8. Cross-Validation

Cross-validation of the current method was examined by measuring the concentrations of posaconazole ranged from 0.1 to 1.4 μg/mL in 20 de-identified patient samples. The results obtained by this method were compared well with results from the reference method (Figure 4). The correlation between the two methods was good; the linear regression statistics indicated an r^2 value of 0.982 ($p < 0.0001$). The linear regression equation for the correlation was $y = 0.986x - 0.017$ with a standard error value of 0.027; where y denotes the current method and x denotes the reference method.

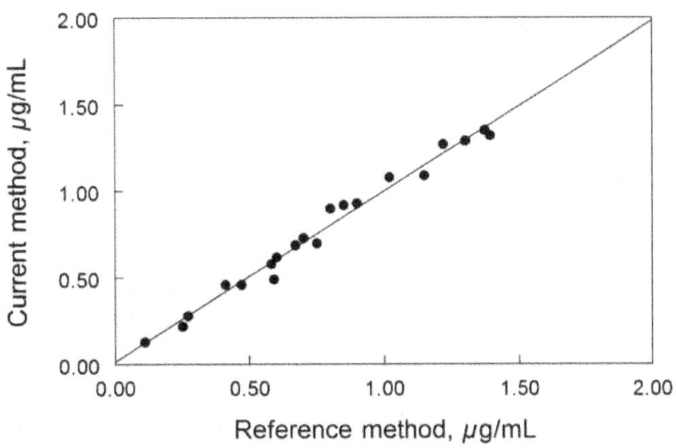

Figure 4. Cross-validation of posaconazole method ($n = 20$).

4. Conclusions

A simple, rapid, and accurate method for the measurement of posaconazole in human plasma/serum has been developed. The current method uses an efficient procedure consisting of a single dilution step, protein precipitation and centrifugation which avoids the use of more complex liquid-liquid extraction or solid-phase extraction procedures and substantially decreases the sample preparation time. The method meets the requirements of high sample throughput in clinical practice. The method has potential cost savings lying in the use of a single dilution step process and a quick chromatographic run to be completed within 8 min. This method can also be readily adopted by clinical and analytical laboratories to support the monitoring of posaconazole concentrations in plasma/serum, to optimize the efficacy of posaconazole therapy and minimize adverse effects.

Acknowledgments: The author wishes to thank the Department of Pathology and Laboratory Medicine, Cincinnati Children's Hospital Medical Center for the internal funding for the work performed in this study.

Conflicts of Interest: The author declares no conflict of interest.

References

1. Schiller, D.; Fung, H. Posaconazole: An extended-spectrum triazole antifungal agent. *Clin. Ther.* **2007**, *29*, 1862–1886. [CrossRef] [PubMed]
2. Rachwalski, E.; Wieczorkiewicz, J.; Scheetz, M. Posaconazole: An oral triazole with an extended spectrum of activity. *Ann. Pharmacother.* **2008**, *42*, 1429–1438. [CrossRef] [PubMed]
3. Li, X.; Brown, N.; Chau, A.; López-Ribot, J.; Ruesga, M.; Quindos, G.; Mendrick, C.; Hare, R.; Loebenberg, D.; DiDomenico, B.; et al. Changes in susceptibility to posaconazole in clinical isolates of Candida albicans. *J. Antimicrob. Chemother.* **2004**, *53*, 74–80. [CrossRef] [PubMed]
4. Walsh, T.; Raad, I.; Patterson, T.; Chandrasekar, P.; Donowitz, G.; Graybill, R.; Greene, R.; Hachem, R.; Hadley, S.; Herbrecht, R.; et al. Treatment of Invasive Aspergillosis with Posaconazole in Patients Who Are Refractory to or Intolerant of Conventional Therapy: An Externally Controlled Trial. *Clin. Infect. Dis.* **2007**, *44*, 2–12. [CrossRef] [PubMed]
5. Torres, H.; Hachem, R.; Chemaly, R.; Kontoyiannis, D.; Raad, I. Posaconazole: A broad-spectrum triazole antifungal. *Lancet Infect. Dis.* **2005**, *5*, 775–785. [CrossRef]
6. Lipp, H. Clinical pharmacodynamics and pharmacokinetics of the antifungal extended-spectrum triazole posaconazole: An overview. *Br. J. Clin. Pharmacol.* **2010**, *70*, 471–480. [CrossRef] [PubMed]
7. Li, Y.; Theuretzbacher, U.; Clancy, C.; Nguyen, M.; Derendorf, H. Pharmacokinetic/pharmacodynamic profile of posaconazole. *Clin. Pharmacokinet.* **2010**, *49*, 379–396. [CrossRef] [PubMed]
8. Kraft, W.; Chang, P.; van Iersel, M.; Waskin, H.; Krishna, G.; Kersemaekers, W. Posaconazole tablet pharmacokinetics: Lack of effect of concomitant medications altering gastric pH and gastric motility in healthy subjects. *Antimicrob. Agents Chemother.* **2014**, *58*, 4020–4025. [CrossRef] [PubMed]
9. Maertens, J.; Cornely, O.; Ullmann, A.; Heinz, W.; Krishna, G.; Patino, H.; Caceres, M.; Kartsonis, N.; Waskin, H.; Robertson, M. Phase 1B study of the pharmacokinetics and safety of posaconazole intravenous solution in patients at risk for invasive fungal disease. *Antimicrob. Agents Chemother.* **2014**, *58*, 3610–3617. [CrossRef] [PubMed]
10. McKeage, K. Posaconazole: A review of the gastro-resistant tablet and intravenous solution in invasive fungal infections. *Drugs* **2015**, *75*, 397–406. [CrossRef] [PubMed]
11. Kersemaekers, W.; van Iersel, T.; Nassander, U.; O'Mara, E.; Waskin, H.; Caceres, M.; van Iersel, M. Pharmacokinetics and safety study of posaconazole intravenous solution administered peripherally to healthy subjects. *Antimicrob. Agents Chemother.* **2015**, *59*, 1246–1251. [CrossRef] [PubMed]
12. Luke, D.; Tomaszewski, K.; Damle, B.; Schlamm, H. Review of the basic and clinical pharmacology of sulfobutylether-beta-cyclodextrin (SBECD). *J. Pharm. Sci.* **2010**, *99*, 3291–3301. [CrossRef] [PubMed]
13. Dolton, M.; Brüggemann, R.; Burger, D.; McLachlan, A. Understanding variability in posaconazole exposure using an integrated population pharmacokinetic analysis. *Antimicrob. Agents Chemother.* **2014**, *58*, 6879–6885. [CrossRef] [PubMed]

14. Dolton, M.; Ray, J.; Marriott, D.; McLachlan, A. Posaconazole exposure-response relationship: Evaluating the utility of therapeutic drug monitoring. *Antimicrob. Agents Chemother.* **2012**, *56*, 2806–2813. [CrossRef] [PubMed]
15. Dolton, M.; Ray, J.; Chen, S.; Ng, K.; Pont, L.; McLachlan, A. Multicenter study of posaconazole therapeutic drug monitoring: Exposure-response relationship and factors affecting concentration. *Antimicrob. Agents Chemother.* **2012**, *56*, 5503–5510. [CrossRef] [PubMed]
16. Seyedmousavi, S.; Mouton, J.; Verweij, P.; Brüggemann, R. Therapeutic drug monitoring of voriconazole and posaconazole for invasive aspergillosis. *Expert Rev. Anti-Infect. Ther.* **2013**, *11*, 931–941. [CrossRef] [PubMed]
17. Ashbee, H.; Barnes, R.; Johnson, E.; Richardson, M.; Gorton, R.; Hope, W. Therapeutic drug monitoring (TDM) of antifungal agents: Guidelines from the British Society for Medical Mycology. *J. Antimicrob. Chemother.* **2013**, *69*, 1162–1176. [CrossRef] [PubMed]
18. Cunliffe, J.; Noren, C.; Hayes, R.; Clement, R.; Shen, J. A high-throughput LC-MS/MS method for the quantitation of posaconazole in human plasma: Implementing fused core silica liquid chromatography. *J. Pharm. Biomed. Anal.* **2009**, *50*, 46–52. [CrossRef] [PubMed]
19. Vogeser, M.; Rieger, C.; Ostermann, H.; Spöhrer, U. A routine method for the quantitation of the novel antimycotic drug posaconazole in plasma using liquid chromatography-tandem mass spectrometry. *Clin. Chem. Lab. Med.* **2009**, *47*, 579–584. [CrossRef] [PubMed]
20. Rochat, B.; Pascual, A.; Pesse, B.; Lamoth, F.; Sanglard, D.; Decosterd, L.; Bille, J.; Marchetti, O. Ultra-performance liquid chromatography mass spectrometry and sensitive bioassay methods for quantitation of posaconazole plasma concentrations after oral dosing. *Antimicrob. Agents Chemother.* **2010**, *54*, 5074–5081. [CrossRef] [PubMed]
21. Alffenaar, J.; Wessels, A.; van Hateren, K.; Greijdanus, B.; Kosterink, J.; Uges, D. Method for therapeutic drug monitoring of azol antifungal drugs in human serum using LC/MS/MS. *J. Chromatogr. B Anal. Technol. Biomed. Life Sci.* **2010**, *878*, 39–44. [CrossRef] [PubMed]
22. Baietto, L.; D'Avolio, A.; Ventimiglia, G.; De Rosa, F.; Siccardi, M.; Simiele, M.; Sciandra, M.; Di Perri, G. Development, validation, and routine application of a high-performance liquid chromatography method coupled with a single mass detector for quantification of itraconazole, voriconazole, and posaconazole in human plasma. *Antimicrob. Agents Chemother.* **2010**, *54*, 3408–3413. [CrossRef] [PubMed]
23. Beste, K.; Burkhardt, O.; Kaever, V. Rapid HPLC-MS/MS method for simultaneous quantitation of four routinely administered triazole antifungals in human plasma. *Clin. Chim. Acta* **2012**, *413*, 240–245. [CrossRef] [PubMed]
24. Baietto, L.; D'Avolio, A.; Marra, C.; Simiele, M.; Cusato, J.; Pace, S.; Ariaudo, A.; De Rosa, F.; Di Perri, G. Development and validation of a new method to simultaneously quantify triazoles in plasma spotted on dry sample spot devices and analysed by HPLC-MS. *J. Antimicrob. Chemother.* **2012**, *67*, 2645–2649. [CrossRef] [PubMed]
25. Chae, H.; Cho, S.; Yu, H.; Cha, K.; Lee, S.; Kim, M.; Kim, Y.; Kim, Y.J.; Kim, H.J.; Lee, D.G. Determination of posaconazole concentration with LC-MS/MS in adult patients with hematologic malignancy. *Clin. Chim. Acta* **2015**, *450*, 220–226. [CrossRef] [PubMed]
26. Chhun, S.; Rey, E.; Tran, A.; Lortholary, O.; Pons, G.; Jullien, V. Simultaneous quantification of voriconazole and posaconazole in human plasma by high-performance liquid chromatography with ultra-violet detection. *J. Chromatogr. B Anal. Technol. Biomed. Life Sci.* **2007**, *852*, 223–228. [CrossRef] [PubMed]
27. Storzinger, D.; Swoboda, S.; Lichtenstern, C.; Müller, C.; Weigand, M.; Hoppe-Tichy, T. Development and validation of a high-performance liquid chromatography assay for posaconazole utilizing solid-phase extraction. *Clin. Chem. Lab. Med.* **2008**, *46*, 1747–1751. [CrossRef] [PubMed]
28. Kahle, K.; Langmann, P.; Schirmer, D.; Lenker, U.; Keller, D.; Helle, A.; Klinker, H.; Heinz, W. Simultaneous determination of voriconazole and posaconazole concentrations in human plasma by high-performance liquid chromatography. *Antimicrob. Agents Chemother.* **2009**, *53*, 3140–3142. [CrossRef] [PubMed]
29. Cendejas-Bueno, E.; Forastiero, A.; Rodriguez-Tudela, J.; Cuenca-Estrella, M.; Gomez-Lopez, A. HPLC/UV or bioassay: Two valid methods for posaconazole quantification in human serum samples. *Clin. Microbiol. Infect.* **2012**, *18*, 1229–1235. [CrossRef] [PubMed]

30. Wissen, C.; Burger, D.; Verweij, P.; Aarnoutse, R.; Brüggemann, R. Simultaneous determination of the azoles voriconazole, posaconazole, isavuconazole, itraconazole and its metabolite hydroxyl-itraconazole in human plasma by reversed phase ultra-performance liquid chromatography with ultraviolet detection. *J. Chromatogr. B Anal. Technol. Biomed. Life Sci.* **2012**, *887–888*, 79–84. [CrossRef] [PubMed]
31. Zhang, M.; Moore, G.; Barclay, M.; Begga, E. A simple high-performance liquid chromatography method for simultaneous determination of three triazole antifungals in human plasma. *Antimicrob. Agents Chemother.* **2013**, *57*, 484–489. [CrossRef] [PubMed]
32. Muller, C.; Arndt, M.; Queckenberg, C.; Cornely, O.; Theisohn, M. HPLC analysis of the antifungal agent posaconazole in patients with haematological diseases. *Mycoses* **2006**, *49* (Suppl. 1), 17–22. [CrossRef] [PubMed]
33. Neubauer, W.; Konig, A.; Bolek, R.; Trittler, R.; Engelhardt, M.; Jung, M.; Kümmerer, K. Determination of the antifungal agent posaconazole in human serum by HPLC with parallel column-switching technique. *J. Chromatogr. B Anal. Technol. Biomed. Life Sci.* **2009**, *877*, 2493–2498. [CrossRef] [PubMed]
34. Buckner, S.; Ceesay, M.; Pagliuca, A.; Morgan, P.; Flanagan, R. Measurement of posaconazole, itraconazole, and hydroxyitraconazole in plasma/serum by high-performance liquid chromatography with fluorescence detection. *Ther. Drug Monit.* **2011**, *33*, 735–741. [CrossRef] [PubMed]
35. Hens, B.; Corsetti, M.; Brouwers, J.; Augustijns, P. Gastrointestinal and systemic monitoring of posaconazole in humans after fasted and fed state administration of a solid dispersion. *J. Pharm. Sci.* **2016**, *105*, 2904–2912. [CrossRef] [PubMed]
36. Remmel, R.; Amoh, K.; Abdel-Monem, M. The disposition and pharmacokinetics of ketoconazole in the rat. *Drug Metab. Dispos.* **1987**, *15*, 735–739. [PubMed]
37. Khashaba, P.; El-Shabouri, S.; Emara, K.; Mohamed, A. Analysis of some antifungal drugs by spectrophotometric and spectrofluorimetric methods in different pharmaceutical dosage forms. *J. Pharm. Biomed. Anal.* **2000**, *22*, 363–376. [CrossRef]
38. Tang, P.; Miles, M.; DeGrauw, A.; Hershey, A.; Pesce, A. HPLC analysis of reduced and oxidized coenzyme Q10 in human plasma. *Clin. Chem.* **2001**, *47*, 256–265. [PubMed]
39. U.S. Department of Health and Human Services; Food and Drug Administration; Center for Drug Evaluation and Research. Guidance for Industry: Bioanalytical Method Validation. 2001.

4

Determination of Carotenoids in Human Serum and Breast Milk Using High Performance Liquid Chromatography Coupled with a Diode Array Detector (HPLC-DAD)

Jing Tan [1,*], Jason Gek Leong Neo [2], Tania Setiawati [3] and Chunyan Zhang [1]

1 Abbott Nutrition Research and Development, 20 Biopolis Way #09-01/02, Centros Building, Singapore 138668, Singapore; chunyan.zhang@abbott.com

2 Department of Chemistry, Faculty of Science, National University of Singapore, Singapore 117543, Singapore; jasonneo20@hotmail.com

3 Division of Chemistry and Biological Chemistry, School of Physical and Mathematical Sciences, Nanyang Technological University, Singapore 637371, Singapore; tania.setiawati@gmail.com

* Correspondence: jtan@abbott.com

Academic Editor: Michael J. Dunphy

Abstract: High performance liquid chromatography (HPLC) coupled with a diode array detector (HPLC-DAD) for the identification and quantification of carotenoids, namely all-*trans* lutein, zeaxanthin, β-cryptoxanthin, α-carotene, and β-carotene, in biological samples such as human serum and breast milk, has been developed and validated. Good chromatography separation was achieved using a binary mobile phase system on a YMC C_{30} column (150 × 2.1 mm, 3 μm) at 30 °C. Owing to the smaller column particle size and diameter of the column, the separation was achieved in 18 min, which is significantly reduced from the typical 30–40 min of other methods. The diode array detector (DAD) acquisition was set at a wavelength of 445 nm; 3D spectra ranging from wavelengths of 240–600 nm were also recorded. Peaks were identified by matching their retention time and spectra with those of standards. Quantification was achieved by internal standard calibration using echinenone as the internal standard. Good linearity was obtained for each compound ($R^2 > 0.9999$). The method quantification limits (MQLs) for serum and breast milk were 10 ng/mL and 5 ng/mL, in matrix, respectively. A spike recovery study and standard reference material (SRM) from the National Institute of Standards and Technology (NIST) 968e analysis has proven that the method has a high degree of accuracy, precision, and robustness. The stability study showed that the carotenoid standard and sample extracts could be stored in a chilled autosampler at 8 °C up to 48 h without being comprised, which provides guidance on re-test time frames. The freeze/thaw process was found to be detrimental to carotenoids, and should always be avoided. Most importantly, UV standardization of the stock standard is to be performed prior to each assay, and simply taking the values on Certificate of Analysis (CoA) for calculation of the standard concentration is not recommended.

Keywords: carotenoids; high performance liquid chromatography; human breast milk; human serum; internal standard

---

1. Introduction

Carotenoids are fat-soluble micronutrients that play an important role in human health. Studies have shown that carotenoids can provide a multitude of health benefits, including provitamin A activities, a role as antioxidants, preventing age-related macular degeneration (AMD), and reducing the risk of cancer [1]. In addition, evidence has associated the presence of carotenoids and the cognitive development of fetuses and infants [2–5]. In order to support clinical and pre-clinical studies, a bioanalysis method has been developed and validated for the accurate determination of key carotenoids and isomers in human breast milk and serum. The target analytes are all-*trans* lutein, zeaxanthin, β-cryptoxanthin, α-carotene, and β-carotene.

High performance liquid chromatography (HPLC), using C_{18} or C_{30} columns, is widely adopted for analyzing carotenoids in biological samples. The problems with this approach are long run times, insufficient resolution of peaks, or tedious sample-preparation steps [6–11]. Recently, a liquid chromatography-tandem mass spectrometry (LC-MS/MS)-based bioanalysis method has been developed [12,13]. However, due to the large sample sizes from clinical or pre-studies, where great sample variation is expected, isotopic-labeled standards are desired for the MS/MS method to minimize potential matrix effects. To the best of our knowledge, not all isotopic standards for the target carotenoids are available, and, thus, an HPLC-based method was finally chosen due to its high reliability.

It is well known that carotenoids are susceptible to degradation [14–17], which poses great challenges for their accurate determination. Therefore, we have conducted stability studies that assessed factors affecting the quantitation of carotenoids, such as freeze/thaw cycles, storage conditions, and stability of the reference standard solution and sample extracts. The findings complement current knowledge and provide guidelines on the proper handling of carotenoids samples and standards.

2. Results and Discussion

2.1. Optimzation of Sample Preparation for Breast Milk Samples

Sample extraction of breast milk was carefully optimized due to the complexity of the sample matrices. Water was added to the samples to adjust the aqueous/organic ratio to prevent the formation of an emulsion [18]. Five percent potassium hydroxide was added for saponification to remove significant matrix interference from the high level of lipids in the samples [18–20]. To minimize the loss of lutein and zeaxanthin, we chose a low-concentration alkali saponification and shortened the saponification duration. Tetrahydrofuran (THF) was added as a modifier to enhance the solubility and improve the resolution [18,20–22]. A small amount of methyl *tert*-butyl ether (MTBE) was added to the final diluent (MTBE:Ethanol with butylated hydroxytoluene (BHT), 2:8, *v*/*v*) to improve the solubility of carotenes. Overall, the saponification and modifier step aimed at improving the method's performance.

2.2. HPLC Separation and Specificity

By using a column with a smaller particle size and column diameter, YMC C_{30}, 150 × 2.1 mm, 3 μm (YMC, Kyoto, Japan), the run time per sample was significantly reduced, from the typical 30–40 min to 18 min. The signal was recorded at a wavelength of 445 nm, which provided better specificity compared to low-range UV wavelengths. Retention time and spectra were compared with those of standards to confirm the identity of the compounds. The spectra found in sample were also matched with those in standards using the library matching function in Chemstation. Typical chromatograms are shown in Figure 1. An example of spectrum-matching of all-*trans*-lutein in standards and samples is demonstrated in Figure 2, where a specific peak is successfully identified as all-*trans*-lutein with a high matching score of 993.402. The peak purities were checked in Chemstation. The purity factors for the five carotenoids in samples were all above 950, which are considered acceptable for such a bioanalysis with high complexity.

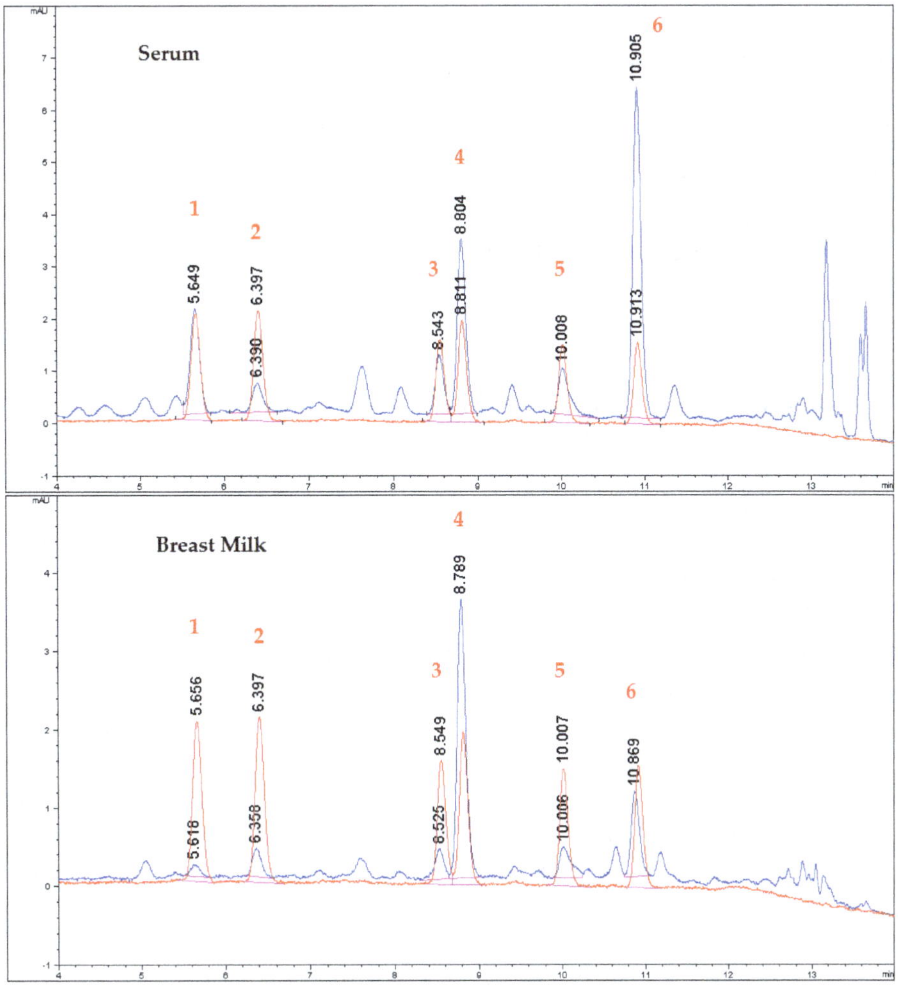

Figure 1. Typical chromatograms of standard (red trace) and sample (blue trace) shown in overlaid pattern. Peaks: (**1**) all-*trans*-lutein; (**2**) zeaxanthin; (**3**) β-cryptoxanthin; (**4**) Internal standard; (**5**) α-carotene; (**6**) β-carotene.

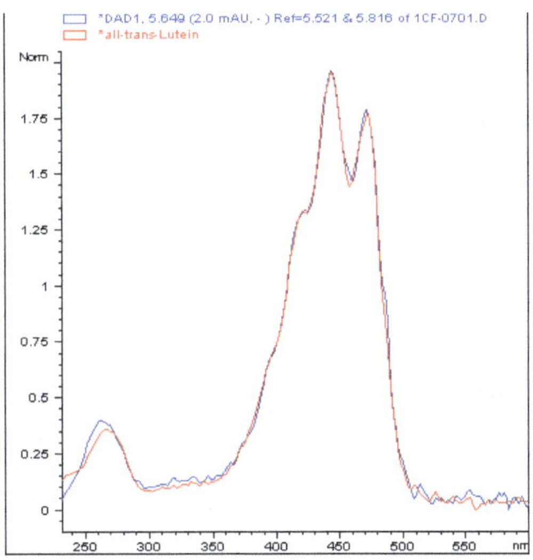

Figure 2. The spectrum of a peak found in a serum sample at retention time of 5.649 is matched against the library, which contains the spectra obtained from carotenoid standards. The peak is identified as all-*trans*-lutein with a matching score of 993.402 and a purity factor of 998.601.

2.3. Linearity

The R^2 obtained were >0.9999 for all compounds, indicating excellent linearity.

2.4. MDL/MQL

The method detection limit (MDL) was determined as the concentration at which the signal to noise ratio (S/N) of the analyte peak equals 3. The method quantification limit (MQL) equals 3 times that of MDL. The MQL obtained in this study for each carotenoid was 10 ng/mL in serum and 5 ng/mL in breast milk.

2.5. Accuracy

For serum samples, Standard Reference Material (SRM) 968e from the National Institute of Standards and Technology (NIST) was analyzed using six determinants at three certified concentrations on different days. The concentrations obtained in the study were then compared with the certified values which can be found on NIST website. It was shown that all results were within the ranges except one sample with lutein concentration (88 ng/mL) lower than the assigned range (90–104 ng/mL) (Table 1). This is expected, as the assigned value for the NIST 968e is for total lutein while this method merely quantifies all-*trans*-lutein.

As there is no reference material available for human breast milk, an over-spike study was carried out: milk samples were spiked with known amounts of carotenoids at two concentrations prior to extraction. The obtained values were later subtracted from the sample blank (without standard addition) and then divided by the known amount and expressed in percentage. The spike was performed at two levels, representing low level (5 ng/mL; at the MQL), which is most likely only found in mature breast milk, and high level (50 ng/mL), which can be observed in both colostrum and mature milk. The recoveries varied from 85% to 105% (Table 2), which meet the requirement of the bioanalysis validation guideline from the US. Food and Drug Administration [23], and demonstrates a high degree of accuracy.

Table 1. National Institute of Standards and Technology (NIST) Standard Reference Material 968e analysis results for serum (unit: ng/mL in serum).

Level I	NIST 968e Certified Value		Day 1 (*n* = 3)			Day 2 (*n* = 3)			Mean [a]	RSD [b]
	Mean	**Range**	**1-1**	**1-2**	**1-3**	**1-4**	**1-5**	**1-6**		
Lutein [c]	67	59–75	74	75	73	72	69	70	72	3.3%
Zeaxanthin	31	26–36	36	33	33	31	31	39	34	9.6%
β-Cryptoxanthin	41	35–47	43	42	42	40	39	41	41	3.7%
α-Carotene	11	6–16	8	9	9	8	8	9	9	5.2%
β-carotene	88	78–98	89	87	88	84	83	88	87	2.8%
Level II			**2-1**	**2-2**	**2-3**	**2-4**	**2-5**	**2-6**		
Lutein	97	90–104	89	93	88	96	91	91	91	3.3%
Zeaxanthin	29	23–35	28	28	27	30	28	25	28	6.3%
β-Cryptoxanthin	40	34–46	40	42	40	41	40	41	41	1.4%
α-Carotene	31	27–35	29	29	29	31	30	31	30	4.1%
β-carotene	203	183–223	215	220	212	221	217	220	218	1.5%
Level III			**3-1**	**3-2**	**3-3**	**3-4**	**3-5**	**3-6**		
Lutein	124	114–134	131	121	117	124	117	119	121	4.2%
Zeaxanthin	29	24–34	34	26	24	31	32	28	29	13.2%
β-Cryptoxanthin	21	17–25	21	20	20	24	22	23	22	7.7%
α-Carotene	15	13–17	17	14	17	14	17	16	16	9.2%
β-carotene	363	325–401	377	364	364	391	380	396	379	3.5%

[a]: Mean of six determinants in the study; [b]: RSD; relative standard deviation of the six determinants in the study; [c]: Reported value in Certificate of Analysis (CoA) is for total lutein. This method quantifies all-*trans*-lutein.

Table 2. Mature milk (MM) and colostrum (CL) over-spike recoveries [a].

	Day 1		Day 2		Day 3			
5 ng/mL in Mature Milk	**MML1 [b]**	**MML2**	**MML3**	**MML4**	**MML5**	**MML6**	**Mean [c]**	**RSD [d]**
Lutein	89%	101%	89%	96%	106%	101%	97%	7.5%
Zeaxanthin	93%	100%	95%	88%	87%	87%	91%	6.0%
β-Cryptoxanthin	107%	93%	81%	89%	93%	90%	92%	9.2%
α-Carotene	92%	106%	91%	85%	104%	109%	98%	9.9%
β-carotene	92%	109%	96%	86%	86%	89%	93%	9.4%
50 ng/mL in Mature Milk	**MMH1**	**MMH2**	**MMH3**	**MMH4**	**MMH5**	**MMH6**	**Mean**	**RSD**
Lutein	91%	90%	92%	93%	88%	88%	91%	6.1%
Zeaxanthin	89%	90%	91%	92%	91%	89%	90%	3.0%
β-Cryptoxanthin	89%	92%	91%	92%	85%	86%	89%	4.1%
α-Carotene	85%	87%	86%	86%	87%	87%	86%	4.6%
β-carotene	89%	94%	89%	90%	97%	99%	93%	5.0%
50 ng/mL in Colostrum	**CLH1**	**CLH2**	**CLH3**	**CLH4**	**CLH5**	**CLH6**	**Mean**	**RSD**
Lutein	91%	92%	100%	104%	102%	105%	99%	2.2%
Zeaxanthin	93%	95%	98%	98%	99%	101%	97%	1.4%
β-Cryptoxanthin	95%	96%	97%	97%	105%	104%	99%	3.4%
α-Carotene	91%	93%	98%	94%	103%	99%	96%	1.0%
β-carotene	92%	90%	95%	97%	103%	101%	96%	4.8%

[a]: Mature milk is pooled milk collected from multiple donors, four-weeks postpartum, and colostrum is pooled milk collected from multiple donors, one-week postpartum; [b]: MML: Mature milk low spike; MMH: Mature milk high spike; CLH: Colostrum high spike; [c]: Mean of six determinants in the study; [d]: RSD; relative standard deviation of the six determinants in the study.

2.6. Repeatability and Intermediate Precision

Injection repeatability was demonstrated by four consecutive injections of WS3. The relative standard deviations (RSDs) of the peak ratios, obtained over the course of method validation, were <3%. It was shown that the RSDs of the SRM results were lower than 15% for all compounds (Table 1), and the RSDs of over-spike recoveries in milk were all within 10% for colostrum and mature milk (Table 2). Therefore, the method is proven to be highly repeatable and precise.

2.7. Storage and Stability

2.7.1. Sample Storage

It has been previously indicated that carotenoids are stable in plasma stored at −70 °C for at least 28 months, or at −20 °C for five months [16]. However, there is an absence of data on long-term storage of breast milk. Assuming a similar behavior of breast milk at ultra-low temperatures, the samples are recommended to be stored at −70 °C for optimal preservation [17].

2.7.2. Stability of Standard Solution and Sample Extracts in HPLC Autosampler

To determine whether overnight automated analysis or re-injection was feasible, the standard solution (ethanol:MTBE, 80:20, *v*/*v*) and sample extracts were kept in an autosampler (chilled to 8 °C), and repeated injections were conducted at time points of 0, 24, and 48 h. The concentrations of the re-injections were calculated using the original calibration curve derived from time 0.

It was found that only mild fluctuations in concentration occurred at 48 h at 8 °C in the dark (Figure 3). The small changes in the concentrations of carotenoids in solution, and extracts, may be due to the protection rendered by BHT, as well as compensation from the internal standard. Thus, if troubleshooting is needed, standard solutions and sample extracts are allowed to be re-injected within 48 h.

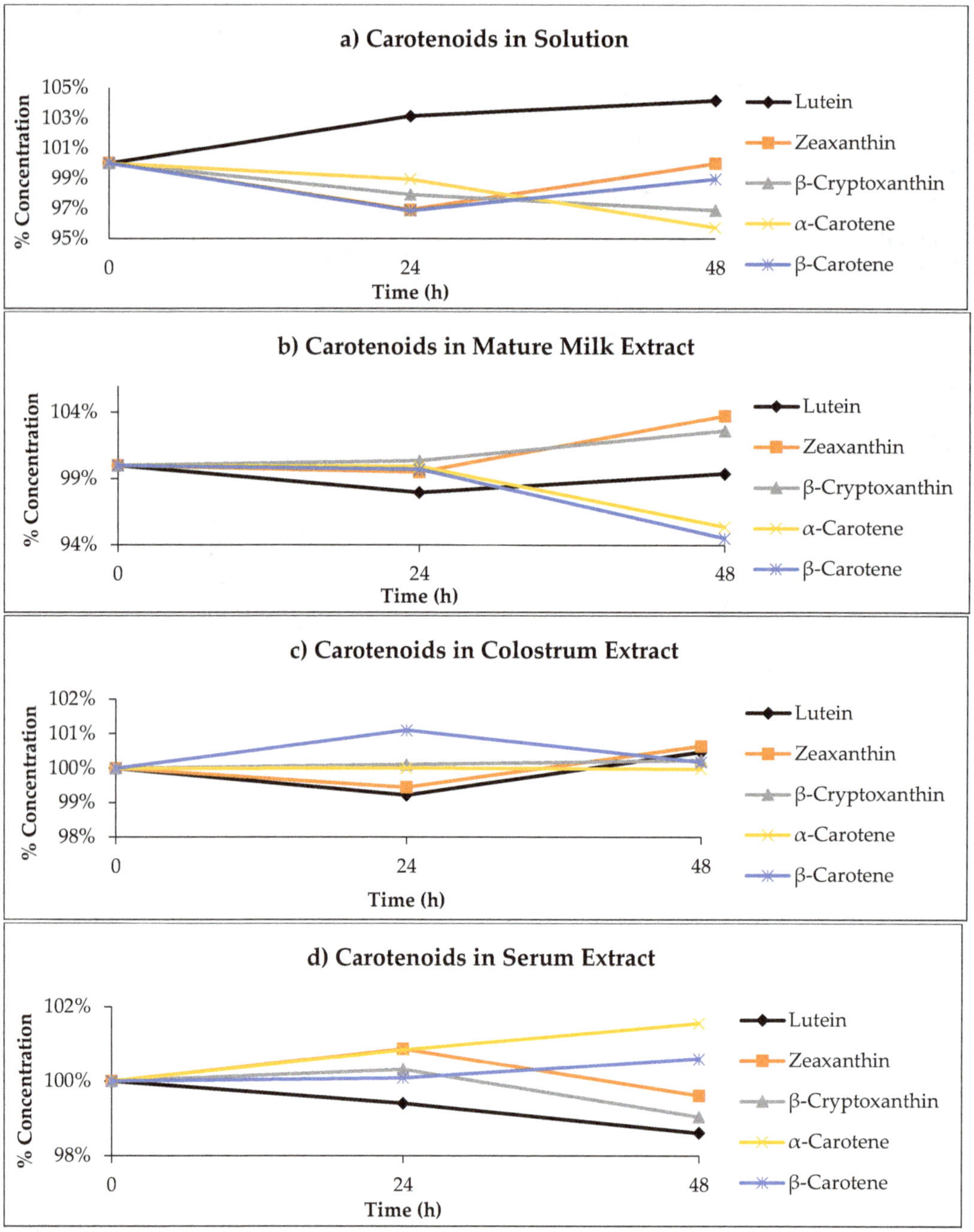

Figure 3. Stability of carotenoids in (**a**) solution (MTBE:Ethanol with BHT, 80:20, *v*/*v*) and in different sample extracts (**b**–**d**) in an autosampler at 8 °C. Testing points at 0, 24, and 48 h. *Y* axis: Concentrations of carotenoids (normalized to percentage). *X* axis: Testing points.

2.7.3. Freeze/Thaw Cycle

The impact of the freeze/thaw cycle on sample integrity was also evaluated. The samples were thawed to room temperature, and an aliquot was taken for analyses. The remaining part was re-frozen at −70 °C. After 24 h, the remaining sample was thawed and then re-tested; this process was repeated three times. It was found that freezing/thawing was detrimental to carotenoids, as all compounds started to degrade after one cycle (Figure 4). Therefore, if the samples have ever been subjected to a freeze/thaw cycle before, the concentrations of these carotenoids may likely be underestimated.

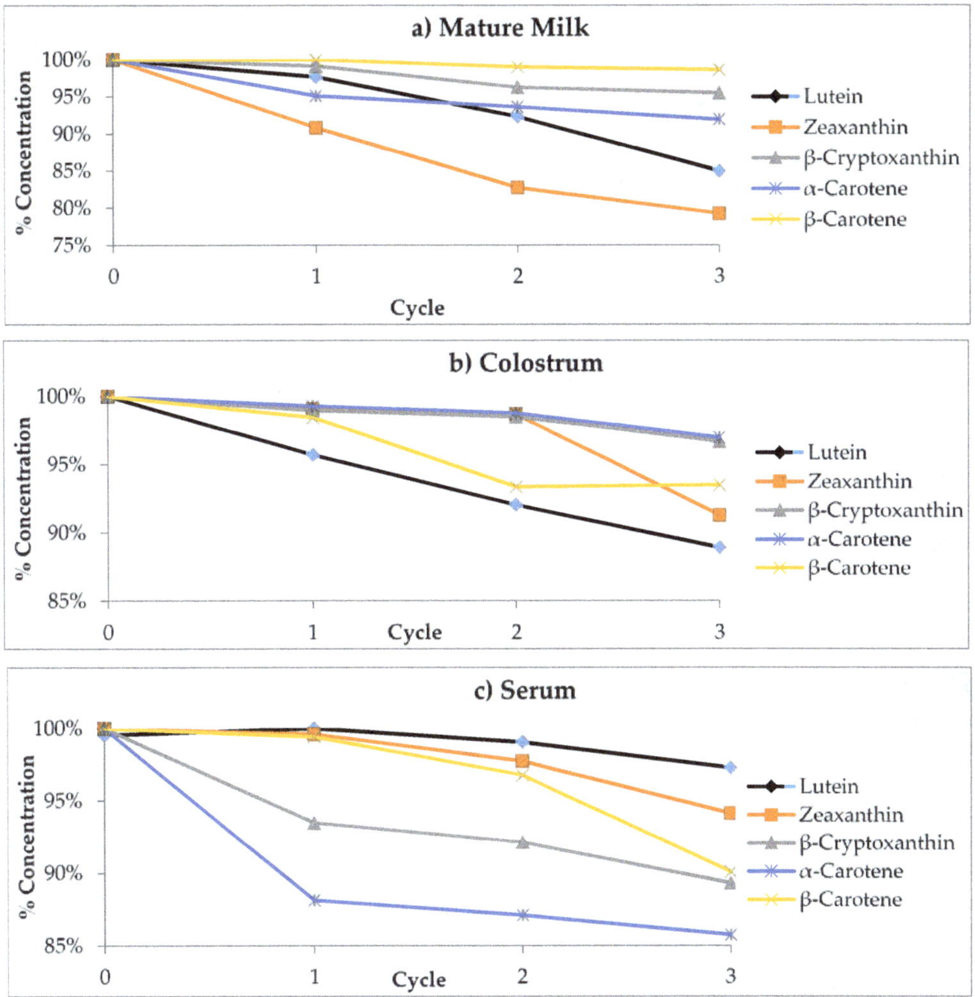

Figure 4. Effect of freeze/thaw on carotenoids in (**a**) mature milk; (**b**) colostrum; and (**c**) serum samples stored at −70 °C. *Y* axis: Concentrations of carotenoids (normalized to percentage). *X* axis: Freeze/thaw cycles.

2.7.4. Stock Solution Stability

Carotenoids are highly susceptible to degradation in solutions; thus, UV standardization of the stock solution, prior to use, is highly recommended to ensure a high accuracy (preparation procedures are detailed in Section 3.2.2). In order to understand the stability of stock solutions at −70 °C (see Section 3.2.1) and to save on effort for checking the concentration every time, UV standardization was performed on aliquot stock solutions, periodically, over a span of 24 days. The trends are shown in Figure 5. It was shown that zeaxanthin degrades rapidly in solutions (up to 40%) during the first week, and α- and β-carotene exhibited high fluctuation patterns, which could be due to precipitation in ultra-low temperatures, which affect the homogeneity of the solution. Therefore, after three readings, the experiment was discontinued for the three compounds. Lutein exhibited 11% degradation and β-cryptoxanthin was stable up to 24 days. Thus, it can be concluded that, except for β-cryptoxanthin, the standard solutions all need to be UV standardized, each time, prior to use for best accuracy.

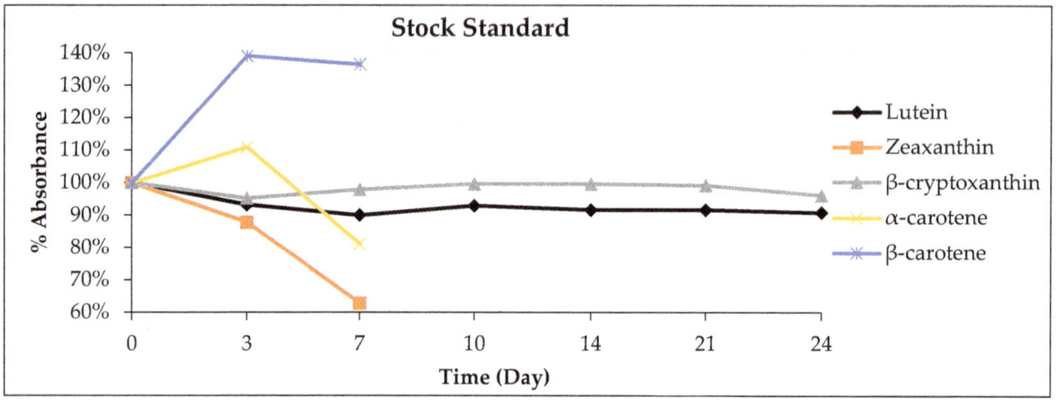

Figure 5. UV absorbance of carotenoids standard stock solutions at different time points. All standard solutions were stored at −70 °C prior to analysis. *Y* axis: Measured absorbance by UV spectrometer compared to the reading from 0-time (normalized to percentage). *X* axis: Testing time point.

3. Materials and Methods

3.1. Chemicals and Reagents

Absolute ethanol was purchased from Merck Millipore (Billerica, MA, USA). Methanol, dichloromethane, methyl *tert*-butyl ether (MTBE), butylated hydroxytoluene (BHT), ammonium acetate, petroleum ether, tetrahydrofuran (THF) with BHT as inhibitor (250 ppb), and potassium hydroxide (ACS grade, ≥85%) were purchased from Sigma-Aldrich (St. Louis, MO, USA). N-hexane was purchased from ThermoFisher Scientific (Waltham, MA, USA). Unless otherwise specified, all reagents were of HPLC grade or higher. Ethanol/BHT solution was prepared by adding 30 mg of BHT into 1 L of absolute ethanol.

All-*trans*-utein (X6250-1MG) and zeaxanthin (14681-1MG-F) were purchased from Sigma-Aldrich (St. Louis, MO, USA) or Chromadex (Irvine, CA, USA) ((ASB-00012453) and (00026504) respectively). β-cryptoxanthin (0055), α-carotene (0007), and echinenone (0283) were purchased from Carotenature (Ostermundigen, Switzerland). β-carotene (00003211) was purchased from Chromadex or United States Pharmacopeia (USP, Rockville, MD, USA) (1065480). Standard Reference Material (SRM) NIST 968e was purchased from the National Institute of Standards and Technology (Gaithersburg, MD, USA).

Human serum and breast milk samples, which were used for spike-recovery studies, were purchased from LeeBio (Maryland Heights, MO, USA).

Carotenoids are light sensitive. All samples and standards were prepared and handled under shielded lighting with a minimum UV cutoff of 385 nm. The standards and samples were tightly wrapped in aluminum foil when being transported through, or into, an area without shielded lighting.

3.2. Standard Preparation

3.2.1. Stock Solution Preparation

Stock solutions (about 100 μg/mL) of lutein, zeaxanthin, and β-cryptoxanthin were individually prepared by dissolving 1 mg of the standard in 1 mL of THF and then diluting to 10 mL with ethanol/BHT solution. Stock solutions (about 100 μg/mL) of α-carotene, β-carotene, and echinenone were individually prepared by dissolving 1 mg of the standard in 7 mL of THF, and then diluting to 10 mL with ethanol/BHT solution. Stock solutions were then distributed in aliquots into amber vials (e.g., 200 μL each), sufficient for one-time use, and then stored at −70 °C.

3.2.2. UV Standardization of Stock Solutions

Unlike the preparation of stable compounds, for carotenoids analyses, concentrations of the standard should not follow the certified values on the Certificate of Analysis (CoA). This is because the standards may be subjected to significant degradation during transportation and/or prolonged storage. Therefore, UV standardization of the stock solution (Section 3.2.1) must be performed to calculate the true working standard concentrations.

The standards in Section 3.2.1 were diluted 100 times to 1 μg/mL. The diluent for lutein, zeaxanthin, and β-cryptoxanthin was the ethanol/BHT solution, while α-carotene and β-carotene were dissolved in hexane for better solubility. The cuvette for UV standardization of α-carotene and β-carotene must be made of quartz, while, for the rest of the compounds, disposable cuvettes made of polystyrene (PS) are acceptable. A λ_{max} of 445, 452, 452, 444, and 452 nm and an absorptivity of 2550, 2540, 2356, 2800, and 2592 dL/(g·cm), for all-*trans*-lutein, zeaxanthin, β-cryptoxanthin, α-carotene, and β-carotene were used, respectively [24]. Spectrum scans needed to be performed to ensure the experimental λ_{max} was within ±2 nm of the reference values.

3.2.3. Working Standard Solution Preparation

The internal standard solution was prepared by dissolving echinenone at a concentration of 0.2 μg/mL in ethanol/BHT solution.

Five-point working standard solutions were freshly prepared at concentrations of 10, 20, 100, 400, and 600 ng/mL with an internal standard of 500 ng/mL in ethanol/BHT solution.

3.3. Sample Preparation

Briefly, samples (100 μL of serum or 200 μL of breast milk) were first spiked with internal standard and then vortexed to mix. Serum samples were then extracted with hexane (0.8 mL) twice. Extracts were dried under a gentle stream of nitrogen at 35 °C, reconstituted with 100 μL reconstitution solution (MTBE:Ethanol/BHT, 80:20, *v*/*v*), centrifuged, and the clear supernatants were transferred into amber autosampler vials with a 250 μL micro-insert. Reagents were added sequentially in breast milk samples: ultrapure water (200 μL), 5% potassium hydroxide (0.4 mL), and THF (0.2 mL), followed by extraction with the extraction solvent (0.4 mL; petroleum ether:hexane:dichloromethane containing BHT 50 mg/L, 4:4:2, *v*/*v*/*v*) twice. The solutions were vortexed for 2 min between additions. The upper organic layer from the extraction was then transferred to a new vial and treated the same way as the extracts from the serum samples.

For quality control, each batch included at least one standard reference material (NIST 968e) for serum and at least one over-spike for breast milk or colostrum. The results of NIST 968e analyses should be within the assigned reference values, and the recovery for the over-spike should be 80–120% of the target value.

3.4. HPLC-DAD Conditions and Methods

The instrument used was an Agilent 1290 Infinity, equipped with a photodiode array detector (DAD) able to scan between 190 and 640 nm (60 mm Max-Light Cartridge Flow Cell). Separation of the carotenoids was achieved on a YMC C_{30} column (150 × 2.1 mm, 3 μm) at 30 °C, with a binary mobile phase system. The autosampler temperature was kept at 8 °C, and the injection volume was 10 μL. Mobile phase A (MPA) was methanol with 1.5% ammonium acetate solution at a ratio of 98:2, *v*/*v*, and mobile phase B (MPB) was a mixture of MTBE, methanol and 1.5% ammonium acetate solution at a ratio of 90:8:2, *v*/*v*/*v*. The gradient started at 100% MPA, which was held for 1.5 min, then ramped to 55% at 9.4 min, then to 5% over 2.0 min, held for 2.0 min, and finally returned to 100% MPA over 0.1 min, and held for 3.0 min. The flow rate was constant at 0.5 mL/min. Carotenoids were detected at 445 nm, with a band width of 2.0 nm (reference wavelength: 560 ± 80 nm). A three-dimensional spectrum, ranging from 240–600 nm, was also recorded. Identification was conducted by matching the peak retention times (R_t) and spectra with those of the pure standards.

Internal standard calibration was used by plotting the peak area ratio of analytes to internal standard (IS) against the concentration ratio of analytes to IS.

As all-*trans*-lutein reference standards may contain significant amounts of impurities (e.g., zeaxanthin or other lutein isomers), such contributions need to be carefully examined.

4. Conclusions

We have successfully developed and validated a method for simultaneously quantifying five major carotenoids in human serum and breast milk. Good chromatography separation was achieved on a YMC C_{30} column with a significantly shorter run time. The method had a high degree of accuracy and precision, and the MQLs were sufficiently low for studies on carotenoids levels in human biological samples. The stability study showed that the carotenoid standard and the sample extracts could be stored in a chilled autosampler at 8 °C without being significantly comprised for 48 h, which provides guidance on re-test time frames. Freeze/thaw was found to be detrimental to carotenoids, and should always be avoided. Most importantly, although the standard/standard solution was stored at ultra-low temperatures, UV standardization of the stock standard should be performed prior to each assay to obtain the true concentrations, and the use of CoA values is not recommended.

Acknowledgments: We are grateful for valuable input from colleagues at Ross Park, Columbus, particularly the prior work done by Karen Schimpf. We would also like to express our thanks to the financial support from Abbott Nutrition Research and Development Singapore.

Author Contributions: Jing Tan and Chunyan Zhang conceived and designed the experiments; Jing Tan, Jason Gek Leong Neo and Tania Setiawati performed the experiments, analyzed the data and wrote the paper.

Conflicts of Interest: The authors declare no conflicts of interest.

References

1. Krinsky, N.I.; Johnson, E.J. Carotenoid actions and their relation to health and disease. *Mol. Asp. Med.* **2005**, *26*, 459–516. [CrossRef] [PubMed]
2. Johnson, E.J. Role of lutein and zeaxanthin in visual and cognitive function throughout the lifespan. *Nutr. Rev.* **2014**, *72*, 605–612. [CrossRef] [PubMed]
3. Johnson, E.J.; Vishwanathan, R.; Johnson, M.A.; Hausman, D.B.; Davey, A.; Scott, T.M.; Green, R.C.; Miller, L.S.; Gearing, M.; Woodard, J.; et al. Relationship between serum and brain carotenoids, α-tocopherol, and retinol concentrations and cognitive performance in the oldest old from the georgia centenarian study. *J. Aging Res.* **2013**, *2013*, 951786. [CrossRef] [PubMed]
4. Vishwanathan, R.; Kuchan, M.J.; Sen, S.; Johnson, E.J. Lutein and preterm infants with decreased concentrations of brain carotenoids. *J. Pediatr. Gastroenterol. Nutr.* **2014**, *59*, 659–665. [CrossRef] [PubMed]
5. Jeon, S.; Neuringer, M.; Johnson, E.E.; Kuchan, M.J.; Pereira, S.L.; Johnson, E.J.; Erdman, J.W. Effect of carotenoid supplemented formula on carotenoid bioaccumulation in tissues of infant rhesus macaques: A pilot study focused on lutein. *Nutrients* **2017**, *9*, 51. [CrossRef] [PubMed]
6. Rajendran, V.; Pu, Y.S.; Chen, B.H. An improved HPLC method for determination of carotenoids in human serum. *J. Chromatogr. B* **2005**, *824*, 99–106. [CrossRef] [PubMed]
7. Khachik, F.; Spangler, C.J.; Smith, J.C.; Canfield, L.M.; Steck, A.; Pfander, H. Identification, quantification, and relative concentrations of carotenoids and their metabolites in human milk and serum. *Anal. Chem.* **1997**, *69*, 1873–1881. [CrossRef] [PubMed]
8. Craft, N.E. Carotenoid reversed-phase high-performance liquid chromatography methods: Reference compendium. *Methods Enzymol.* **1992**, *213*, 185–205. [PubMed]
9. Giuliano, A.R.; Neilson, E.M.; Kelly, B.E.; Canfield, L.M. Simultaneous quantitation and separation of carotenoids and retinol in human milk by high-performance liquid chromatography. *Methods Enzymol.* **1992**, *213*, 391–399. [PubMed]
10. Liu, Y.; Xu, M.J.; Canfield, L.M. Enzymatic hydrolysis, extraction, and quantitation of retinol and major carotenoids in mature human milk. *J. Nutr. Biochem.* **1998**, *9*, 178–183. [CrossRef]

11. Chauveau-Duriot, B.; Doreau, M.; Noziere, P.; Graulet, B. Simultaneous quantification of carotenoids, retinol, and tocopherols in forages, bovine plasma, and milk: Validation of a novel uplc method. *Anal. Bioanal. Chem.* **2010**, *397*, 777–790. [CrossRef] [PubMed]
12. Hrvolová, B.; Martínez-Huélamo, M.; Colmán-Martínez, M.; Hurtado-Barroso, S.; Lamuela-Raventós, R.; Kalina, J. Development of an advanced HPLC–MS/MS method for the determination of carotenoids and fat-soluble vitamins in human plasma. *Int. J. Mol. Sci.* **2016**, *17*, 1719. [CrossRef] [PubMed]
13. Nimalaratne, C.; Sun, C.; Wu, J.; Curtis, J.M.; Schieber, A. Quantification of selected fat soluble vitamins and carotenoids in infant formula and dietary supplements using fast liquid chromatography coupled with tandem mass spectrometry. *Food Res. Int.* **2014**, *66*, 69–77. [CrossRef]
14. Henry, L.K.; Catignani, G.L.; Schwartz, S.J. Oxidative degradation kinetics of lycopene, lutein, and 9-cis and all-trans β-carotene. *J. Am. Oil Chem. Soc.* **1998**, *75*, 823–829. [CrossRef]
15. Henry, L.K.; Puspitasari-Nienaber, N.L.; Jarén-Galán, M.; van Breemen, R.B.; Catignani, G.L.; Schwartz, S.J. Effects of ozone and oxygen on the degradation of carotenoids in an aqueous model system. *J. Agric. Food Chem.* **2000**, *48*, 5008–5013. [CrossRef] [PubMed]
16. Craft, N.E.; Brown, E.D.; Smith, J.C., Jr. Effects of storage and handling conditions on concentrations of individual carotenoids, retinol, and tocopherol in plasma. *Clin. Chem.* **1988**, *34*, 44–48. [PubMed]
17. Comstock, G.W.; Alberg, A.J.; Helzlsouer, K.J. Reported effects of long-term freezer storage on concentrations of retinol, beta-carotene, and alpha-tocopherol in serum or plasma summarized. *Clin. Chem.* **1993**, *39*, 1075–1078. [PubMed]
18. Rodriguez-Amaya, D.B. Structures and analysis of carotenoid molecules. *Sub-Cell. Biochem.* **2016**, *79*, 71–108.
19. Schweigert, F.J.; Hurtienne, A.; Bathe, K. Improved extraction procedure for carotenoids from human milk. *Int. J. Vitam. Nutr. Res.* **2000**, *70*, 79–83. [CrossRef] [PubMed]
20. Rodriguez-Amaya, D.B. Structures and analysis of carotenoid molecules. In *Carotenoids in Nature: Biosynthesis, Regulation and Function*; Stange, C., Ed.; Springer International Publishing: Cham, Switzerland, 2016; p. 457.
21. Craft, N.E.; Soares, J.H. Relative solubility, stability, and absorptivity of lutein and β-carotene in organic solvents. *J. Agric. Food Chem.* **1992**, *40*, 431–434. [CrossRef]
22. Amorim-Carrilho, K.T.; Cepeda, A.; Fente, C.; Regal, P. Review of methods for analysis of carotenoids. *TrAC Trends Anal. Chem.* **2014**, *56*, 49–73. [CrossRef]
23. *Guidance for Industry Bioanalytical Method Validation (Draft)*; USA Food and Drug Administration, 2013. Available online: https://www.fda.gov/downloads/Drugs/Guidances/ucm368107.pdf (accessed on 13 May 2017).
24. Thomas, J.B.; Duewer, D.L.; Mugenya, I.O.; Phinney, K.W.; Sander, L.C.; Sharpless, K.E.; Sniegoski, L.T.; Tai, S.S.; Welch, M.J.; Yen, J.H. Preparation and value assignment of standard reference material 968e fat-soluble vitamins, carotenoids, and cholesterol in human serum. *Anal. Bioanal. Chem.* **2012**, *402*, 749–762. [CrossRef] [PubMed]

5

A Novel Protocol to Monitor Trace Levels of Selected Polycyclic Aromatic Hydrocarbons in Environmental Water Using Fabric Phase Sorptive Extraction Followed by High Performance Liquid Chromatography-Fluorescence Detection

Shivender Singh Saini [1], Abuzar Kabir [2,*], Avasarala Lakshmi Jagannadha Rao [1], Ashok Kumar Malik [1,*] and Kenneth G. Furton [2]

1 Department of Chemistry, Punjabi University, Patiala 147002, India; shiv_algal2007@yahoo.com (S.S.S.); aljrao@gmail.com (A.L.J.R.)

2 International Forensic Research Institute, Department of Chemistry and Biochemistry, Florida International University, Miami, FL 33193, USA; furtonk@fiu.edu

* Correspondence: akabir@fiu.edu (A.K.); malik_chem2002@yahoo.co.uk (A.K.M.); (A.K.)

Academic Editor: Victoria F. Samanidou

Abstract: Fabric phase sorptive extraction (FPSE) combines the advanced material properties of sol–gel derived microextraction sorbents and the flexibility and permeability of fabric to create a robust, simple and green sample preparation device. It simultaneously improves the extraction sensitivity and the speed of the extraction by incorporating high volumes of sponge-like, porous sol–gel hybrid inorganic–organic sorbents into permeable fabric substrates that are capable of extracting target analytes directly from both simple and complex aqueous sample matrices. For the first time, this technique was applied to the trace-level determination of selected polycyclic aromatic hydrocarbons (PAHs) in environmental water samples using a non-polar sol–gel C18 coated FPSE media. Several extraction parameters were optimized to improve the extraction efficiency and to achieve a high detection sensitivity. Validation tests of spiked samples showed good linearity for four selected PAHs (R^2 = 0.9983–0.9997) over a wide range of concentrations (0.010–10 ng/mL). Limits of detection (LODs) and quantification (LOQs) were measured at pg/mL levels; 0.1–1 pg/mL and 0.3–3 pg/mL, respectively. Inter- and intra-day precision tests showed variations of 1.1%–4.1% for four selected PAHs. Average absolute recovery values were in the range of 88.1%–90.5% with relative standard deviations below 5%, surpassing the values predicted by the recovery prediction model. Finally, the developed FPSE-HPLC-FLD protocol was applied to analyze 8 environmental water samples. Out of four selected PAHs, fluoranthene (Flu) and phenanthrene (Phen) were the most frequently detected in four samples, at concentrations of 5.6–7.7 ng/mL and 4.1–11 ng/mL, respectively, followed by anthracene (Anth) and pyrene (Pyr) in two samples. The newly developed FPSE-HPLC-FLD protocol is simple, green, fast and economical, with adequate sensitivity for trace levels of four selected PAHs and seems to be promising for routine monitoring of water quality and safety.

Keywords: fabric phase sorptive extraction (FPSE); polycyclic aromatic hydrocarbons (PAHs); persistent pollutants; green analytical chemistry (GAC); environmental water; sorptive microextraction

1. Introduction

Polycyclic aromatic hydrocarbons are toxic, carcinogenic and mutagenic organic compounds, and can exert immunologic and reproductive effects. They are introduced into the environment from both natural and anthropogenic sources, typically formed as unintentional byproducts of the incomplete combustion of organic matter. PAHs are ubiquitous in nature and the most damaging pollutants with regard to the ecosystem [1–5]. Although they have low water solubility and exist in the environment at very low concentrations, PAHs are strongly bioaccumulative (e.g., by fish) and can, therefore, pass up the food chain to top predators, including humans [6,7]. Natural inputs such as forest and prairie fires, volcanic eruptions and anthropogenic inputs, such as oil spills, waste incineration, coke and asphalt production, oil refining, aluminum production, urban runoff, emissions from combustion and industrial processes are the main sources of PAHs in the environment [8–10]. The elevated concentrations of PAHs in the environment, together with their ecological toxicity and health risk for humans, have spawned numerous environmental studies [11–14]. Due to their environmental concern, PAHs are included in the USEPA (United States Environmental Protection Agency) and in the EU (European Union) priority lists of pollutants [15]. Human exposure to PAHs occurs mainly by direct inhalation of polluted air and tobacco smoke, dietary intake of smoked food stuffs, direct intake of and contact with contaminated water, direct contact with contaminated soil and dermal contact with soot, tars and oils [16]. PAHs enter the aquatic environment through run off from contaminated roads or sealed parking lots [17], urban and industrial waste water discharge, direct spillage and wet and dry deposition of atmospheric contaminants [18]. Therefore, trace determination of PAHs in water matrices [19] is considered to be a valuable tool in risk assessment, remediation and management of PAHs in the environment.

In this study, four selected small PAHS including anthracene (a linear three ring PAH), phenanthrene (a fused three ring PAH), pyrene (a fused four ring PAH), and fluoranthene (a non-alternant four ring PAH) (Figure 1).

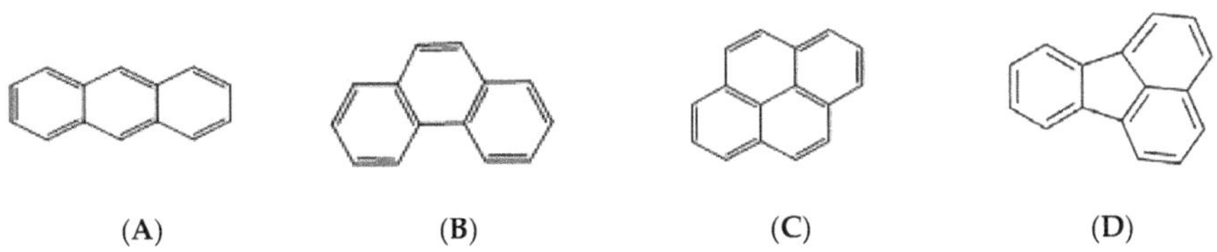

Figure 1. Chemical structures of four selected polycyclic aromatic hydrocarbons (PAHs): (**A**) Anthracene; (**B**) Phenanthrene; (**C**) Pyrene; (**D**) Fluoranthene.

Anthracene is used as a raw material in the industrial production of hydrogen peroxide and in the manufacturing of alizarin and anthraquinone dyes for cotton fibers [20]. Phenanthrene, a tricyclic aromatic hydrocarbon, has a "K-region" and a "bay-region", where the main carcinogenic species can be formed and is therefore commonly used as a model substrate for carcinogenic studies [21]. Fluoranthene, a non-alternant PAH, contains a five-member ring and is a persistent environmental organic pollutant structurally similar to other environmentally important compounds such as fluorene, dibenzothiophene, acenaphthylene, carbazole, dibenzofuran, and dibenzodioxin, i.e., compounds of [22,23]. Pyrene occurs at relatively high concentrations in PAH mixtures and is one of the most highly concentrated PAHs detected in drinking water [24].

Until now, numerous protocols comprising gas chromatography (GC) and high-performance liquid chromatography (HPLC) have been successfully developed for the determination of PAHs in the environment [25–29]. However, since PAHs are generally present at trace levels in the environment and are accompanied by diversified matrices, they cannot be directly handled by analytical instruments. This means that sample pre-treatment is required, with the aim of concentrating the PAHs, as well as eliminating or decreasing interference [30]. For the isolation of PAHs in environmental matrices, various sample pre-treatment techniques have been proposed, including solid-phase extraction (SPE) [31,32], liquid-liquid extraction (LLE) [33], miniaturized homogeneous liquid-liquid extraction (MHLLE) [34], solid-phase microextraction (SPME) [35,36] and single drop microextraction (SDME) [37,38]. Although these techniques are very useful, they possess some serious drawbacks. For example, LLE and SPE are laborious, time consuming, utilize high volumes of organic solvents and sample solutions, often incur significant loss of analytes and have poor reproducibility. The shortcomings of SPME include high cost, fragility of the fiber, low thermal and chemical stability, swelling if exposed to organic solvents, limited lifetime and sample carry-over problems [39,40]. Breaking of the organic drop due to fast stirring, reduced extraction rate due to air bubble formation, time-consumption and non-equilibrium are the major drawbacks associated with SDME [41].

Nowadays, a number of unique materials have been synthesized and applied as sorbents for PAH extraction, including multi-walled carbon nanotubes [32], magnetic nanoparticles [42], and metal-organic frameworks [43]. However, the synthesis process of such materials is rather complex and often consumes large amounts of organic solvents and time. Therefore, new, versatile and high-performance adsorbents with a simple preparation process are still highly desirable.

Kabir et.al., in 2014 integrated the rich surface chemistry of cellulose cotton fabric substrates and sol–gel technology to develop a novel sorptive microextraction technique called fabric phase sorptive extraction (FPSE) The inherent advantages of the synthesized sorptive material include (1) a flexible hydrophobic/hydrophilic fabric substrate that can be bent, twisted and squeezed to insert directly into unmodified samples; (2) the ability to extract target analytes directly from a raw sample matrix i.e., environmental water, whole milk, whole blood, urine, saliva containing proteins, lipids, particulates, biomasses or debris without any sample pre-treatment; (3) a high loading capacity with unique and tunable selectivity; (4) the possibility of using any organic or aqueous-organic solvent mixture for elution/solvent back-extraction; (5) the capability to extract polar, nonpolar, acidic, and basic compounds; (6) the absence of a requirement for solvent evaporation and analyte reconstitution; (7) the fact that its operational simplicity meets green analytical chemistry and economic criteria as well. One major advantage of FPSE is its ability to immobilize highly polar polymers in a sol–gel hybrid organic–inorganic network, chemically bonded to the fabric substrate, resulting in a microextraction material that can efficiently extract both polar and nonpolar analytes directly from an aqueous sample matrix. This has been used in the analysis of a wide variety of analytes in environmental and biological samples [44–47].

PAHs are inherently nonpolar, hydrophobic compounds which do not ionize in water. Intuitively, the selective extraction of highly nonpolar analytes, such as PAHs, from an aqueous sample matrix would be facilitated by a hydrophobic sorbent as the extraction phase material [48] and consequently sol–gel C18 coated FPSE media containing long hydrophobic C18 chains has been evaluated and successfully applied in this case.

Herein, we describe the design and preparation of sol–gel C18 coated FPSE media and the development of a novel application protocol for the efficient extraction of trace amounts of four selected small PAHs of high environmental interest from aqueous solutions. The hydrophilic cellulose fabric substrate incorporated in the core of the adsorbent aids in the extraction kinetics by attracting water molecules containing PAHs towards its surface for a successful sorbent-analyte interaction, resulting in the trapping of the analyte on the FPSE media. Our strategy offers the following advantages in water analysis: (1) highly efficient analysis of PAHs (to ppt level concentrations); (2) simultaneous determination of four different PAHs in water using just one adsorbent; (3) facile regeneration of the adsorbent; (4) no solvent evaporation or analyte reconstitution is needed and (5) the FPSE strategy can be extended to design a range of adsorbents for sensitive determination of related pollutant compounds. This study serves the analytical and environmental community by providing a better and superior pathway for effective extraction and determination of trace PAHs in water matrices of domestic and environmental interest. In addition, HPLC-FLD was used for chromatographic separation of four PAHs with the advantage of a shorter analysis time and superior resolution and sensitivity. Finally, and most importantly, this is the first manuscript presenting the application of FPSE for the trace analysis of polycyclic aromatic hydrocarbons in aqueous media.

2. Materials and Methods

2.1. Choice of Target PAHs

The four selected small, low molecular weight PAHs are in the list of 16 priority PAHs designated by USEPA [1] and are source markers in water with diagnostic ratios for discriminating petrogenic from pyrolytic sources (fluoranthene/pyrene) [27], a model substrate for cancer studies (phenanthrene) [21], and the PAHs most detected in the highest concentrations in drinking water (fluoranthene, phenanthrene, pyrene, and anthracene) [49]. In addition to the feasibility of HPLC-FLD analysis, the commercial availability of PAHs for analytical standards was also considered.

2.2. Chemicals and Materials

The Muslin 100% cotton cellulose substrate for the creation of sol–gel C18 coated FPSE media was purchased from Jo-Ann Fabric (Miami, FL, USA). The precursors for sol–gel synthesis, octadecyl trimethoxysilane, tetramethyl orthosilicate, trifluoroacetic acid (TFA) and organic solvents such as acetone and dichloromethane were purchased from Sigma-Aldrich (St. Louis, MO, USA). Sodium hydroxide and hydrochloric acid were purchased from Thermo Fisher Scientific (Milwaukee, WI, USA).

Certified analytical standards of anthracene (Anth), fluoranthene (Flu), pyrene (Pyr), and phenanthrene (Phen) were purchased from Sigma-Aldrich (Darmstadt, Germany). HPLC grade methanol, acetonitrile and water were purchased from Sigma-Aldrich (Darmstadt, Germany) and filtered through a 0.22 μm filter before use. Stock solutions of each PAH were made at a concentration of 1 mg/mL, and a standard mixture of all four PAHs (1 mg/mL each) was prepared in methanol to achieve a response at the comparable level in HPLC-FLD. Working solutions were freshly prepared by diluting the mixed standard solution with HPLC-grade water to the required concentrations. All standards and working solutions were stored at 4 °C. All other reagents were of analytical grade.

2.3. Instrumentation

HPLC-FLD analysis of four PAHs was performed with a Dionex P680 HPLC pump (Germering, Germany) equipped with a Dionex Ultimate 3000 Fluorescence detector and the Chromeleon chromatography management software (Dionex, Germering, Germany). An Acentis Express reverse-phase C18 column (10 cm × 4.6 mm, particle size 5 μm, Supelco, Darmstadt, Germany) maintained at 25 °C was used for separation. Ultrasonic degassing was performed with an ultrasonic bath (Sarthak Scientific Services, Panchkula, India). The mobile phase consisted of water (solvent A) and acetonitrile (solvent B) and the flow rate was 1 mL/min. The isocratic elution mobile phase was set as follows: 15% A and 85% B with a total chromatographic run time of 10 min at λ_{ex} = 260 nm and λ_{em} = 420 nm using a fluorescent detector. The injection volume was 20 μL.

2.4. Water Sample Collection

The water samples selected for the investigation included two river water samples collected from the Chakki river (Pathankot, Punjab, India), two rain water samples and two bore-well drinking water samples collected from the Punjabi university campus (Patiala, Punjab, India), two metal factory wastewater samples were collected from a chemical factory (Patiala, India). Before the experiments were performed, all the water samples were filtered through Whatman filter paper and then through 0.22 μm micropore membranes and stored at 4 °C in a refrigerator.

2.5. Preparation of Sol–Gel C18 Coated FPSE Media

Due to the presence of starch and other finishing chemicals on the commercially available cellulose fabric used in apparel making, the substrate required a thorough cleaning. At the same time, a chemical treatment was applied to the substrate that maximizes the number of available hydroxyl groups which are required to effectively bind the sol–gel network via a condensation reaction. Preliminary cleaning of the substrate was accomplished by immersing a 100 cm^2 section of cotton cellulose fabric in deionized water for 30 min under constant sonication, followed by multiple rinsing steps with excess deionized (DI) water. For the surface chemical activation, the fabric was immersed in 1M NaOH solution for an hour under constant sonication. This step was followed by thorough rinsing with DI water and treatment with a 0.1M HCl solution for an hour to neutralize any residual NaOH which might still be on the fabric surface. Finally, the cleaned and activated cellulose substrate was dried and stored in an air-tight container until it was used for sol–gel C18 coating.

The sol solution used to create the sol–gel C18 coated FPSE media was prepared by sequentially mixing the sol–gel precursors methyl trimethoxysilane (MTMS); solvents methylene chloride and acetone; organically modified sol–gel precursor, octadecyl trimethoxysilane; sol–gel catalyst, trifluoroacetic acid (TFA) and water. In order to obtain a uniform sol solution for coating the fabric substrate, the molar ratio between methyl trimethoxysilane precursor:methylene chloride:acetone:trifluoroacetic acid:water was maintained at 1:2.33:1.94:0.5:0.20. The molar ratio between methyl trimethoxysilane and octadecyl trimethoxysilane was maintained at 1:0.38. To ensure that all the sol solution ingredients mixed homogeneously, the sol solution was vigorously vortexed for 3 min after adding each of the ingredients. The final sol solution was centrifuged for 5 min, followed by collection of the supernatant into a clean 3 oz. amber-colored glass reaction bottle. The solution was then sonicated for 10 min to remove trapped gaseous molecules from the sol solution.

The cleaned and pretreated cellulose fabric substrate was then gently immersed into the sol solution to initiate the substrate coating via a dip-coating process. The substrates were kept in the appropriate sol solutions for 2 h. During this surface coating period, a three-dimensional sol–gel network chemically bonded to the substrate was formed. At the end of the coating period, the sol solutions were expelled from the reaction bottle, the coated fabrics were dried in a desiccator and finally the sol–gel sorbent coated FPSE media was conditioned/aged in a home-made thermal conditioning device built inside a gas chromatography oven with continuous helium gas flow at 50 °C for 24 h. After conditioning/ageing, the sol–gel C18 coated FPSE media were cleaned sequentially with dichloromethane and methanol. Finally, the FPSE media were dried in the presence of continuous helium gas flow at 50 °C for 1 h. The FPSE media were then cut into 2.5 cm × 2.0 cm pieces, the typical application size for FPSE, and were stored in an airtight glass container for future use.

2.6. Fabric Phase Sorptive Extraction Procedure

Firstly, the sol–gel C18-coated FPSE media was conditioned in a mixture of 1 mL methanol and 1 mL acetonitrile for 5 min and then rinsed with 2 mL deionized water to remove any residual organic solvent. The FPSE procedure is shown in Figure 2.

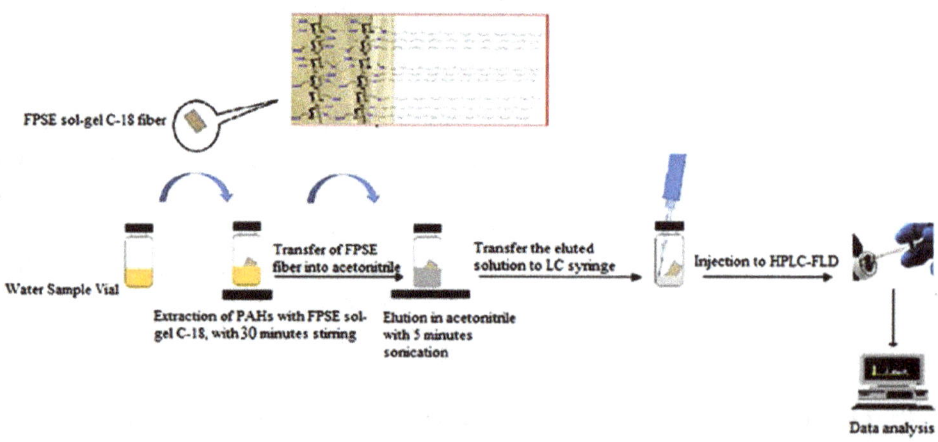

Figure 2. Chemical structure of sol–gel C18 coated fabric phase sorptive extraction (FPSE) media and the schematic representation of FPSE protocol.

The extraction process was performed quickly in a 20 mL glass vial containing 10 mL of the aqueous sample solution. A piece of sol–gel C18 coated FPSE media was directly immersed in the sample solution for 30 min under constant stirring at 1000 rpm at room temperature. After extraction from the sample solution, the PAHs were eluted from the sol–gel C18-coated FPSE media in 300 μL of acetonitrile with sonication for 5 min. Finally, 20 μL of this solution was injected directly into the HPLC-FLD system for analysis.

3. Results and Discussion

3.1. Selection of Fabric Phase Sorptive Extraction Sorbent Chemistry

Taking the nonpolar, hydrophobic characteristics of the selected PAHs into consideration (log K_{OW} values between 4.45 and 5.16), it is obvious that a nonpolar sorbent would provide the best affinity towards the PAHs to selectively isolate them from complex environmental water sample matrix. As such, octadecyl trimethoxysilane was selected as the source of the C18 pendant group in the sol–gel silica network during the sol solution design. C18 has long been known as a nonpolar sorbent in solid phase extraction and in nonpolar stationary phase in reversed phase liquid chromatography. It is worth mentioning that the selectivity of commercially available C18 sorbent is substantially different from that of sol–gel C18 sorbent and often the latter demonstrates unique selectivity towards a wide range of analytes including polar, moderately polar and nonpolar analytes due the presence of residual surface silanol groups in the sol–gel C18 sorbent matrix.

Unlike the substrate used in C18 sorbent (silica particles), the substrate used in FPSE plays an active role in determining the overall polarity and selectivity of the FPSE media. A fabric substrate made of 100% cotton cellulose was selected as the support for sol–gel C18 coating because the hydrophilic nature of the substrate can cause water molecules to come close to the extraction device during extraction so that the requisite interactions between the C18 pendant groups and the PAHs can result in successful extraction of PAHs onto the FPSE media. The permeable structure of the fabric support also plays an important role as a pseudo-solid phase extraction disk and facilitates rapid analyte extraction due to the continuous diffusion of water through the FPSE media when under the influence of magnetic stirring during the extraction.

3.2. Characterization of Sol–Gel C18 Coated FPSE Media

3.2.1. Scanning Electron Microscopy

Figure 3 represents the scanning electron micrographs (SEM) of (a) the uncoated Muslin cotton (100% cellulose) substrate at 100× magnification; (b) sol–gel C18 coated fabric phase sorptive extraction media at 100× magnification; and (c) sol–gel C18 coated fabric phase sorptive extraction media at 500× magnification.

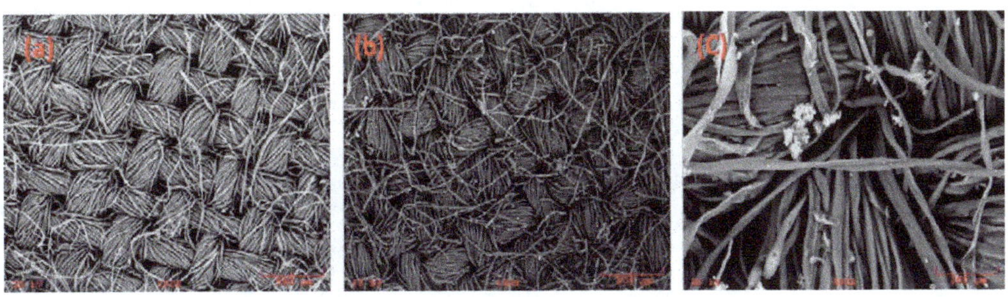

Figure 3. Scanning electron microscopy images of (**a**) uncoated cotton (100%) cellulose fabric substrate at 100× magnification; (**b**) uniformly coated sol–gel C18 sorbent coating on fabric matrix at 100× magnification; (**c**) sol–gel C18 coated FPSE media at 500× magnification.

The extraction efficiency of FPSE primarily depends on the sponge-like porous architecture of the sol–gel sorbent coating (for faster analyte diffusion) and the permeability of the fabric substrate that mimics a solid phase extraction disk. The highly porous sorbent coating as well as the permeability and hydrophilicity of the cellulose fabric synergistically allow the aqueous sample matrix to flow through the FPSE medium, leading to rapid interaction between the sorbent and the analytes. Consequently, the analytes are adsorbed on the sorbent with high efficiency in a short period of time. As such, it is important to study the surface morphology of the FPSE media before and after the coating to ensure that the pores of the fabric substrate are well maintained even after the sol–gel coating.

The SEM images demonstrated that the microstructures and the pores of the cellulose substrate were well preserved even after the sol–gel C18 sorbent coating. This flow-through extraction mechanism is only exploited in solid phase extraction (SPE), and is totally absent in solid phase microextraction and related techniques (such as stir bar sorptive extraction, thin film microextraction, etc.) due to the impermeable nature of the substrate used in these microextraction techniques. The flow-through extraction system consequently helps to achieve faster extraction equilibrium. The enlarged image of the sol–gel C18-coated FPSE media demonstrates that C18 coatings are homogeneously distributed on the fabric substrate surface while maintaining the pores of the substrate.

3.2.2. Fourier-Transform Infrared Spectroscopy (FT-IR)

Figure 4 illustrates FT-IR spectra representing (a) uncoated Muslin cotton (100% cellulose) fabric; (b) octadecyl trimethoxysilane; (c) sol–gel C18 coated fabric phase sorptive extraction media.

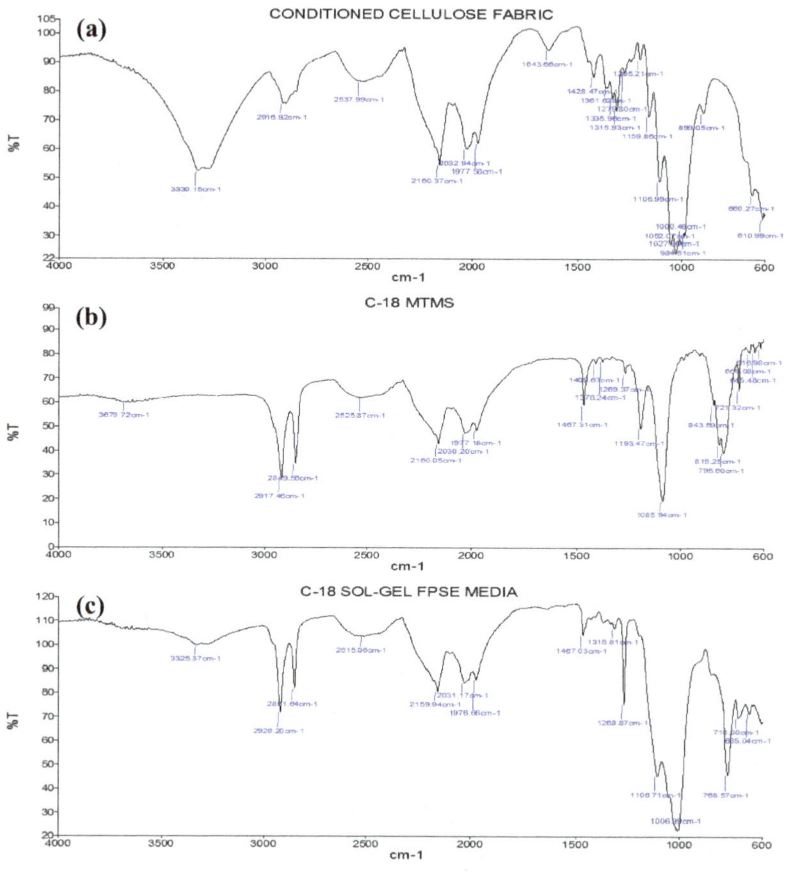

Figure 4. FT-IR spectra of (**a**) uncoated cellulose substrate; (**b**) octadecyl trimethoxysilane; and (**c**) sol–gel C18 coated FPSE medium.

The FT-IR spectra of the uncoated Muslin cotton 100% cellulose fabric demonstrates characteristic absorption bands at ~3330 cm^{-1}, 2916 cm^{-1}, 1315 cm^{-1}, and 1028 cm^{-1} that correspond to O–H, C–H,

C–O vibration and C–H bending vibration, respectively [50]. The characteristic peaks of octadecyl trimethoxysilane appear at 2917 cm^{-1}, and 2849 cm^{-1} which correspond to asymmetric and symmetric vibrations of $-CH_2-$, $-CH_2-$ groups, respectively [51]. The strong peak at 1467 cm^{-1} and 1085 cm^{-1} are due to the vibration absorption of Si–O–C, and Si–O–Si, respectively. The strong peak at 815 cm^{-1} is assigned to Si–C bonds and the peak at 795 cm^{-1} is assigned to the vibration of $(-CH_2-)_n$ $(n \geq 4)$ [52]. The characteristic peaks of sol–gel C18-coated FPSE media appeared at 2890 cm^{-1} and 2851 cm^{-1} which represent symmetric vibration of $-CH_2-$ and asymmetric vibration of $-CH_3$, respectively. The same characteristic peaks are also seen in octadecyl trimethoxysilane spectra. In addition, the presence of ~1467 cm^{-1}, ~1268 cm^{-1}, ~1977 cm^{-1} in both the sol–gel C18 coated FPSE media and octadecyl trimethoxysilane strongly suggests the successful integration of octadecyl moieties into the sol–gel network. The substantial reduction of the O–H stretching vibrations (at 3325 cm^{-1}) in sol–gel C18 coated FPSE media compared to uncoated cotton (100% cellulose) indicates the chemical integration of the sol–gel C18 network to the cellulose structure via condensation. Due to the chemical integration of sol–gel sorbent to the substrate surface, the resulting FPSE media offers a remarkably superior thermal, solvent and chemical stability than its commercial counterparts such as SPME, SBSE etc.

3.3. Optimization of the FPSE Procedure

In order to achieve accurate and sensitive chromatographic quantification of the trace PAHs in the water samples, the optimum conditions for using sol–gel C18 coated FPSE media were investigated. Several conditions affecting the extraction efficiency were optimized, including extraction time, sample volume, eluting solvent, elution time, volume of organic modifier, and salt concentration. Optimization experiments were performed using a standard aqueous solution of PAHs containing 0.10 µg/mL each of four PAHs to ensure a comparable level of response to each compound.

3.3.1. Optimization of Sample Volume

To obtain high enrichment factors and high recoveries for all PAHs, the initial sample volume should be as large as possible. Therefore, different volumes (5 mL, 10 mL, 15 mL, 20 mL) of an aqueous solution were investigated. It was found that the highest extraction efficiency was obtained with a sample volume of 10 mL, as shown in Figure 5a. As the sample volume was increased up to 15 mL, recovery increased and after 15 up to 20 mL no obvious change was observed. After 20 mL, recovery decreased up to 50 mL, inferring that the extraction efficiency was insufficient at volumes above 20 mL. Therefore, the initial sample volume was set at 15 mL for future FPSE protocols.

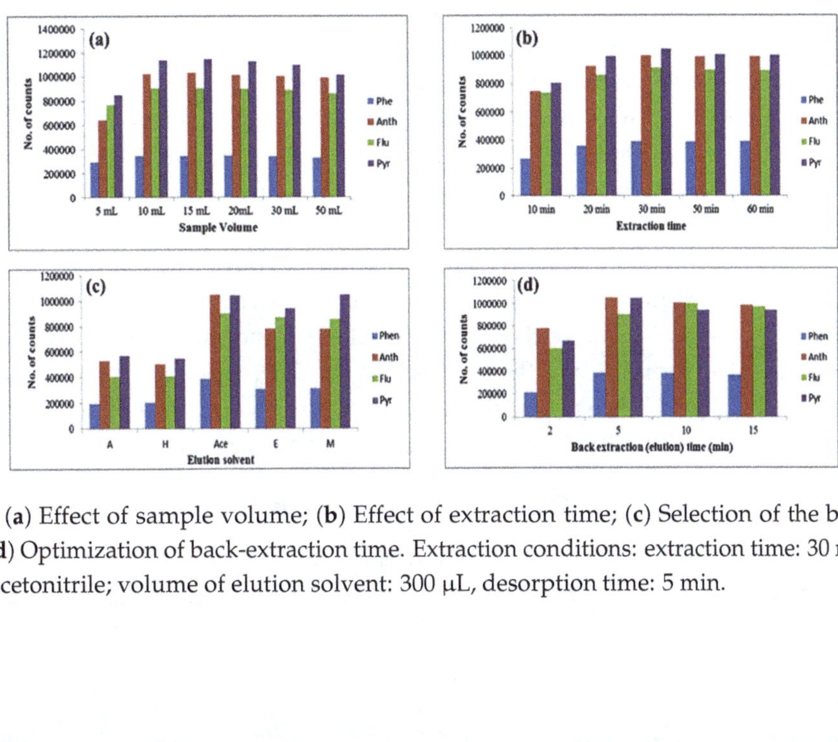

Figure 5. (**a**) Effect of sample volume; (**b**) Effect of extraction time; (**c**) Selection of the best elution solvent; (**d**) Optimization of back-extraction time. Extraction conditions: extraction time: 30 min; eluent solvent: acetonitrile; volume of elution solvent: 300 µL, desorption time: 5 min.

3.3.2. Optimization of Extraction Time

Another key factor affecting extraction efficiency is the extraction time and it so it was imperative that this factor be thoroughly investigated. The extraction time was set from 10 to 60 min. As shown in Figure 5b, recoveries of all four PAHs increased with extraction time as it was increased from 10 to 30 min, and then remained unchanged even when the time was increased up to 60 min, implying that the extraction equilibrium was achieved at about 30 min. Thirty minutes was therefore selected as the extraction time of FPSE protocol.

3.3.3. Optimization of Desorption Solvent and Time

As far as the FPSE protocol is concerned, PAH desorption from the sol–gel C18 coated FPSE media can significantly affect the sensitivity of PAH extraction. Therefore, choice of an appropriate elution solvent plays a key role in the process. Having considered the properties of PAHs, acetone (A), n-hexane (H), acetonitrile (Ace), ethanol (E) and methanol (M) were selected as potential elution solvents in this experiment. As shown in Figure 5c, acetone and n-hexane were poor eluents for PAHs. Methanol yielded the highest recovery for Pyr, whereas acetonitrile was preferable for recovery of all four PAHs. Therefore, acetonitrile was chosen as the elution solvent and was used for further studies. Furthermore, the elution efficiency also relies on the volume of the elution solvent. As shown in Supplementary Figure S1, for most of the PAHs, the recovery increased as the eluent volume increased from 100 μL to 300 μL, but remains nearly unchanged if the volume is further increased from 300 μL to 700 μL, after which dilution causes a decrease in recovery. To get maximum recovery with minimum solvent use, 300 μL of acetonitrile was selected for desorption.

In order to resolve any possible carry-over problems and to avoid any loss of PAHs, the sonication desorption time was further optimized. The process of desorption was carried out in an ultrasonic bath with desorption times of 1, 3, 5 and 10 min. The results shown in Figure 5d prove that the peak areas of PAHs increased as the desorption time increased from 1 min to 5 min, but remained unchanged as the desorption time was increased further. Therefore, 5 min was sufficient to achieve maximum desorption.

3.3.4. Effect of Salt Concentration and Organic Modifiers

Salt ions in the sample might also affect FPSE by competitive interaction between the salting-in and the salting-out effect. The salting-out effect causes the analyte to enhance its partition onto the sol–gel C18-coated FPSE media by decreasing its solubility in water, while the salting-in effect has the opposite effect. Thus, the effect of the addition of salt to the samples was investigated and is shown in Supplementary Figure S2. No obvious change was observed for the recoveries with KCl at 0–0.15 M, indicating that salt ion addition does not affect extraction efficiency. This is probably due to the low polarity of all four PAHs. Thus, no salt was added in the following experiments.

Addition of an organic modifier, such as methanol, might promote the extraction efficiency of FPSE for PAHs by preventing the FPSE C18 carbon chains from cross-linking and thus retaining the ability to contact the target analytes completely. It was found that methanol addition (0–3 mL) did not cause any changes in extraction efficiency as shown in Supplementary Figure S3. This was probably because of the fact that FPSE media coated with C18 hydrophobic long chains spreads out well in the form of a uniform film on both sides of cellulose fabric substrate, due to strong chemical bonding between the sol–gel sorbent and the fabric substrate. Therefore, no organic modifier is needed. This indicates that the sol–gel C18 coated FPSE media is stable in various solutions, and the extraction efficiency is independent of the salinity and the presence of an organic modifier.

3.4. Regeneration and Reusability of Sol–Gel C18 Coated FPSE Media

The durability of sol–gel C18 coated FPSE media was also investigated by extracting PAHs from a water sample 30 times using the same FPSE media. The FPSE media was regenerated with sonication in 10 mL acetonitrile for 15 min. The extraction efficiency of the sol–gel C18 coated FPSE media was almost unchanged after 30 extraction procedures the results of which are shown in Supplementary Figure S4. These results indicate that sol–gel C18 coated FPSE media can be repeatedly used for extraction.

3.5. Analytical Performance

All data were subjected to strict quality control procedures. The linearities of the chromatographic responses of the protocol were tested with calibration standards at seven concentration levels ranging from 0.01 to 10 ng/mL. Good linearities were observed for the four target PAHs, with all the correlation coefficients (R^2) above 0.99. Limits of quantification (LOQs) (signal-to-noise ratio = 10) and limits of detection (LODs) (signal-to-noise ratio = 3) of the four target compounds are shown in Table 1.

Table 1. Linear range, linearity curve, correlation coefficients, limits of detection (LODs) and limits of quantification (LOQs) for the determination of PAHs (n = 5).

Analyte	Linear Range (ng/mL)	Linearity Curve	R^2	LOD (pg/mL)	LOQ (pg/mL)
Phen	0.010–10	$y = 34{,}591x + 772$	0.9997	1	3.33
Anth	0.010–10	$y = 104{,}873x + 2147$	0.9997	0.1	0.33
Flu	0.010–10	$y = 90{,}258x + 247$	0.9987	0.7	2.33
Pyr	0.010–10	$y = 114{,}921x + 497$	0.9983	0.4	1.33

The accuracy, precision and recovery studies were performed on 15 mL deionized organic-free water spiked with (low, medium, and high concentration levels) 0.01, 0.5 and 5 ng of each PAH per mL (five replicates). Mean recoveries ranged from 88.1% to 90.5% for all the target compounds. The mixture of standards was analyzed five times within a day. The relative standard deviations (RSD) of concentrations for all the targets ranged from 1.1% to 4.1% (Table 2), demonstrating the high precision of the analytical protocol. In order to ensure the accuracy of the analysis, all samples were replicated five times, and the final concentration averages were used.

Table 2. Precision, accuracy and recovery of four selected PAHs in spiked deionized organic-free water at (low, medium, and high concentration levels) (n = 5).

Analyte	Phen			Anth			Flu			Pyr		
Precision (RSD%)	0.01	0.5	5	0.01	0.5	5	0.01	0.5	5	0.01	0.5	5
Intra-day	5.6	4.7	3.6	1.9	2.8	2.2	2.1	2.8	1.8	1.9	2.1	1.1
Inter-day	4.9	5.01	4.1	2.2	2.3	2.5	1.8	2.9	1.9	3.1	2.7	2.0
Accuracy (%)												
Intra-day	88	87	90	90	92	91	91.1	90.1	92.1	90.8	91.6	92.8
Inter-day	86	89	89.4	91	90	89.3	91	90	92	90	89	91
Recovery (%)	88.1	88	88.4	88.6	88	88.1	92	91	90.5	90.7	91	90.1
(RSD%)	(2.9)	(4.1)	(3.8)	(3.1)	(3.2)	(2.4)	(2.2)	(2.8)	(1.8)	(3.2)	(4.2)	(1.4)

3.6. Mathematical Model for Predicting Extraction Efficiency (Absolute Recovery, %)

A mathematical model was created for sol–gel C18 coated FPSE media that can be used as a predictive tool to assess the extraction efficiency (expressed as the absolute recovery) of analytes on sol–gel C18-coated FPSE media using their log K_{ow} values. A carefully chosen test mixture was created to develop the mathematical model, consisting of 10 compounds representing different polarity and functionality (log K_{ow} values ranging from 0.3 to 5.07). The selected test compounds included piperonal (PIP), phenol (PHE), furfuryl alcohol (FA), benzodioxole (BDO), naphthalene (NAP), 4-nitrotoluene (4NT), 9-anthracene methanol (9AM), 1,2,4,5-tetramethyl benzene (TMB), triclosan (TCL) and diethylstilbestrol (DESB). The extraction recovery value of each compound from an aqueous solution of the test mixture was determined and the values were plotted against their log K_{ow} values to obtain a second order mathematical model: Extraction efficiency (absolute recovery, %) = −2.274875 + 20.816015 × log K_{ow} − 4.1478973 × (log K_{ow} − 2.737)^2. This second order mathematical model can be used to predict the extraction efficiency of sol–gel C18-coated FPSE media for a given analyte using its log K_{ow} value. A graphical representation of the model is shown in Figure 6.

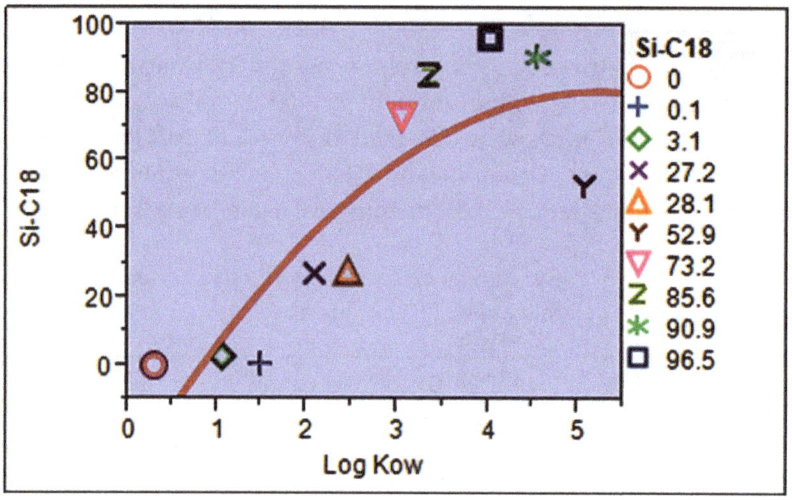

Figure 6. Correlation curve between absolute recovery and logarithmic values of octanol-coefficients of analytes with wide polarity range.

The predicted recovery values for four PAHs using the mathematical model and their actual recovery values are given in Table 3.

Table 3. Data demonstrating model predicted absolute recovery (%) and actual recovery (%) obtained by FPSE–HPLC–FLD method.

Compound	Log K_{ow}	Expected Recovery (%)	Actual Recovery (%)
Phen	4.46	78.30	88.4
Anth	4.45	78.24	88.1
Flu	5.16	80.85	90.5
Pyr	4.88	80.32	90.1

For all the target PAHs, the actual extraction recovery values were found to be higher than the predicted values obtained from the model. This was attributed to the strong hydrophobic interactions between the PAHs and the sol–gel C18 sorbent, in addition to efficient trapping of the PAHs on the sol-gel C18 extraction media with high primary contact surface area, high loading of sol–gel nanocomposite sorbent (in the form of ultrathin film), as well as the permeability of the extraction device that mimics a solid phase extraction disk (characterized by exhaustive extraction).

3.7. Application to Real Water Samples

Subsequently, the FPSE-HPLC-FLD protocol developed was applied for PAH determination in real environmental water samples: bore-well water, river water, rain water and factory wastewater. Quintuplicate analyses were performed and the concentrations found for the target PAHs are summarized in Table 3. In the four types of environmental water samples, the target four PAHs were not detected in bore-well water and rain water. Phen and Flu were detected in two river water samples and all four target PAHs were found in each of the two metal-fabrication factory wastewater samples (Table 4).

Table 4. The four selected PAHs concentrations detected in real water samples (n = 5).

Sample	Phen (ng/mL)	Anth (ng/mL)	Flu (ng/mL)	Pyr (ng/mL)
A [a]	n.d	n.d	n.d	n.d
B [a]	n.d	n.d	n.d	n.d
C [b]	7.8 ± 2.6	n.d	5.6 ± 3.3	n.d
D [b]	8.8 ± 2.9	n.d	6.8 ± 3.4	n.d
E [c]	n.d	n.d	n.d	n.d
F [c]	n.d	n.d	n.d	n.d
G [d]	11 ± 3.1	7.8 ± 2.5	7.5 ± 3.8	5.8 ± 4.1
H [d]	4.1± 3.2	3.7 ± 3.1	7.7 ± 3.3	3.0 ± 2.8

[a]: bore-well water; [b]: river water; [c]: rain water; [d]: factory waste water; n.d: not detected.

The two river water samples, collected from two different locations in the Chakki River in Pathankot were analyzed. At one outlet point of metal factory wastewater, Phen, Anth, Flu, Pyr were detected at 11 ng/mL, 7.8 ng/mL, 7.5 ng/mL and 5.8 ng/mL, respectively, while at second outlet point, Phen, Anth, Flu, Pyr were detected at 4.1 ng/mL, 3.7 ng/mL, 7.7 ng/mL and 3.0 ng/mL, respectively, (shown in Table 3); suggesting that the PAHs came mainly from the wastewater. These results indicated that the method could be successfully applied to PAH analysis in real water samples. The typical chromatograms of a river water sample and a metal factory waste water sample are presented in Figure 7.

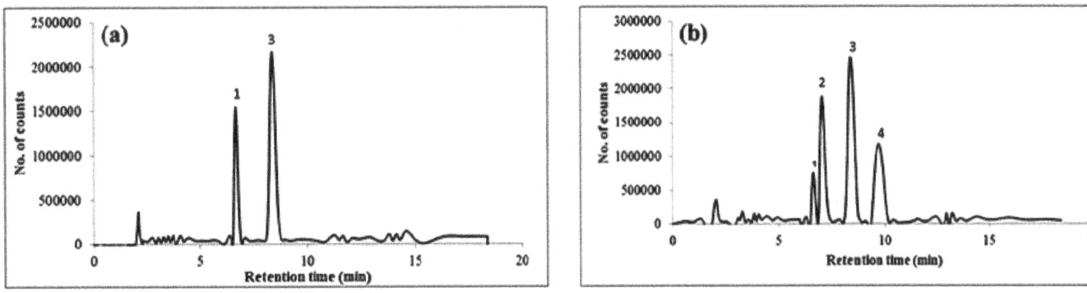

Figure 7. Chromatogram representing analysis of real environmental samples: (**a**) Chakki river water sample 2; (**b**) factory wastewater sample 2.

3.8. Comparison of Sol–Gel C18 Coated FPSE Media with Other Sorbent Materials

The performance of the new analytical protocol with the sol–gel C18-coated FPSE media as the adsorbent material was also compared with other materials reported in the literature (as shown in Table 5).

Table 5. Comparisons of the analytical performance of the method developed with reported methods in the literature.

Method	Sorbent Material	Sorbent Preparation Time (Hours)	LOD (pg/mL)	References
MSPE-HPLC-FLD	TPA-functionalized MNPs	37.5	0.04–3.75	[53]
μ-SPE-GC-MS	Functionalized graphene sheet	29.5	0.8–3.9	[54]
SPE-HPLC-FLD	Cotton fiber	0.5	0.1–2	[55]
MIPs-SPE-GC-MS	Imprinted sol-gel adsorbent	38	5.2–12.6	[56]
FPSE-HPLC-FLD	Sol-gel C18 (Cellulose)	24	0.1–1	Current work

Abbreviations: MIPs: molecularly imprinted polymers; MSPE: magnetic solid-phase extraction; TPA-functionalized MNPs: triphenylamine functionalized magnetic microspheres. HPLC-FLD: high performance liquid chromatography with fluorescence detector; GC-MS: gas chromatography mass spectrometry.

The time required in sorbent preparation and the LOD values were compared. The LOD values achieved in the present research are lower than those reported in the literature. Furthermore, sol–gel C18-coated FPSE media as a sorbent has obvious advantages: preparation of sorbents often requires several days to complete the synthesis process, involves dozens or hundreds of milliliters of organic solvents, is time consuming and may cause environment pollution, whereas sol–gel C18-coated FPSE media, an advanced inorganic–organic hybrid material with tunable porosity, selectivity, thermal and chemical stability, high reproducibility and solvent resistance, requires less organic solvents, is fast and is a green analytical endeavor. Therefore, the proposed FPSE-HPLC-FLD protocol is proven to be a green, convenient, efficient and reliable method for the pre-concentration of trace PAHs from water samples.

4. Conclusions

In summary, for the first time, sol–gel C18-coated FPSE media was directly and successfully applied as a selective adsorbent for PAH analysis in real environmental water samples. The LODs obtained for four target PAH compounds using the novel analytical protocol were 0.1–1 pg/mL. Compared to traditional SPE methods, the sol–gel C18-coated FPSE media is simple to prepare and regenerate for recurring usage, thereby meeting the need for rapid analysis. In addition, the costs of the preparation of the sol–gel C18-coated FPSE media are economical and organic solvent consumption is minimal. Sol–gel C18-coated FPSE media can also be regenerated and reused more than 30 times. The FPSE media is also degradable and therefore more environment friendly. Utilizing the unique attributes of FPSE, a novel, simple, efficient, fast, sensitive, green, economical and reliable FPSE-HPLC-FLD protocol is presented for trace level determination of four environmentally important PAHs. This protocol is suitable for many applications in health-related water contamination studies and offers new analytical capabilities for water quality assurance, particularly for routine monitoring of the presence of PAHs in water samples in order to ensure water quality, safety and consumer protection.

Supplementary Materials: The following are available online at http://www.mdpi.com/2297-8739/4/2/22/s1, Figure S1: Effect of eluent solvent volume: Extraction conditions: Sample volume, 15 mL; extraction time: 30 min; eluent solvent: acetonitrile; sonication desorption time 5 min. Figure S2: Effect of salt concentration: Extraction conditions: sample volume, 15 mL; extraction time: 30 min; eluent solvent, acetonitrile; volume of elution solvent, 300 μL desorption time, 5 min. Figure S3: Effect of organic modifier volume: Extraction conditions: sample volume, 15 mL; extraction time: 30 min; eluent solvent, acetonitrile; volume of elution solvent, 300 μL desorption time, 5 min. Figure S4: Regeneration and reusability of sol–gel C18 coated FPSE media. Extraction conditions: sample volume, 15 mL; extraction time: 30 min; eluent solvent: acetonitrile; volume of elution solvent: 300 μL desorption time, 5 min.

Acknowledgments: This study was financially supported by University Grants Commission (UGC), New Delhi, India.

Author Contributions: Shivender Singh Saini and Abuzar Kabir conceived and designed the experiments; Shivender Singh Saini and Abuzar Kabir performed the experiments; Shivender Singh Saini and Abuzar Kabir analyzed the data; Avasarala Lakshmi Jagannadha Rao, Ashok Kumar Malik and Kenneth G. Furton contributed reagents/materials/analysis tools; Shivender Singh Saini and Abuzar Kabir wrote the paper. Avasarala Lakshmi Jagannadha Rao, Ashok Kumar Malik and Kenneth G. Furton reviewed and edited the paper.

Conflicts of Interest: The authors declare no conflict of interest.

References

1. Agency for Toxic Substances and Drug Registry (ATSDR). *Toxicological Profile for Polycyclic Aromatic Hydrocarbons (PAHs) Update*; Agency for Toxic Substances and Drug Registry (ATSDR): Washington, DC, USA, 1995; Chapter 1, pp. 1–9.
2. Agency for Toxic Substances and Drug Registry (ATSDR). *Toxicological Profile for Polycyclic Aromatic Hydrocarbons (PAHs) Update*; Agency for Toxic Substances and Drug Registry (ATSDR): Washington, DC, USA, 1995; Chapter 2, pp. 11–207.
3. Bostrom, C.E.; Gerde, P.; Hanberg, A.; Jernstrom, B.; Johansson, C.; Kyrklund, T.; Rannug, A.; Tornqvist, M.; Victorin, K.; Westerholm, R. Cancer risk assessment, indicators, and guidelines for polycyclic aromatic hydrocarbons in the ambient air. *Environ. Health Perspect.* **2002**, *110*, 451–488. [CrossRef] [PubMed]

4. White, K.L. An overview of immunotoxicology and carcinogenic polycyclic aromatic hydrocarbons. *Environ. Carcinog. Rev.* **1986**, *4*, 163–202. [CrossRef]
5. Srogi, K. Monitoring of environmental exposure to polycyclic aromatic hydrocarbons: A review. *Environ. Chem. Lett.* **2007**, *5*, 169–195. [CrossRef]
6. Beyer, J.; Jonsson, G.; Porte, C.; Krah, M.M.; Ariese, F. Analytical methods for determining metabolites of polycyclic aromatic hydrocarbon (PAH) pollutants in fish bile: A review. *Environ. Toxicol. Pharmacol.* **2010**, *30*, 224–244. [CrossRef] [PubMed]
7. Van der Oost, R.; Beyer, J.; Vermeulen, N.P.E. Fish bioaccumulation and biomarkers in environmental risk assessment: A review. *Environ. Toxicol. Pharmacol.* **2003**, *13*, 57–149. [CrossRef]
8. Liu, L.Y.; Wang, J.Z.; Wei, G.L.; Guan, Y.F.; Wong, C.S.; Zeng, E.Y. Sediment records of polycyclic aromatic hydrocarbons (PAHs) in the continental shelf of China: Implications for evolving anthropogenic impacts. *Environ. Sci. Technol.* **2012**, *46*, 6497–6504. [CrossRef] [PubMed]
9. Baek, S.O.; Field, R.A.; Goldstone, M.E.; Kirk, P.W.; Lester, J.N.; Perry, R. A review of atmospheric polycyclic aromatic hydrocarbons: Sources, fate and behavior. *Water Air Soil Pollut.* **1991**, *60*, 279–300. [CrossRef]
10. Chen, H.Y.; Teng, Y.G.; Wang, J.S.; Song, L.T.; Zuo, R. Source apportionment of sediment PAHs in the Pearl River Delta region (China) using nonnegative matrix factorization analysis with effective weighted variance solution. *Sci. Total Environ.* **2013**, *444*, 401–408. [CrossRef] [PubMed]
11. Weinstein, J.E.; Crawford, K.D.; Garner, T.R.; Flemming, A.J. Screening level ecological and human health risk assessment of polycyclic aromatic hydrocarbons in stormwater detention pond sediments of Coastal South Carolina, USA. *J. Hazard. Mater.* **2010**, *178*, 906–916. [CrossRef] [PubMed]
12. Guo, J.F.; Guo, Q.Z.; Yan, G.P. Determination of polycyclic aromatic hydrocarbons in water samples by hollow fiber extraction coupled with GC-MS. *Anal. Methods* **2015**, *7*, 1071–1075. [CrossRef]
13. Shamsipur, M.; Hashemi, B. Extraction and determination of polycyclic aromatic hydrocarbons in water samples using stir bar sorptive extraction (SBSE) combined with dispersive liquid-liquid microextraction based on the solidification of floating organic drop (DLLME-SFO) followed by HPLC-UV. *RSC Adv.* **2015**, *5*, 20339–20345.
14. Zhou, Q.; Gao, Y. Determination of polycyclic aromatic hydrocarbons in water samples by temperature-controlled ionic liquid dispersive liquid-liquid microextraction combined with high performance liquid chromatography. *Anal. Methods* **2014**, *6*, 2553–2559. [CrossRef]
15. Manoli, E.; Samara, C. Polycyclic aromatic hydrocarbons in natural waters: Sources, occurrence and analysis. *TrAC Trends Anal. Chem.* **1999**, *18*, 417–428. [CrossRef]
16. Pánková, K. IACR Monographs on the Evaluation of the Carcinogenicity Risk of Chemical to Humans. *Biol. Plant.* **1986**, *28*, 354. [CrossRef]
17. Mahler, B.J.; Van Metre, P.C.; Bashara, T.J.; Wilson, J.T.; Johns, D.A. Parking lot sealcoat: An unrecognized source of urban polycyclic aromatic hydrocarbons. *Environ. Sci. Technol.* **2005**, *39*, 5560–5566. [CrossRef] [PubMed]
18. Dickhut, R.M.; Canuel, E.A.; Gustafson, K.E.; Liu, K.; Arzayus, K.M.; Walker, S.E.; Edgecombe, G.; Gaylor, M.O.; MacDonald, E.H. Automotive sources of carcinogenic polycyclic aromatic hydrocarbons associated with particulate matter in the Chesapeake Bay region. *Environ. Sci. Technol.* **2000**, *34*, 4635–4640. [CrossRef]
19. Sicre, M.A.; Marty, J.C.; Saliot, A.; Aparicio, X.; Grimalt, J.; Albaiges, J. Aliphatic and aromatic hydrocarbons in different sized aerosols over the Mediterranean Sea: Occurrence and origin. *Atmos. Environ.* **1967**, *21*, 2247–2259. [CrossRef]
20. Fermin, L.; Valley, C.; Alonso, E.J.; Olmo, M.D.; Vilchez, J.L. Determination of ultra-traces of anthracene in water samples by solid-phase spectrofluorometry. *Anal. Sci.* **1993**, *9*, 117–120.
21. Arbabi, M.; Nasseri, S.; Mesdaghinia, A.R.; Rezaie, S.; Naddafi, K.; Omrani, G.H.; Yunesian, M. Survey on physical, chemical and microbiological characteristics of PAH-contaminated soils in Iran. *Iran. J. Environ. Health Sci. Eng.* **2004**, *1*, 26–33.
22. Kanaly, R.; Harayama, S. Biodegradation of high-molecular-weight polycyclic aromatic hydrocarbons by bacteria. *J. Bacteriol.* **2000**, *182*, 2059–2067. [CrossRef] [PubMed]
23. Kweon, O.; Kim, S.J.; Jones, R.C.; Freeman, J.P.; Adjei, M.D.; Edmondson, R.D.; Cerniglia, C.E. A polyomic approach to elucidate the fluoranthene-degradative pathway in *Mycobacterium vanbaalenii* PYR-1. *J. Bacteriol.* **2007**, *189*, 4635–4647. [CrossRef] [PubMed]

24. Zhang, X.J.; Shi, Z.; Lyv, J.X.; He, X.; Englert, N.A.; Zhang, S.Y. Pyrene is a novel constitutive androstane receptor (CAR) activator and causes hepatotoxicity by CAR. *Toxicol. Sci.* **2015**, *147*, 436–445. [CrossRef] [PubMed]
25. Hollosi, L.; Wenzl, T. Development and optimisation of a dopant assisted liquid chromatographic-atmospheric pressure photo ionisation-tandem mass spectrometric method for the determination of 15 + 1 EU priority PAHs in edible oils. *J. Chromatogr. A* **2011**, *1218*, 23–31. [CrossRef] [PubMed]
26. Purcaro, G.; Moret, S.; Bucar-Miklavcic, M.; Conte, L.S. Ultra-high performance liquid chromatographic method for the determination of polycyclic aromatic hydrocarbons in a passive environmental sampler. *J. Sep. Sci.* **2012**, *35*, 922–928. [CrossRef] [PubMed]
27. Zhao, B.; Zhang, S.; Zhou, Y.; He, D.; Li, Y.; Ren, M.; Xu, Z.; Fang, J. Characterization and quantification of PAH atmospheric pollution from a large petrochemical complex in Guangzhou: GC–MS/MS analysis. *Microchem. J.* **2015**, *119*, 140–144. [CrossRef]
28. Mirivel, G.; Riffault, V.; Galloo, J.C. Simultaneous determination by ultra-performance liquid chromatography-atmospheric pressure chemical ionization time-of-flight mass spectrometry of nitrated and oxygenated PAHs found in air and soot particles. *Anal. Bioanal. Chem.* **2010**, *397*, 243–256. [CrossRef] [PubMed]
29. Fujiwara, F.; Guinez, M.; Cerutti, S.; Smichowski, P. UHPLC-(+)APCI-MS/MS determination of oxygenated and nitrated polycyclic aromatic hydrocarbons in airborne particulate matter and tree barks collected in Buenos Aires city. *Microchem. J.* **2014**, *116*, 118–124. [CrossRef]
30. Cruz-Vera, M.; Lucena, R.; Ardenas, S.C.; Valcarcel, M. One-step in-syringe ionic liquid-based dispersive liquid-liquid microextraction. *J. Chromatogr. A* **2009**, *1216*, 6459–6465. [CrossRef] [PubMed]
31. Wang, W.D.; Huang, Y.M.; Shu, W.Q.; Cao, J. Multiwalled carbon nanotubes as adsorbents of solid-phase extraction for determination of polycyclic aromatic hydrocarbons in environmental waters coupled with high-performance liquid chromatography. *J. Chromatogr. A* **2007**, *1173*, 27–36. [CrossRef] [PubMed]
32. Ma, J.; Xiao, R.; Li, J.; Yu, J.; Zhang, Y.; Chen, L. Determination of 16 polycyclic aromatic hydrocarbons in environmental water samples by solid-phase extraction using multi-walled carbon nanotubes as adsorbent coupled with gas chromatography-mass spectrometry. *J. Chromatogr. A* **2010**, *1217*, 5462–5469. [CrossRef] [PubMed]
33. Brum, D.M.; Cassella, R.J.; Pereira Netto, A.D. Multivariate optimization of a liquid-liquid extraction of the EPA-PAHs from natural contaminated waters prior to determination by liquid chromatography with fluorescence detection. *Talanta* **2008**, *74*, 1392–1399. [CrossRef] [PubMed]
34. Shamsipur, M.; Hassan, J. A novel miniaturized homogenous liquid-liquid solvent extraction-high performance liquid chromatographic-fluorescence method for determination of ultra traces of polycyclic aromatic hydrocarbons in sediment samples. *J. Chromatogr. A* **2010**, *1217*, 4877–4882. [CrossRef] [PubMed]
35. Elisabeth, R.; Rajasekhar, B. Optimization and validation of solid phase micro-extraction (SPME) method for analysis of polycyclic aromatic hydrocarbons in rainwater and stormwater. *Phys. Chem. Earth* **2009**, *34*, 857–865.
36. Bagheri, H.; Babanezhad, E.; Es-haghi, A. An aniline-based fiber coating for solid phase microextraction of polycyclic aromatic hydrocarbons from water followed by gas chromatography-mass spectrometry. *J. Chromatogr. A* **2007**, *1152*, 168–174. [CrossRef] [PubMed]
37. Wu, Y.L.; Xia, L.B.; Chen, R.; Hu, B. Headspace single drop microextraction combined with HPLC for the determination of trace polycyclic aromatic hydrocarbons in environmental samples. *Talanta* **2008**, *74*, 470–477. [CrossRef] [PubMed]
38. Yao, C.; Twu, P.; Anderson, J.L. Headspace single drop microextraction using micellar ionic liquid extraction solvents. *Chromatographia* **2010**, *72*, 393–402. [CrossRef]
39. Djozan, D.J.; Assadi, Y. Modified pencil lead as a new fiber for solid-phase microextraction. *Chromatographia* **2004**, *60*, 313–317. [CrossRef]
40. Djozan, D.; Assadi, Y.; Haddadi, S.H. Anodized aluminum wire as a solid-phase microextraction fiber. *Anal. Chem.* **2001**, *73*, 4054–4058.
41. Bai, D.S.; Li, J.H.; Chen, S.B.; Chen, B.H. A novel cloud-point extraction process for preconcentrating selected polycyclic aromatic hydrocarbons in aqueous solution. *Environ. Sci. Technol.* **2001**, *35*, 3936–3940. [CrossRef] [PubMed]

42. Heidari, H.; Razmi, H.; Jouyban, A. Preparation and characterization of ceramic/carbon coated Fe_3O_4 magnetic nanoparticle nanocomposite as a solid-phase microextraction adsorbent. *J. Chromatogr. A* **2012**, *1245*, 1–7. [CrossRef] [PubMed]
43. Yang, S.; Chen, C.; Yan, Z.; Cai, Q.; Yao, S. Evaluation of metal-organic framework 5 as a new SPE material for the determination of polycyclic aromatic hydrocarbons in environmental waters. *J. Sep. Sci.* **2013**, *36*, 1283–1290. [CrossRef] [PubMed]
44. Kabir, A.; Furton, K.G. Fabric Phase Sorptive Extractors. U.S. Patent 9,283,544, 15 March 2016.
45. Racamonde, I.; Rodit, R.; Quintana, J.B.; Sieira, J.B.; Kabir, A.; Furton, K.G.; Cela, R. Fabric phase sorptive extraction: A new sorptive microextraction technique for the determination of non-steroidal anti-inflammatory drugs from environmental water samples. *Anal. Chim. Acta* **2015**, *865*, 22–30. [CrossRef] [PubMed]
46. Rodríguez-Gomez, R.; Roldan-Pijuan, M.; Lucena, R.; Cárdenas, S.; Zafra-Gómez, A.; Ballesteros, O.; Valcárcel, M. Stir-membrane solid-liquid-liquid microextraction for the determination of parabens in human breast milk samples by ultra high performance liquid chromatography-tandem mass spectrometry. *J. Chromatogr. A* **2014**, *1354*, 26–33. [CrossRef] [PubMed]
47. Samanidou, V.; Kabir, A.; Galanopoulos, L.D.; Furton, K.G. Fast extraction of amphenicols residues from raw milk using novel fabric phase sorptive extraction followed by high-performance liquid chromatography-diode array detection. *Anal. Chim. Acta* **2015**, *855*, 41–50. [CrossRef] [PubMed]
48. Kayillo, S.; Dennis, G.R.; Shalliker, R.A. An assessment of the retention behaviour of polycyclic aromatic hydrocarbons on reversed phase stationary phases: Selectivity and retention on C18 and phenyl-type surfaces. *J. Chromatogr. A* **2006**, *1126*, 283–297. [CrossRef] [PubMed]
49. World Health Organization. *Guidelines for Drinking-Water Quality*, 2nd ed.; World Health Organization: Geneva, Switzerland, 1997.
50. Kumar, R.; Heena, G.; Malik, A.K.; Kabir, A.; Furton, K.G. Efficient analysis of selected estrogens using fabric phase sorptive extraction and high performance liquid chromatography-fluorescence detection. *J. Chromatogr. A* **2014**, *1359*, 16–25. [CrossRef] [PubMed]
51. Brambilla, R.; Pires, G.P.; dos Santos, J.H.Z.; Miranda, M.S.L.; Chornik, B. Octadecylsilane-modified silicas prepared by grafting and sol–gel methods. *J. Electron Spectrosc. Relat. Phenom.* **2007**, *156*, 413–420. [CrossRef]
52. Pan, H.; Wang, X.D.; Xiao, S.S.; Yu, L.G.; Zhang, Z.J. Preparation and characterization of TiO_2 nanoparticles surface-modified by octadecyltrimethoxysilane. *Indian J. Eng. Mater. Sci.* **2013**, *20*, 561–567.
53. Long, Y.; Chen, Y.; Yang, F.; Chen, C.; Pan, D.; Cai, Q.; Yao, S. Triphenylamine-functionalized magnetic microparticles as a new adsorbent coupled with high performance liquid chromatography for the analysis of trace polycyclic aromatic hydrocarbons in aqueous samples. *Analyst* **2012**, *137*, 2716–2722. [CrossRef] [PubMed]
54. Zhang, H.; Low, W.P.; Lee, H.K. Evaluation of sulfonated graphene sheets as sorbent for micro-solid-phase extraction combined with gas chromatography-mass spectrometry. *J. Chromatogr. A* **2012**, *1233*, 16–21. [CrossRef] [PubMed]
55. Song, X.; Li, J.; Xu, S.; Ying, R.; Ma, J.; Liao, C.; Liu, D.; Yu, J.; Chen, L. Determination of 16 polycyclic aromatic hydrocarbons in seawater using molecularly imprinted solid-phase extraction coupled with gas chromatography-mass spectrometry. *Talanta* **2012**, *99*, 75–82. [CrossRef] [PubMed]
56. Wang, J.; Liu, S.; Chen, C.; Zou, Y.; Hu, H.; Cai, Q.; Yao, S. Natural cotton fibers as adsorbent for solid phase extraction of polycyclic aromatic hydrocarbons in water samples. *Analyst* **2014**, *139*, 3593–3599. [CrossRef] [PubMed]

6

Determination of the Biocide Econea® in Artificial Seawater by Solid Phase Extraction and High Performance Liquid Chromatography Mass Spectrometry

Robert A. Downs [1], John R. Dean [1], Adrian Downer [2] and Justin J. Perry [1,*]

1 Department of Applied Sciences, Northumbria University, Newcastle upon Tyne NE1 8ST, UK; robert.a.downs@googlemail.com (R.A.D.); John.Dean@northumbria.ac.uk (J.R.D.)

2 AkzoNobel, Stoneygate Lane, Felling Gateshead, Tyne & Wear NE10 0JY, UK; Adrian.Downer@akzonobel.com

* Correspondence: Justin.Perry@northumbria.ac.uk

Abstract: Econea®, or 4-bromo-2-(4-chlorophenyl)-5-(trifluoromethyl)-1H-pyrrole-3-carbonitrile, is a new type of environmentally friendly anti-fouling compound used in the immersed coatings of commercial sea going vessels. This paper reports the development of a new analytical method to directly detect the active biocide Econea® in an artificial sea water matrix using liquid chromatography mass spectrometry. The developed method is both rapid and sensitive, with a limit of detection of 0.05 μg/L and a limit of quantitation of 0.17 μg/L in artificial seawater. The subsequent developed method was then applied to investigate the biocide's release from a commercially available Econea® containing paint immersed in artificial sea water over a 45-day period. It was found that the average release rate of Econea® from this paint was 4.3 ± 0.6 μg cm^{-2} d^{-1}.

Keywords: Econea®; 4-bromo-2-(4-chlorophenyl)-5-(trifluoromethyl)-1H-pyrrole-3-carbonitrile; paint; artificial seawater; LC-MS

---

1. Introduction

The anti-fouling of the immersed hull of commercial boats, ships and yachts is an important economic consideration for transporting products and people across the earth's maritime environment. Fouling is the build-up on the immersed hull of marine organisms such as barnacles, molluscs, slime and weed. Their adherence to the hull severely impacts its hydrodynamic efficiency with a consequent effect on performance and fuel efficiency [1]. To prevent this disadvantageous build-up, the immersed surface needs to be regularly treated with anti-fouling products. Historically, a range of anti-fouling products have been used that contain metal-based biocides (including copper, organotin and zinc), though these can lead to the build-up of elemental residues, and are increasingly the target of environmental legislation [2]. More recently, alternative technologies have been proposed that do not use metals, with one approach being biodegradable—and hence environmentally friendly—alternative biocides [3]. An example of this is Econea®, or 4-bromo-2-(4-chlorophenyl)-5-(trifluoromethyl)-1H-pyrrole-3-carbonitrile (see Scheme 1, compound **A**). Whilst the initial purpose of this biocide was to protect growing plants from infestation and attack by insects and plant mites [4], Econea® is also sold in anti-fouling paints by several different paint manufacturers [5]. Typically, paints are formulated using between 0.5 wt % and 9.9 wt % Econea® based on the total weight of the dry mass composition. When formulated at this level, excellent activity is observed against barnacles at 2 wt % [6]. Unlike metal-centred biocides, Econea® does not accumulate in the marine environment, due to its rapid degradation in sea water (3 h and 15 h at 25 °C and 10 °C, respectively) [5] to 3-bromo-5-(4-chlorophenyl)-4-cyano-1H-pyrrole-3-carboxylic acid (BCCPCA) (see Scheme 1, compound **B**).

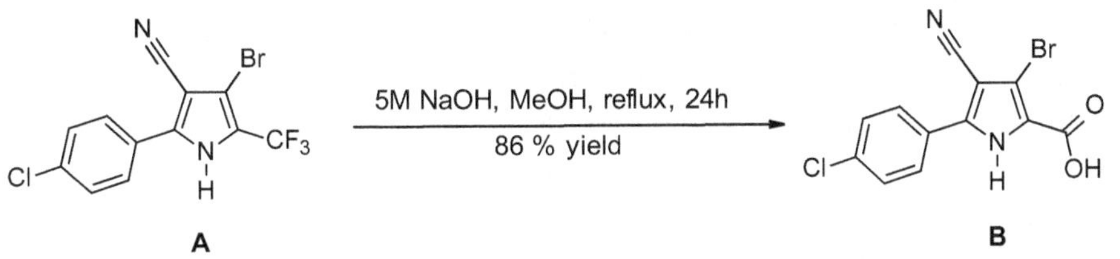

Scheme 1. Hydroylsis of Econea® (or 4-bromo-2-(4-chlorophenyl)-5-(trifluoromethyl)-1H-pyrrole-3-carbonitrile) (**A**) to give 3-bromo-5-(4-chlorophenyl)-4-cyano-1H-pyrrole-3-carboxylic acid (BCCPCA) (**B**).

An ISO method for the indirect detection and quantification of Econea® in sea water samples has been reported [7]. The method is based on the use of high performance liquid chromatography with ultraviolet detection (HPLC-UV) at 280 nm to detect a hydrolysis breakdown product of Econea®, 3-bromo-5-(4-chlorophenyl)-4-cyano-1H-pyrrole-3-carboxylic acid (BCCPCA) (Scheme 1, compound **B**). As a seawater sample may contain a mixture of Econea®, as well as the breakdown product, in this method any Econea® in the sample is converted to BCCPCA, and the method quantitates this compound. The ISO method requires the samples to be degraded in a thermostatically controlled cabinet at 50 ± 5 °C for between 4 h and a maximum of 24 h. This method allows a level of quantification of BCCPCA of 1.87 μg/L.

As part of a research programme to develop novel marine coatings with enhanced controlled release characteristics, a sensitive method for detection of Econea® in sea water, in preference to BCCPCA, was sought. It has been shown previously that solid phase extraction followed by high performance liquid chromatography-electrospray ionisation-mass spectrometry (HPLC-ESI-MS) is a highly sensitive method for detecting and quantitating trace amounts of biocidal compounds in sea water [8]; work has also been reported using the more sensitive technique of LC-MS/MS [9]. This paper outlines a HPLC-ESI-MS method for the determination of Econea® in artificial seawater and its application to follow its release over time from a commercially available anti-fouling paint, and a range of polymer coatings bearing different dispersed amounts of Econea®. The commercially available paint is an anti-fouling paint that uses an ablative action, and contains two biocidal compounds, Econea®, and zinc pyrithione, as well as number of other components that aid its performance in other aspects of its utility [10].

2. Materials and Methods

2.1. Chemicals and Reagents

Analytical standard chemicals were purchased as follows: Econea® (PESTANAL®, Sigma-Aldrich, Gillingham, UK); acetonitrile and ammonium acetate (HPLC grade, Fisher Scientific, Loughbrough, UK); acetic acid (VWR, Lutterworth, UK); Tetra Marine SeaSalt (Amazon, Berkshire, UK). The sample of a commercial available Econea® containing paint was supplied by AkzoNobel, Gateshead, UK. Deionised water was prepared from distilled water purified using a Milli-Q-system (18 MΩ cm^{-1}) (Merck Millipore, Darmstadt, Germany).

2.2. Instrumentation

High performance liquid chromatography was performed on a Thermo Surveyor HPLC system (Thermo Fisher Scientific, Hemel Hempstead, UK) equipped with a degasser, quaternary pump and autosampler. Separation was achieved on a Phenomenex Luna C18 150 mm × 2.00 mm i.d. × 3 μm column with a 10 μL injection volume. The chromatography was performed using an isocratic separation, 75% (A): 25% (B) at 200 μL min^{-1}. Elutant (A) was acetonitrile acidified with 1% (v/v) acetic acid and Elutant B consisted of filtered water with acetonitrile 5% (v/v) and 1% (v/v) acetic acid. The LC was coupled to an electrospray ion-trap mass spectrometer (MS) (Thermo LCQ Advantage) and was operated in negative ion mode. The ion source conditions were optimised using the automatic optimisation function of the MS software (LCQ Tune). This was done by introducing a 10 μg/mL solution of Econea®, in acetonitrile, by direct infusion into the mass spectrometer at a flow rate of 20 μL/min and tuning on the most abundant ion (M+2) at m/z 349. The electrospray ion source was operated at 280 °C. Other source conditions were sheath flow rate 30 (arbitrary units); ionisation spray voltage 4.5 kV; capillary voltage −4.00 V; tube lens offset 10.00 V; second octapole offset 11.00 V, first octapole offset 1.75 V and inter-octapole lens 16.00 V.

^{1}H-NMR spectra were obtained using a JEOL ECS 400 NMR spectrometer operating at a frequency of 400 MHz, using 8–32 scans, a relaxation delay of 5 s, and a flip angle of 45° (5 μs pulse). Spectra were Fourier transformed typically into 32,000 data points using standard exponential window with a line broadening factor of 0.2 Hz. ^{13}C spectra were obtained at a frequency of 100.53 MHz, from 128 to 1048 scans, a relaxation delay of 2 s, and a flip angle of 30° (2.7 μs pulse). Spectra were Fourier transformed typically into 64,000 data points using standard exponential window with a line-broadening factor of 0.5 Hz. All high resolution mass spectra (HRMS) were obtained from the EPSRC UK National Mass Spectrometry Service Centre, Swansea, UK.

2.3. Sample Preparation: Solid Phase Extraction

To investigate, and develop, a suitable solid phase extraction (SPE) protocol, a broad range of hydrophobic, hydrophilic, cation exchange and polymeric cartridges were compared. The highest recoveries were obtained with a C8 cartridge. Off-line SPE was performed by passing 10 mL of artificial sea water sample through a TELOS C8 (EC) (200 mg/3 mL) cartridge from Kinesis (St Neots, UK). The procedure was as follows. The SPE cartridge was washed with 6 mL acetonitrile, followed by 6 mL of filtered water containing 5% v/v acetonitrile. The cartridge was then conditioned with 6 mL acetonitrile followed by 12 mL of filtered water ensuring that the stationary phase did not run dry. Then, 10 mL of the aqueous sample containing Econea® was pipetted from the volumetric flask into a beaker. This sample was passed through the SPE cartridge, and the beaker was washed with approximately 3 × 3 mL of filtered water which was also passed through the SPE cartridge ensuring that the stationary phase did not run dry until the last aliquot. A vacuum was applied for at least 30 s to remove any residual water from the cartridge. The biocide was then eluted from the stationary phase using 8 mL of acetonitrile and quantitatively transferred to a 10 mL volumetric flask. Finally, 50 μL of internal standard of concentration 50 μg/L, which was dissolved in acetonitrile, was also added to the volumetric flask prior to making up to volume with acetonitrile. An aliquot of this solution was then transferred to a 2 mL autosampler vial prior to analysis.

2.4. Artificial Seawater Preparation

Artificial seawater was prepared from the sea water salt mix, 40 g of sea water salts were added to 1 L of distilled water, contents of the beaker were stirred until the sea water salts were dissolved. Any particulates were removed by vacuum through a Qualitative Filter Paper (Fisherbrand FILTER 1 QL100 240 mm). This provided artificial seawater (with the heavy metal component omitted) as per the appropriate ASTM [11].

2.5. Sampling under Controlled Release Conditions

A commercially available paint was supplied by AkzoNobel Ltd which is designed specifically for controlled release of Econea® into sea water. In addition, a solution (in butyl acetate and toluene) of a methacrylate polymer prepared from butyl methacrylate and isobornyl methacrylate using 2,2′-azobisisobutyronitrile as catalyst initiator was polymerised [12] with a starved feed methodology to disfavour polymer block formation [13]. From this polymer solution were created three samples of dissolved polymer containing either 2, 15 or 38 percent composition by weight Econea® (as calculated from the mass of dried polymer product) by addition of powdered Econea® to the solution.

For all four polymer solutions, a 7 cm × 7 cm square coating was drawn down on to a glass panel using a 300 μm draw down bar. The panels were then dried for three days at room temperature in a low airflow extraction rack. The glass panel was then placed on an aluminium support in a beaker containing 250 mL of artificial sea water. The beaker was stirred using a stirrer hotplate calibrated to rotate at 240 rpm at room temperature. Each panel was prepared in triplicate. The volume of sample removed from the beaker is shown in Table 1.

Table 1. Average leach rates of Econea® from synthetic mixtures and a commercial coating.

Coating	Sampling Regime from Panels		Average Leach Rate/μg cm^{-2} d^{-1}
	Volume of Sample/mL	Volume of Eluant/mL	
2 wt % Econea®	100	10	0.4 ± 0.1
15 wt % Econea®	100	10	2.4 ± 0.2
38 wt % Econea®	10	10	7.9 ± 0.7
Commercial coating	10	10	4.3 ± 0.6

The leach rate was determined against a standard calibration graph using standards of Econea®, over the concentration range 0.5–100 μg/L, using 6 data points at concentrations of 0.5, 1, 5, 25, 50 and 100 μg/L. The limit of detection (LOD) and limit of quantitation (LOQ) was determined based on 3 or 10× standard deviation of the blank, respectively.

2.6. Synthesis of BCCPCA

To methanol (200 mL), Econea® (3.70 g, 1.1 mmol) and sodium hydroxide (5 M, 200 mL, 1000 mmol) were added. The mixture was refluxed for 24 h. Methanol was removed from the mixture using a rotary evaporator, and the solution diluted with water (200 mL). Hydrochloric acid (37%) was added until a solid precipitated as the pH of the solution reached pH 1. The solid was washed with water then dried to give BCCPCA (Scheme 1, compound **B**) (2.9533 g, 86% yield). m.p. 184–186 °C; ^{1}H-NMR (400 MHz; DMSO) δ 7.83 (d, J = 8.7 Hz, 2H, ArH) 7.63 (d, J = 8.7Hz, 2H, ArH). ^{13}C-NMR: (100 MHz, DMSO): δ 160.5 (C=O). The HRMS for $[M + H]^+$ was calculated as 324.9374, 325.9408, 326.9351, 327.9385, 328,9324; found 324.9383, 325.9413, 326.9356, 327.9387, 328.9325 (See Supplementary Information).

2.7. Synthesis of Internal Standard: Methyl Ester of BCCPCA

To 20 mL methanol, Econea® (3.57 g, 1.02 mmol) and powdered sodium hydroxide (0.83 g, 2.08 mmol) was added. The mixture was refluxed for 44 h. Sodium hydroxide (0.81 g, 2.03 mmol) and 5.51 g of methanol were then added to the reaction flask. After stirring for 3 days, hydrochloric acid (2 M; 20 mL, 40 mmol) was added until the pH was 1. The resulting solid was collected and washed with water to give an off white solid. Analysis by ^{1}H-NMR showed the solid to be a mixture of the trimethoxymethyl (see Scheme 2, compound **C**) and the methyl ester (see Scheme 2, compound **D**).

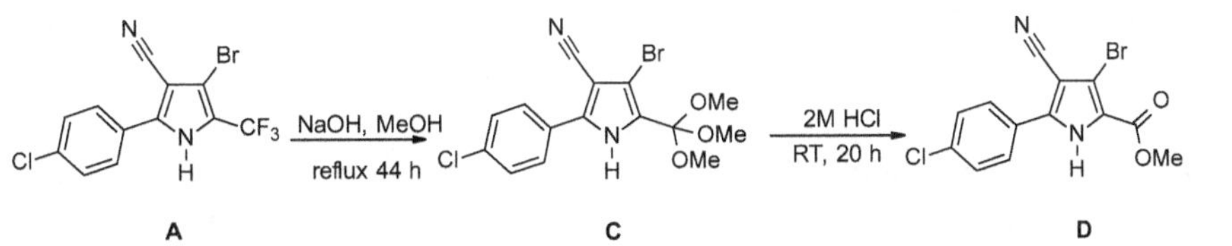

Scheme 2. Synthesis of internal standard, the methyl ester of BCCPCA (Compound D). Compound **A**, Econea® (or 4-bromo-2-(4-chlorophenyl)-5-(trifluoromethyl)-1H-pyrrole-3-carbonitrile); Compound **C**, tri(methoxy)methyl of BCCPCA (i.e., 3-bromo-5-(4-chlorophenyl)-4-cyano-1H-pyrrole-3-carboxylic acid); and Compound **D**, methyl ester of BCCPCA (i.e., 3-bromo-5-(4-chlorophenyl)-4-cyano-1H-pyrrole-3-carboxylic acid).

The crude white solid was added to a round bottom flask, to which hydrochloric acid (2 M; 80 mL, 160 mmol) was added, and then stirred overnight; the resulting solid was collected and washed with water. Analysis by ^{1}H-NMR showed the solid to have converted completely to the methyl ester (see Scheme 2, compound **D**). m.p. 165–167 °C; ^{1}H-NMR (400 MHz; DMSO) δ 7.83 (d, *J* = 8.7 Hz, 2H, ArH) 7.64 (d, *J* = 8.7Hz, 2H, ArH) 3.87 (s, 1H, OCH_3). ^{13}C-NMR: (100 MHz, DMSO): δ 159.5 (C=O), 52.6 (OCH_3); HRMS for $[M + H]^+$ calculated 338.9530, 339.9564, 340.9508, 341.9542, 342.9481; found 338.9533, 339.9563, 340.9506, 341.9537, 342.9476.

3. Results and Discussion

The analytical performance of the new method is shown in Table 2. The experimentally determined limit of quantitation (i.e., 0.17 μg/L) is of the order of 10× more sensitive than the ISO method (LOQ 1.87 μg/L) for the breakdown product of Econea®; this is not unexpected, given that this new method uses HPLC-MS. An example chromatogram of a sample containing 50 μg/L of Econea® and 50 μg/L internal standard is shown in Figure 1. Excellent baseline separation was observed for the internal standard and the Econea® peak.

Table 2. Analytical data for the determination of Econea®.

Compound	Calibration Range (μg/L)	Number of Data Points	Equation (y = mx + c)	Correlation coefficient, R^2	LOD (μg/L)	LOQ (μg/L)
Econea®	0.5–100	6	y = 0.0193x + 0.0269	0.9994	0.05	0.17

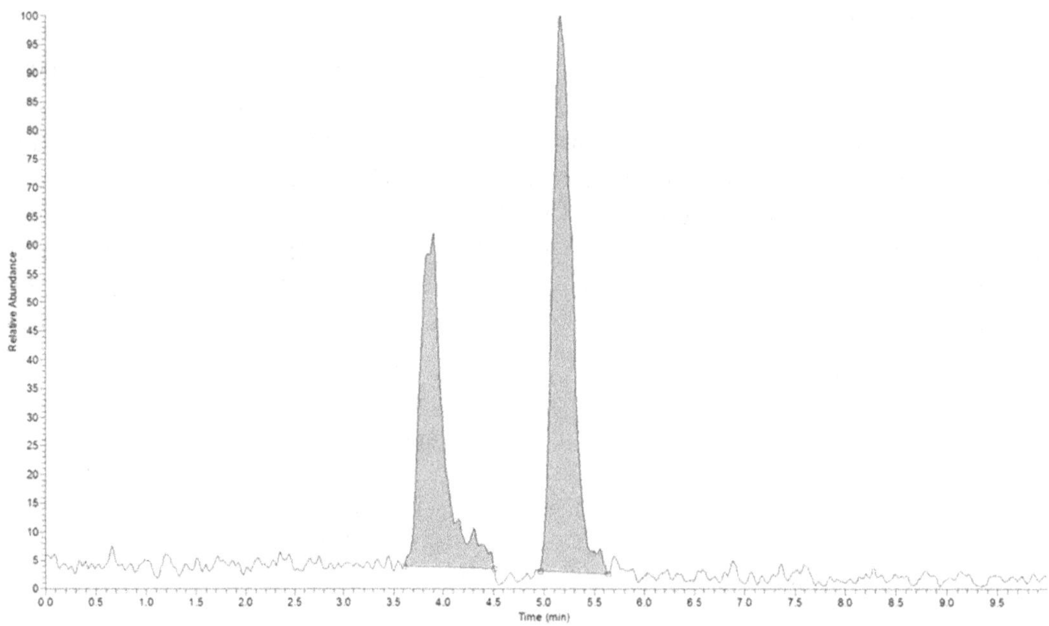

Figure 1. Example chromatogram of sample containing 50 μg/L internal standard (Tr = 3.95 min) and 50 μg/L Econea® (Tr = 5.25 min).

Subsequently, the leach rate of Econea® from the commercial paint was observed over a period of 45 days; samples were taken on days 1, 3, 7, 10, 14, 21, 24, 28, 31, 35, 38, 42 and 45 (as prescribed in ISO/CD15181-1) [14]. Based on the measured concentration of Econea® in the artificial seawater samples, the leach rate, R, (μg cm^{-2} d^{-1}) from the painted test plates was calculated:

$$R = \frac{C_{Econea} \times v \times 24}{t \times A} \quad (1)$$

where C_{Econea} is the concentration of Econea® released into the measuring container in μg/L; 24 is the number of hours per day; v is the volume in litres of artificial sea water (0.25 L); t is the time the test panel is immersed artificial sea water in hours (1 h); and, A is the surface area of the coating applied to the panel in cm^2 (49 cm^2). The results, displayed in Figure 2, show the variation over time; an average leach rate for the commercial paint of approximately 4.3 ± 0.6 μg cm^{-2} d^{-1} was observed for the duration of the study.

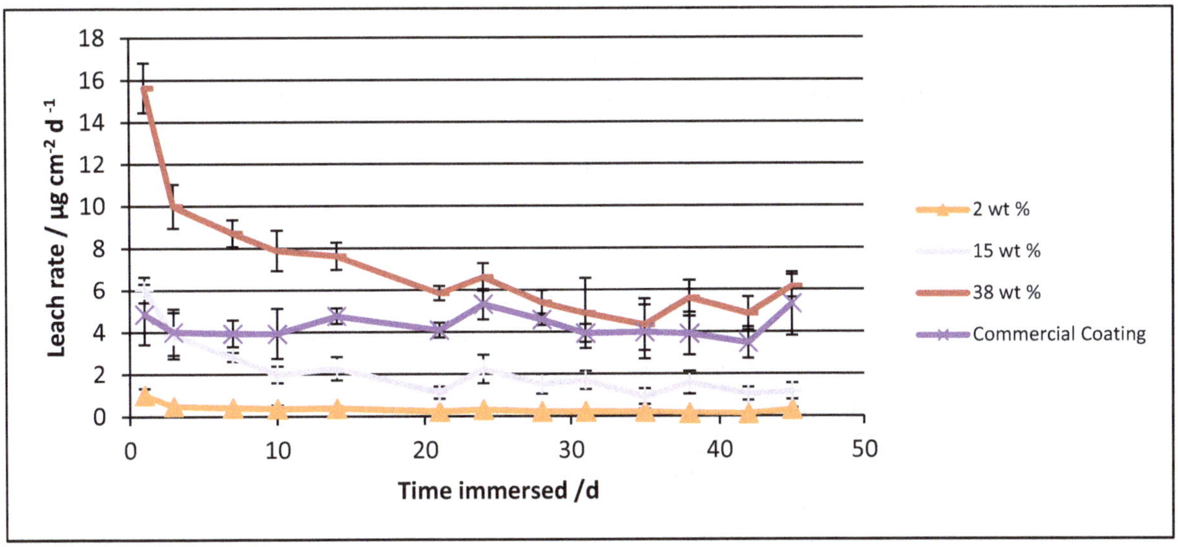

Figure 2. The leach rate of Econea®, from synthetic standards and a commercial coating, in artificial sea water. [The number of replicates per synthetic standard/commercial coating is 3. The error bars represent ± standard deviation].

4. Conclusions

The developed analytical method provides a reliable, robust and sensitive approach for the direct detection of Econea® from complex matrix samples despite its short half-life of 3 h in sea water. LC-MS combined with solid phase extraction is capable of operating with a detection limit of 0.05 μg/L in an artificial sea water sample. It should be noted that this method does not quantitate BCCPCA so whilst it enables the accurate determination of the level of Econea® in solution it cannot be used to make determinations of total leached biocide in the aqueous phase. Unlike the ISO method, which requires a dedicated rig that provides controlled release from coatings painted on rotating cylinders, this method used standard laboratory equipment. Additionally, this method is also applicable to a smaller coated area, and hence a smaller amount of biocide and host polymer need be prepared.

Acknowledgments: The Engineering and Physical Sciences Research Council (EPSRC) and AkzoNobel are gratefully acknowledged for funding this research. We also acknowledge the EPSRC UK National Mass Spectrometry Service Centre, Swansea for providing mass spectrometric analysis.

Author Contributions: R.A.D., J.R.D. and J.J.P. conceived and designed the experiments with input from A.D.; R.A.D. performed the experiments; R.A.D., J.R.D. and J.J.P. analysed the data; A.D. contributed knowledge of the compound and paint; all authors contributed to the writing of the paper.

Conflicts of Interest: The authors (R.A.D., J.R.D. and J.J.P.) declare no conflict of interest. A.D. is employed by a company that produces paint containing the biocide.

References

1. Woods Hole Oceanographic Institution United States (1952). "Chapter 2: Ship Resistance" in Marine Fouling and Its Prevention. Massachusetts Navy Department Bureau of Ships (United States Naval Institute). Available online: http://hdl.handle.net/1912/191 (accessed on 1 July 2017).
2. Omae, I. Organotin Antifouling Paints and their Alternatives. *Appl. Organomet. Chem.* **2003**, *17*, 81–105. [CrossRef]
3. Chambers, L.D.; Stokes, K.R.; Walsh, F.C.; Wood, R.J.K. Modern approaches to marine antifouling coatings. *Surf. Coat. Technol.* **2006**, *201*, 3642–3652. [CrossRef]
4. Lovell, J.B. Insecticidal and Synergistic Miticidal Compositions. U.S. Patent 5,187,184, 16 February 1993.
5. JANNSEN PMP. Antifouling Downloads. Available online: http://www.janssenpmp.com/antifouling.html (accessed on 17 November 2017).
6. Kramer, J.P.; Vos, M. Light-and Bright-Coloured Antifouling Paints. WO Patent 1998012269 A1, 26 March 1998.
7. *Paints and Varnishes—Determination of Release Rate of Biocides from Antifouling Paints—Determination of Econea Release Rate by Quantitation of Its Degradation Product in the Extract*; ISO 15181-6; International Standards Organisation: Geneva, Switzerland, 2008.
8. Bassarab, P.; Williams, D.; Dean, J.R.; Ludkin, E.; Perry, J.J. Determination of quaternary ammonium compounds in seawater samples by solid-phase extraction and liquid chromatography-mass spectrometry. *J. Chromatogr. A* **2011**, *1218*, 673–677. [CrossRef] [PubMed]
9. Oliveira, I.B.; Schönenberger, R.; Barroso, C.M.; Suter, M.J.-F. LC-MS/MS determination of tralopyril in water samples. *Chemosphere* **2016**, *145*, 445–449. [CrossRef] [PubMed]
10. Hellio, C.; Yebra, D.M. *Advances in Marine Antifouling Coatings and Technologies*; Woodhead Publishing Limited: Cambridge, UK, 2009.
11. *Standard Practice for the Preparation of Substitute Ocean Water*; ASTMD1141; ASTM International: West Conshohocken, PA, USA, 2013.
12. Aranaz, I.; Carrasco, S.; Tardajos, M.G.; Elvira, C.; Reinecke, H.; López, D.; Gallardo, A.; Daniel, L. Singular thermosensitivity of polymethyl methacrylate/poly-N-isopropylacrylamide conetworks prepared by a facile synthetic route. *Polym. Chem.* **2011**, *2*, 709–713. [CrossRef]
13. Cao, G.-P.; Zhu, Z.-N.; Zhang, M.-H.; Yuan, W.-K. Kinetics of butylacrylate polymerization in a starved feed reactor. *J. Appl. Polym. Sci.* **2004**, *93*, 1519–1525. [CrossRef]
14. *Paints and Varnishes—Determination of Release Rate of Biocides from Antifouling Paints—General Method for Extraction of Biocides*; ISO 15181-1; International Standards Organisation: Geneva, Switzerland, 2007.

7

Determination of 2-Thioxo-3-pyrrolidinecarbaldehyde in Salted Radish Root (Takuan-zuke) by High-Performance Liquid Chromatography with Fluorescence Detection after Pre-Column Derivatization Using 4-(*N*,*N*-dimethylaminosulfonyl)-7-hydrazino-2,1,3-benzoxadiazole

Taito Kobayashi [1], Kei Kumakura [1], Wataru Kobayashi [1], Asaka Takahashi [2] and Hiroki Matsuoka [1,*]

[1] Department of Health and Nutrition, Takasaki University of Health and Welfare, 37-1 Nakaorui-machi, Takasaki-shi, Gunma 370-0033, Japan; 1610501@takasaki-u.ac.jp (T.K.); kumakura@takasaki-u.ac.jp (K.K.); 1710302@takasaki-u.ac.jp (W.K.)

[2] Higashinihon College of Nutrition and Pharmaceutical, 1098-1 Koyahara-machi, Maebashi-shi, Gunma 379-2184, Japan; asaka@ns.yamasaki.ac.jp

* Correspondence: matsuoka@takasaki-u.ac.jp

Abstract: 2-thioxo-3-pyrrolidinecarbaldehyde (TPC) is an important intermediate in the yellowing of Japanese pickles "takuan-zuke". TPC has been reported to have antibacterial activity against bacteria causing food poisoning and microorganisms associated with the development of caries, as well as various physiological functions such as antimutagenicity. However, since TPC has high reactivity, robust quantitative analysis is difficult with the extraction method or pre-column derivatization method using 2,4-dinitrophenylhydrazine. In this study, a high-performance liquid chromatography (HPLC) method involving labeling with 4-(*N*,*N*-dimethylaminosulfonyl)-7-hydrazino-2,1,3-benzoxadiazole (DBD-H) was developed for the determination of the level of TPC in takuan-zuke. DBD-TPC was successfully stabilized by adding a buffer solution to the reaction solution, which established continuous analysis by HPLC using an autosampler. The DBD-TPC calibration curve was linear in the range of 0.25–250 nmol/mL (final concentration) and showed a wide dynamic range. The lower limit of detection was 0.205 nmol/mL in TPC standard. The proposed method was successfully applied to the measurement of TPC in daikon-oroshi. The results reveal the possibility of determining the variation of TPC level in processed foods containing radish. We concluded that the proposed method is useful for evaluating the quality of processed radish products.

Keywords: 2-thioxo-3-pyrrolidinecarbaldehyde; 4-(*N*,*N*-dimethylaminosulfonyl)-7-hydrazino-2,1,3-benzoxadiazole; daikon-oroshi; takuan-zuke; aldehyde analysis

---

1. Introduction

Daikon (*Raphanus sativus* L.) belongs to the plant family Brassicaceae and is among the most common cruciferous vegetables in Japan. Cruciferous vegetables produce distinctive pungent flavored compounds called isothiocyanates (ITCs) upon the destruction of their cells. 4-Methylthio-3-butenyl isothiocyanate (MTB-ITC) is most abundant (>90%) in white radish [1,2]. Studies on the physiological function of MTB-ITC have revealed various effects including antimutagenicity and cancer prevention [3–5]. However, MTB-ITC is unstable compared with other ITCs in the aqueous phase and is converted to 2-thioxo-3-pyrrolidinecarbaldehyde (TPC) [6]. Consequently MTB-ITC exerts complex effects, including those through degradation products, such as TPC. TPC has been reported to have antibacterial activities against bacteria causative of food poisoning and microorganisms associated with the development of caries, as well as a range of functions such as antimutagenicity [7–10]. In addition, TPC is an important intermediate in the yellowing of takuan-zuke pickle (i.e., salted radish root), which is a popular Japanese pickle [11,12], and is converted to a more stable form, 1-(2-thioxopyrrolidine-3-yl)-1,2,3,4-tetrahydro-β-carboline-3-carboxylic acid (TPCC), in the salting process by reacting with tryptophan [13].

The establishment of a method for measuring TPC is important for evaluating not only the quality of takuan-zuke but also that of other processed radish products. However, since TPC has high reactivity, it is difficult to analyze it quantitatively in a robust manner. Therefore, HPLC analysis following pre-column derivatization was carried out, focusing on the aldehyde group of TPC. In preliminary experiments, a derivatization method using 2,4-dinitrophenylhydrazine (DNPH) reagent, which is widely used for aldehyde analysis, was examined. It was difficult to analyze at low concentrations because DNPH-TPC has low stability at 4 °C and it contains many impurities derived from reagents. Therefore, a method of fluorescence derivatization analysis by 4-(*N*,*N*-dimethylaminosulfonyl)-7-hydrazino-2,1,3-benzoxadiazole (DBD-H), used for high-selectivity and -sensitivity analysis of aldehyde, was investigated [14].

In this study, we aimed to establish optimal conditions for a highly sensitive and stable analytical HPLC method for derivatizing TPC contained in takuan-zuke with DBD-H (Figure 1).

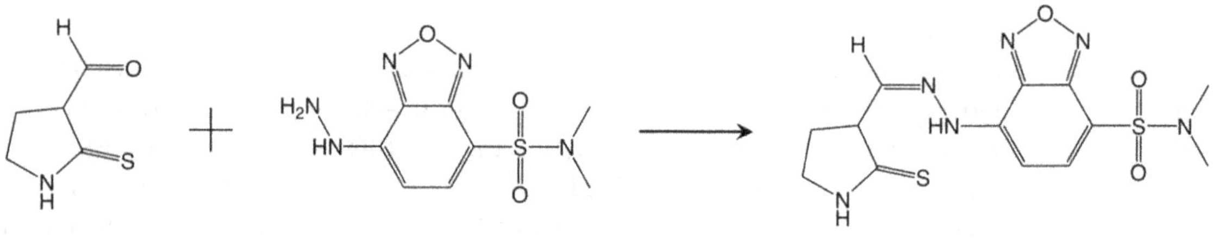

Figure 1. Fluorescence labeling reaction of 2-thioxo-3-pyrrolidinecarbaldehyde (TPC) with DBD-H.

2. Materials and Methods

2.1. Materials and Reagents

Radish roots were purchased from a market in Gunma, Japan. TPC was prepared by previously described methods [15]. 4-(*N*,*N*-dimethylaminosulfonyl)-7-hydrazino-2,1,3-benzoxadiazole (DBD-H) was purchased from Tokyo Chemical Industry (Tokyo, Japan). Trifluoroacetic acid (TFA), acetonitrile, and anisaldehyde, were purchased from Kanto Chemical (Tokyo, Japan).

2.2. Fluorescent Labeling with DBD-H

2.2.1. TPC Standard Solution

In brief, 250 μL of 0.1% DBD in acetonitrile, 250 μL of 0.2 mM anisaldehyde in acetonitrile, and 250 μL of 0.1–1.0% TFA in 20% (*v*/*v*) acetonitrile were added to 250 μL of 0.1 mM TPC in acetonitrile. The reaction mixture was shaken and incubated at 25 °C for 0–120 min. After ice-cooling for 5 min, the mixture was centrifuged at 20,630× *g* for 1 min. To 100 μL of the upper layer, 25 μL of 500 mM McIlvaine buffer (pH 4–6) and 25 mg of NaCl were added. The mixture was shaken to separate the acetonitrile phase from the aqueous phase, and 5 μL of acetonitrile phase was injected into the HPLC system.

2.2.2. Takuan-zuke Sample

In brief, 250 μL of 0.1% DBD in acetonitrile, 250 μL of 0.2 mM anisaldehyde in acetonitrile, 250 μL of acetonitrile, and 250 μL of 1.0% TFA in 20% (v/v) acetonitrile were added to 10–50 mg of takuan-zuke lyophilized powder. The reaction mixture was analyzed in accordance with the method described above.

2.3. Analysis of TPC and MTB-ITC Contained in Daikon-oroshi (Grated Radish)

2.3.1. Preparation of Grated Radish Juice

Fresh radish was grated with a radish grater and then immediately filtered with a tea strainer to obtain squeezed juice. This procedure was carried out under ice-cold conditions. The sample was prepared using a 100-mm long portion (total length 300 mm) obtained from the tip of the radish root.

2.3.2. Fluorescent Labeling of TPC from Grated Radish Juice with DBD-H

In brief, 250 μL of 0.1% DBD in acetonitrile, 250 μL of 0.2 mM anisaldehyde in acetonitrile, 250 μL of 1.0% TFA in acetonitrile, 50 μL of acetonitrile, and 150 μL H_2O were added to 50 μL of grated radish juice. The reaction mixture was analyzed in accordance with the method described above.

2.3.3. Gas Chromatography Analysis of MTB-ITC in Grated Radish Juice

In brief, 200 μL of 0.5 mM allyl isothiocyanate in ethyl acetate was added to 200 μL of grated radish juice. The mixture was shaken vigorously to dissolve released MTB-ITC into ethyl acetate. After centrifugation (20,630× g, 1 min), the layer of ethyl acetate was recovered into a vial and then dehydrated with anhydrous sodium sulfate. The ethyl acetate solution containing isothiocyanates was analyzed by gas chromatography (GC).

2.4. HPLC System and Conditions

The HPLC system consisted of an Agilent 1200 pump (Agilent Technologies, Palo Alto, CA, USA), a 1200 auto sampler with a 100 μL loop, 1100 column compartment, 1260 photodiode array detector, and 1100 fluorescence detector with an 8 μL flow cell. Chromatographic separation was performed on a Poroshell HPH-C18 (100 × 3.0 mm ø, 2.7 μm; Agilent Technologies). These analyzes can be separated by a general reversed phase column (ex. Poroshell EC-C18; Agilent Technologies). The flow rate was 0.85 mL/min. The column temperature was set to 40 °C. Elution was achieved using a gradient of two eluents: H_2O as eluent A and acetonitrile as eluent B without an ion pair additive and/or buffer. The gradient program was as follows: 25% B rising to 73% B at 5.5 min, further rising to 100% B at 5.6 min, and remaining at 100% B to 5.99 min. Finally, the separation column was equilibrated using 25% B from 5.99 to 8.0 min. Fluorescence detection was performed with exciting at 450 nm and emitting at 565 nm as reported previously [14]. Anisaldehyde was used as internal standard.

2.5. GC System and Conditions

Gas chromatography was carried out on an Agilent Technologies 7890A GC system (Agilent Technologies). Chromatographic separation was performed on a DB-23 (30 m × 0.25 mm ø, film thickness 0.25 μm; Agilent Technologies). The carrier gas was helium with a flow rate of 1 mL/min. The column temperature was programmed as follows: 70 °C held for 1 min, increased to 200 °C at a rate of 15 °C/min, then ramped to 250 °C at a rate of 30 °C/min, and held for 1 min, giving a total run time of 10.3 min. Detection was carried out with a flame ionization detector. Allyl isothiocyanate was used as an internal standard.

3. Results and Discussion

3.1. Effects of Acetonitrile Concentration and Reaction Time on Extraction and Reaction Efficiency

We investigated the optimization conditions for TPC analysis by an aldehyde derivatization method with DBD-H. Figure 2 shows a typical chromatogram obtained from the standard solution of TPC after labeling with DBD-H.

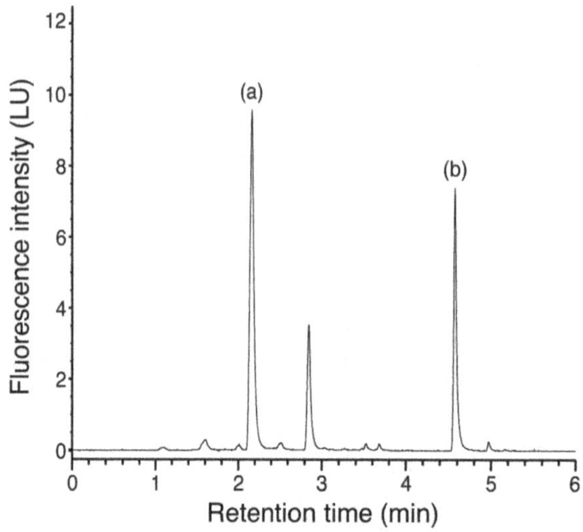

Figure 2. Chromatogram of standard solution of TPC (0.1 mM). This result was obtained by the analysis of the standard solution of 0.1mM TPC (**a**); and 0.2 mM anisaldehyde (**b**) using HPLC.

Uzu et al. reported that the reactivity of DBD-H increased in proportion to the concentration of acetonitrile [14]. The reactions of TPC and anisaldehyde (I.S.) with DBD-H in 50–100% acetonitrile were investigated. Reaction conditions were 0.025% DBD-H and 0.025% TFA for 60 to 120 min at 25 °C. As shown in Figure 3A, the derivative reaction lasting for 2 h occurred at a sufficient level for all acetonitrile concentrations, but the reaction for 1 h was insufficient for 50% to 60% acetonitrile.

Next, we investigated solvent conditions suitable for extraction and reaction using takuan-zuke samples, TPC, and MTB-ITC standard solution. The MTB-ITC standard was used to confirm TPC formation during the reaction. The effect of acetonitrile concentration is shown in Figure 3B. The extraction and reaction efficiency were constant in the range of 50–90% acetonitrile for the TPC standard solution and in the range of 70–90% for the takuan-zuke sample. Conversion of MTB-ITC to TPC was observed during a derivatization reaction with acetonitrile at a concentration of 70% or less. It was suggested that TPC was generated from residual MTB-ITC when the acetonitrile concentration was less than 80%, and that extraction and reaction efficiency decreased above an acetonitrile concentration of 90%; thus, 80% acetonitrile was selected as the optimal condition for subsequent analyses.

3.2. Effects of Trifluoroacetic Acid Concentration on Reaction Efficiency of DBD-H and Stability of DBD-TPC

Uzu et al. reported that DBD-H becomes more reactive with increasing TFA concentration and reacts not only with aldehyde groups but also with ketones. Figure 4 shows the DBD-TPC reaction efficiency in the presence of 0.025–0.25% TFA. Derivatized efficiency increased in proportion to the TFA concentration; the reaction with 0.25% TFA was completed by 60 min.

The area ratio was calculated from the 0.05 mM internal standard and the TPC area value.

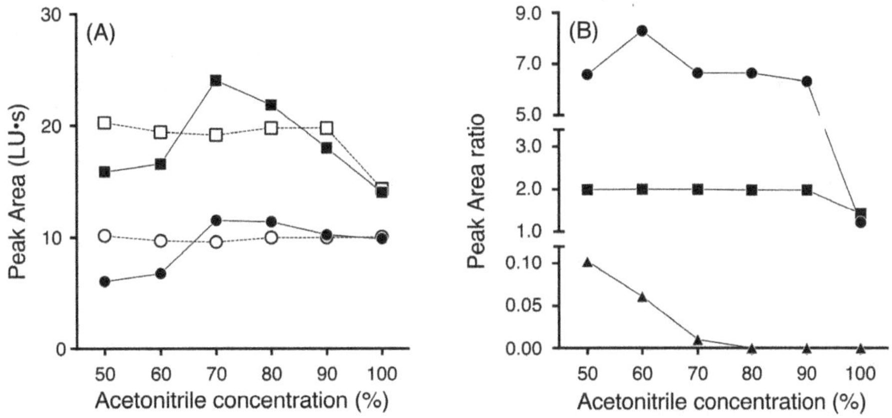

Figure 3. Effects of acetonitrile concentrations on the reaction efficiency of DBD-H. The effect of acetonitrile concentration and reaction time on the DBD-H derivatization reaction (**A**). Symbols: ○ 0.05 mM anisaldehyde and □ 0.025 mM TPC reacted for 60 min; ● 0.05 mM anisaldehyde and ■ 0.025 mM TPC reacted for 120 min. The effect of acetonitrile concentration on extraction and DBD-H derivatization reaction (**B**). Symbols: ● 10 mg of takuan-zuke, ■ 0.025 mM TPC, ▲ 0.025 mM MTB-ITC.

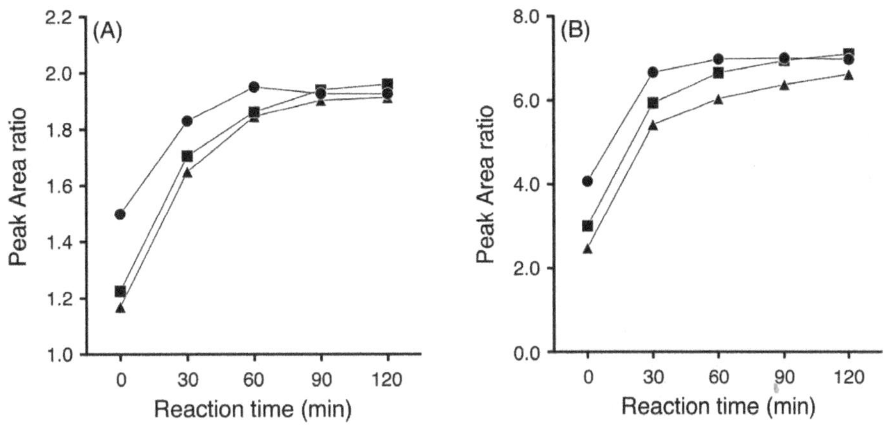

Figure 4. Effects of trifluoroacetic acid (TFA) concentrations on the area ratio of TPC. 0.1 mM TPC standard solution (**A**); and 10 mg of takuan-zuke sample (**B**). Symbols: ● 0.25% TFA, ■ 0.05% TFA, ▲ 0.025% TFA.

The level of DBD-TPC decreased over time during storage at 4 °C for 16 h, while that of DBD-anisaldehyde was stable. Matsuoka et al. reported that TPC is stable around pH 5 [9]. It was assumed that the stability of DBD-TPC was related to the structure of TPC under low pH. It was suggested that the pH buffering ability of takuan-zuke sample influences the stability of DBD-TPC.

The area ratio was calculated from the 0.05 mM internal standard and the TPC area value.

3.3. *Effect of pH on Stability of DBD-TPC*

The effect of pH on the stability of DBD-TPC at 4 °C was confirmed by adding a McIlvaine buffer of pH 4, 5, and 6 to the derivatized sample immediately after the reaction. As shown in Figure 5, the DBD-TPC residual proportion in the TPC standard solution was 86% when no buffer was added. In contrast, DBD-TPC with buffer was stable. Similar results were obtained with the takuan-zuke sample. From these results, DBD-TPC was stabilized by adding pH 5–6 buffer.

The optimal reaction conditions for TPC derivatization were set as 80% acetonitrile solvent containing 0.25% TFA for 60 min at 25 °C. Thereafter, McIlvaine buffer solution (pH 5) was added to the reaction mixture and analysis was performed with an HPLC system equipped with a thermostated autosampler and a fluorescence detector.

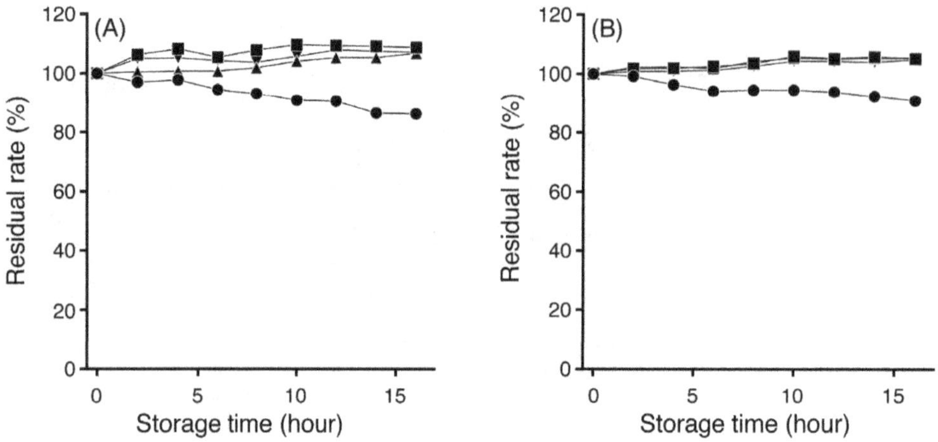

Figure 5. Effects of buffer pH on the stability of TPC. (**A**) 0.1 mM TPC standard solution; and (**B**) 10 mg of takuan-zuke sample. Symbols: ● No addition of buffer, ■ pH4 McIlvaine buffer, ▲ pH5 McIlvaine buffer, ▼ pH6 McIlvaine buffer. The residual ratio was represented by the peak area ratio when the peak area value immediately after the reaction was taken as 100%.

3.4. Verification of Sensitivity

Figure 6 shows the linearity of DBD-TPC under optimized reaction conditions. The calibration curve of TPC was linear ($y = 1.34x - 0.06$, $r^2 = 1.000$) in the range of 0.25 to 250 nmol/mL (final concentration), with a wide dynamic range. The detection limit (signal-to-noise ratio = 3) of the proposed method was 0.205 nmol/mL.

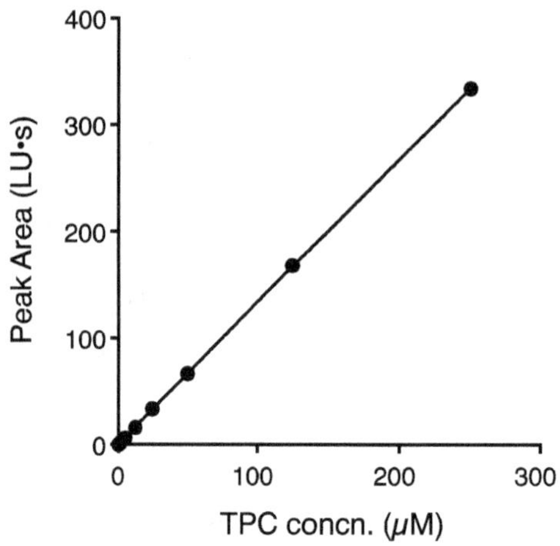

Figure 6. Linearity of TPC concentration in DBD-H fluorescent labeling.

3.5. Quantification of TPC and MTB-ITC in Grated Radish Juice

Time-dependent changes in TPC and MTB-ITC in grated radish juice were analyzed (Figure 7). MTB-ITC was rapidly formed during radish grinding and decreased in a time-dependent manner from the maximum concentration of 213.0 nmol/50 μL of grated juice. The TPC in grated radish juice increased with time; the concentration was 0.5–160.0 nmol/50 μL. The results showed that about 75% of MTB-ITC was converted to TPC in only 120 min. In addition, the cumulative value of MTB-ITC and TPC after 120 min was 182.4 nmol/50 μL, which was 85% of the initial value of MTB-ITC. A strong negative correlation between TPC and MTB-ITC was revealed ($y = -1.20\,x + 211$, $r^2 = 0.997$) in linear regression analysis. These results revealed that TPC is the major degradation product of MTB-ITC. The findings of this study show that this analytical method can be applied not only to takuan-zuke, but also to processed radish products, with potential utility for evaluating the quality of such items.

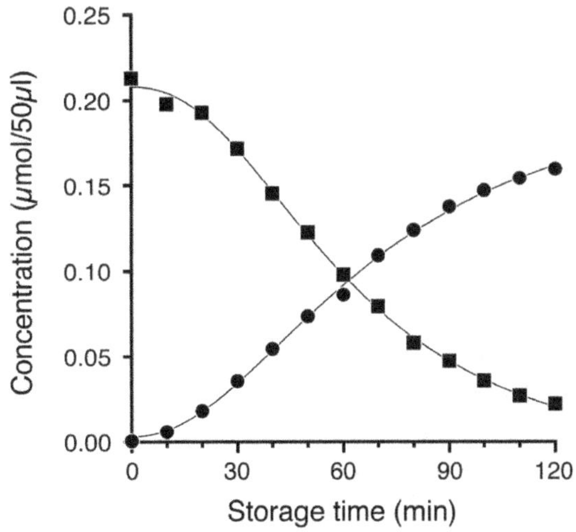

Figure 7. Changes of TPC and MTB-ITC in grated radish juice. Symbols: ● TPC (HPLC method), ■ MTB-ITC (GC method).

4. Conclusions

In this study, we developed a simple method to quantify TPC in takuan-zuke using HPLC by pre-column derivatization with DBD-H. In this method, extraction and fluorescence derivatization were performed in parallel to reduce the loss of TPC and enable stable analysis. This approach enabled the determination of the variation of TPC amounts in processed foods using radish. We considered this method to be useful for evaluating the quality of processed radish products.

Acknowledgments: The authors would like to thank the language editing service Enago for the English language review.

Author Contributions: The experimental design was constructed and supervised by T.K., K.K., W.K., A.T., and H.M. The production of samples was performed by T.K., K.K., W.K., A.T., and H.T. Instrumental analysis were performed by T.K., K.K., and W.K. The manuscript was drafted and written by T.K., and H.M.

Conflicts of Interest: The authors declare no conflict of interest.

References

1. Friis, P.A.; Kjaer, A. 4-methylthio-3-butenyl isothiocyanate, the pungent principle of radish root. *Acta Chem. Scand.* **1966**, *20*, 698–705. [CrossRef]
2. Kjr, A.; Madsen, J.O.; Maeda, Y.; Ozawa, Y.; Uda, Y. Volatiles in distillates of fresh radish of Japanese and Kenyan origin. *Agric. Biol. Chem.* **1978**, *42*, 1715–1721.
3. Nakamura, Y.; Iwahashi, T.; Tanaka, A.; Koutani, J.; Matsuo, T.; Okamoto, S.; Sato, K.; Ohtsuki, K. 4-(methylthio)-3-butenyl isothiocyanate, a principal antimutagen in daikon (*Raphanus sativus*; Japanese white radish). *J. Agric. Food Chem.* **2001**, *49*, 5755–5760. [CrossRef] [PubMed]
4. Okamura, T.; Umemura, T.; Inoue, T.; Tasaki, M.; Ishii, Y.; Nakamura, Y.; Park, E.Y.; Sato, K.; Matsuo, T.; Okamoto, S.; et al. Chemopreventive effects of 4-methylthio-3-butenyl isothiocyanate (Raphasatin) but not curcumin against pancreatic carcinogenesis in hamsters. *J. Agric. Food Chem.* **2013**, *61*, 2103–2108. [CrossRef] [PubMed]
5. Yamasaki, M.; Omi, Y.; Fujii, N.; Ozaki, A.; Nakama, A.; Sakakibara, Y.; Suiko, M.; Nishiyama, K. Mustard oil in "Shibori Daikon" a variety of Japanese radish, selectively inhibits the proliferation of H-ras-transformed 3Y1 cells. *Biosci. Biotechnol. Biochem.* **2009**, *73*, 2217–2221. [CrossRef] [PubMed]
6. Uda, Y.; Ozawa, Y.; Ohshima, T.; Kawakishi, S. Identification of enolated 2-thioxo-3-pyrrolidinecarbaldehyde, a new degradation product of 4-methylthio-3-butenyl isothiocyanate. *Agric. Biol. Chem.* **1990**, *54*, 613–617.
7. Hashimoto, K.; Yanagi, K.; Fukushima, K.; Uda, Y. Effect of 3-hydroxymethylene-2-thioxopyrrolidine on growth of two species of mutans streptococci and their in vitro plaque formation. *Int. J. Antimicrob. Agents* **2001**, *17*, 97–102. [CrossRef]

8. Uda, Y.; Matsuoka, H.; Shima, H.; Kumagami, H.; Maeda, Y. Antimicrobial activity of water-soluble products derived from radish mustard oil and identification of an active component therein. *Nippon Shokuhin Kougyo Gakkaishi* **1993**, *40*, 801–806. [CrossRef]
9. Matsuoka, H.; Takahashi, A.; Yanagi, K.; Uda, Y. Antimicrobial action of 2-thioxo-3-pyrrolidinecarbardehyde, a major thiolactam compound generated from the pungent principle of radish in an aqueous medium. *Food Sci. Technol. Int. Tokyo* **1997**, *3*, 353–356. [CrossRef]
10. Uda, Y.; Hayashi, H.; Takahashi, A.; Shimizu, A. Mutagenic and antimutagenic property of 3-hydroxymethylene-2-thioxopyrrolidine, a major product generating from pungent principle of radish. *LWT-Food Sci. Technol.* **2000**, *33*, 37–43. [CrossRef]
11. Ozawa, Y.; Uda, Y.; Kawakishi, S. Generation of β-carboline derivative, the yellowish precursor of processed radish roots, from 4-methylthio-3-butenyl isothiocyanate and L-tryptophan. *Agric. Biol. Chem.* **1990**, *54*, 1849–1851.
12. Matsuoka, H.; Takahashi, A.; Ozawa, Y.; Yamada, Y.; Uda, Y.; Kawakishi, S. 2-[3-(2-thioxopyrrolidin-3-ylidene)methyl]-tryptophan, a novel yellow pigment in salted radish roots. *Biosci. Biotechnol. Biochem.* **2002**, *66*, 1450–1454. [CrossRef] [PubMed]
13. Takahashi, A.; Yamada, T.; Uchiyama, Y.; Hayashi, S.; Kumakura, K.; Takahashi, H.; Kimura, N.; Matsuoka, H. Generation of the antioxidant yellow pigment derived from 4-methylthio-3-butenyl isothiocyanate in salted radish roots (takuan-zuke). *Biosci. Biotechnol. Biochem.* **2015**, *79*, 1512–1517. [CrossRef] [PubMed]
14. Uzu, S.; Kanda, S.; Imai, K.; Nakashima, K.; Akiyama, S. Fluorogenic reagents: 4-aminosulphonyl-7-hydrazino-2,1,3-benzoxadiazole, 4-(*N*,*N*-dimethylaminosulphonyl)-7-hydrazino-2,1,3-benzoxadiazole and 4-hydrazino-7-nitro-2,1,3-benzoxadiazole hydrazine for aldehydes and ketones. *Analyst* **1990**, *115*, 1477–1482. [CrossRef]
15. Matsuoka, H.; Uda, Y.; Mitani, K.; Yoneyama, K. Easy preparation method for 2-thioxopyrrolidine derivatives including 3-hydroxymethylene-2-thioxopyrrolidine, an antimicrobial degradation product of radish pungent principle, via (*E*,*Z*)-4-methoxy-3-butenyl isothiocyanate. *Biosci. Biotechnol. Biochem.* **1996**, *60*, 914–915. [CrossRef] [PubMed]

8

Comparison of Separation of Seed Oil Triglycerides Containing Isomeric Conjugated Octadecatrienoic Acid Moieties by Reversed-Phase HPLC

Anh Van Nguyen [1], Victor Deineka [1,*], Lumila Deineka [1] and Anh Vu Thi Ngoc [2]

1 Institute of Engineering Technologies and Natural Sciences of Belgorod National Research University, Pobeda str. 85, Belgorod 308015, Russia; vananhkhoahoasp2@gmail.com (A.V.N.); deyneka@bsu.edu.ru (L.D.)

2 Pharmaceutical Department of of Buon Ma Thuot University, 298 Ha Huy Tap, Buon Ma Thuot, Daclak 630000, Vietnam; vuanh0000s@gmail.com

* Correspondence: deineka@bsu.edu.ru

Abstract: Relative retention analysis and increment approach were applied for the comparison of triglycerides (TGs) retention of a broad set of plant seed oils with isomeric conjugated octadecatrienoic acids (CLnA) by reversed-phase HPLC for "propanol-2-acetonitrile" mobile phases and Kromasil 100-5C18 stationary phase with diode array detection (DAD) and mass spectrometric (MS) detection. The subjects of investigation were TGs of seed oils: *Calendula officinalis, Catalpa ovata, Jacaranda mimosifolia, Centranthus ruber, Momordica charantia, Trichosanthes anguina, Punica granatum, Thladiantha dubia, Valeriana officinalis,* and *Vernicia montana.* It was found that a sequence of elution of TGs of the same types is the same without any inversions for full range of mobile phase compositions: punicic ($C18:3^{9Z11E13Z}$) < jacaric ($C18:3^{8Z10E12Z}$) < catalpic ($C18:3^{9E11E13Z}$) < α-eleostearic ($C18:3^{9Z11E13E}$) < calendic ($C18:3^{8E10E12Z}$) < β-eleostearic ($C18:3^{9E11E13E}$) < all-E calendic ($C18:3^{8E10E12E}$) acids. TGs and fatty acid compositions were calculated for all oil samples. Regularities of solute retentions as a function of isomeric conjugated octadecatrienoic acid moiety structure are discussed. Thus, it was proven that it is possible to differentiate TGs of complex composition with moieties of all natural CLnA by retention control accomplished by electronic spectra comparison, even though there are only three types of electronic-vibration spectra for seven isomeric CLnA.

Keywords: reversed-phase HPLC; seed oil; conjugated octadecatrienoic acids; increment approach; relative retention analysis

---

1. Introduction

Among a great variety of plant seed oils, oils with conjugated fatty acid moieties are of special interest because of their high biological activity [1–7]. One type of these acids includes isomeric octadecadienoic acids found mostly in the meat and dairy products derived from ruminants (conjugated linoleic acids, CLA) [8]. Mainly natural octadecatrienoic acids (conjugated linolenic acids, CLnA) [2] are synthesized in some plant seeds, being the substances under investigation in current paper. Other types of acid with more complex structure also known, for example, parinaric (octadecatetraenoic acid), some hydroxylated acids (α-kamlolenic acid) and acids with acetylenic bonds in conjugation [9].

CLnA in native seed oils are mainly in the form of triacylglycerols (TGs). Currently, there exist two ways for oil analysis. The first way implies transmethylation procedure of initial oils to obtain methyl esters of the fatty acids suitable for subsequent analysis by gas chromatography [10]. This method permits determination the oil main fatty acids as well as minor components. However, it should be taken into account that during transesterification some labile acids may undergo transformation or be lost entirely. Moreover, the procedure leads to loss of information about fatty acids distribution among TGs species as a hardly falsifiable parameter. According to an alternative method, non-modified TGs are analyzed by HPLC separations.

The most commonly used method explores reversed-phase HPLC for TGs separation [11]. In this case, some problems arise concerning solutes resolution especially in the case of some TGs pairs [12]. Indeed, if only four types of fatty acids compose the oil, the number of resulting possible different types of TGs is 20 (without specifying the position of the acid moieties in TG). Some of these solutes are easily separated but, for another pair, retentions are close and base-line separation is almost unachievable. Thus, results of calculation of fatty acid composition of the oil by the second method may have some errors.

Another common problem is connected with solutes detection, but TGs with conjugated acid substituents are easily detectable using conventional spectrophotometric detection, while diode-array detectors are favorable due to a possibility of electronic spectra registration being the orthogonal (to retention times) properties for solute identification. Mass-spectrometric detection is highly desirable orthogonal method of solute properties control and for seed oils method APCI mass-spectrometry (atmospheric pressure chemical ionization) was developed [13].

It is commonly known that solute retention coincidence is not a proof of solute identity. However, there exist strict regularities in TGs retention. Equivalent carbon numbers (ECN) connect solute retentions with that of synthetic TGs composed by saturated fatty acids substituents [14,15]. ECN depend upon structure of all moieties, are additive and so their retention is predetermined. Thus, ECN is really an orthogonal property of the solute and so it may be used for tentative solute identification. Meanwhile ECN depends upon mobile phase composition for a given chromatographic system complicating the method utilization. Development of relative retention analysis [16] was directed to escape the dependence upon mobile phase composition. In this approach, retention of solute has two parametric dependence:

$$\log \boldsymbol{k}(i) = a_0 + a_1 \log \boldsymbol{k}(\boldsymbol{A}) \quad (1)$$

where $\boldsymbol{k}(i)$ is capacity factors of solute, while $k(\boldsymbol{A})$ is that of reference solute $\boldsymbol{A}$.

Increment approach implies to use group contribution factor to calculate the solute retention alteration for the exchange of two fatty acids in pair of TGs is the other two moieties remain unchanged:

$$\Delta(i \rightarrow j) = \log \boldsymbol{k}(j) - \log \boldsymbol{k}(i) \quad (2)$$

The increment $\Delta(i \rightarrow j)$ does not depend upon nature of the remaining two unchanged substituents, but it depends upon mobile phase composition. Equation (1) describes the retention of solutes i in a broad composition of given mobile phase system and given stationary phase. For non-polar solutes, the equation may be transferred upon solute separation for other stationary reversed phases trademarks.

The aim of the present investigation was to compare retentions of TGs of oils with isomeric conjugated octadecatrienoic acids in reversed-phase HPLC to elucidate the possibilities of solute identification.

2. Materials and Methods

2.1. Plant Seed Material

Oils were extracted for plant seeds: *Calendula officinalis, Catalpa ovata, Momordica charantia* and *Punica granatum* grown in Belgorod; *Thladiantha dubia* and *Valeriana officinalis*, bought in Belgorod market for gardeners; and *Trichosanthes anguina, Jacaranda mimosifolia* and *Vernicia montana*, grown in Vietnam in 2016.

2.2. Oil Extraction and Purification

Oils were extracted from 2 g of seeds in porcelain mortar by 5 portions (for exhausting extraction) each of 20 mL of hexane at room temperature. The portions were combined and the solvent was withdrawn on the rotary evaporator at 30 °C.

The liquid residue was dissolved in hexane to prepare solution with oil concentration of 10 mg/mL.

The oil was refined by solid phase extraction on silica (in syringe cartridges), the silica being checked for the absence of catalytic activity. The oil was washed from sorbent with dichloromethane, solvent was evaporated and the oil was stored in a refrigerator at 4 °C.

2.3. Chromatography

For reversed-phase high-performance liquid chromatograpy (RP HPLC) we used an Agilent 1260 Infinity chromatography system with diode array and mass spectrometric detectors (Agilent Technologies, 76337 Waidbronn, Germany). The following chromatographic columns were applied: Kromasil 100-5C18 4.6 × 250 mm, and Kromasil 110-3.5C18 2.1 × 150 mm (for mass spectrometric detection).

Mass spectrometric detection was carried out in a mixed mode: atmospheric pressure chemical ionization and electrospray ionization under conventional conditions at a fragmentor voltage of 50 V; signals of positively charged ions were recorded.

2.4. Calculations and Designations

All experiments were carried out in an isocratic mode; chromatograms were recorded, stored, and processed using a specialized ChemStation software products. Non-resolved peaks were handled by Magicplot Student 2.7.2 software.

TGs were denoted in a conventional way, indicating radicals of acids with letters without specifying of their position in the molecule. The letter designations of the acid substituents: L, linoleic acid; O, oleic acid; P, palmitic acid; and S, stearic acids. For conjugated octadecatrienoic acid designations, see text. The formula L_2O means TG with two substituents of linoleic acid and one substituent of oleic acid.

The column void time (t_M) was calculated by the retention times of a series of TGs, assuming that the retention factors (k) in the series of $X_3 \rightarrow X_2L \rightarrow XL_2$ increase by the same value of logarithmic units of retention or:

$$\frac{t_R(X_3) - t_M}{t_R(X_2L) - t_M} = \frac{t_R(X_2L) - t_M}{t_R(XL_2) - t_M} \quad (3)$$

Mole fraction of TGs, α(TGs), was calculated taking into account peak area (S_i) on the chromatogram and number (n_i) of conjugated octadecatrienoic acid substituents in it:

$$\propto (TG_i) = \frac{\frac{S_i}{n_i}}{\sum_i \frac{S_i}{n_i}} \cdot 100\% \quad (4)$$

For acid j mole fraction in the oil calculation, mole fractions of all TGs were used taking into account the numbers of the acid substituents (n_{ij}), in each TGs:

$$\propto (Acid_j) = \frac{\sum_i \propto (TG_i) \cdot n_{ij}}{\sum_j \sum_i \propto (TG_i) \cdot n_{ij}} \cdot 100\% \quad (5)$$

3. Results

3.1. UV Spectroscopic Properties of Conjugated Octadecatrienoic Acids

Theoretically, there exist many isomers for octadecatrienoic acids with different position of double bonds in a carbon atom chain as well as with different *cis-trans* configuration of these bonds.

Only restricted number of the natural isomers was found in plant sources. First, in the isomers, the middle C=C-bond has only *trans*-configuration [9]. Moreover, only two positional isomers are synthesized in plant seeds: octadeca-9,11,13-trienoic and octadeca-8,10,12-trienoic acids [17]. Thus, only three types of chromophores may be responsible for the appearance of UV-spectra due to conjugation of the three C=C-bonds (Figure 1).

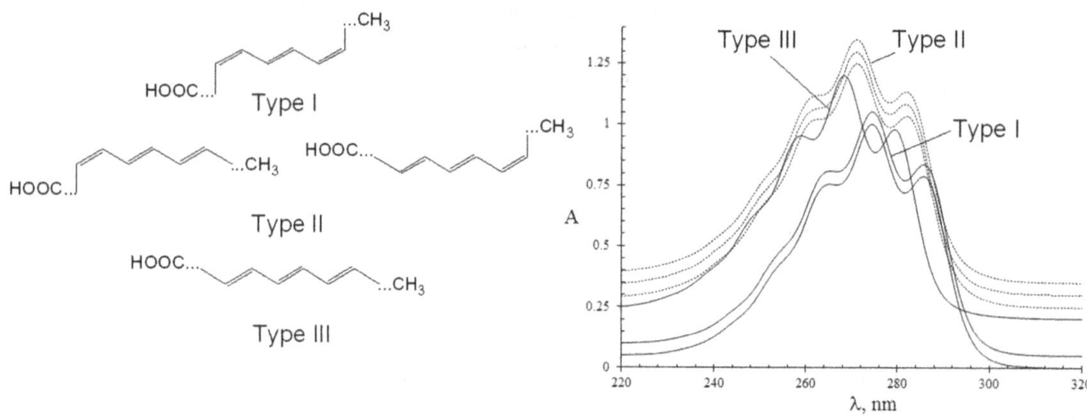

Figure 1. Three types of cis/trans configurations in natural conjugated octadecatrienoic fatty acids and their electronic spectra: Type I, punicic and jacaric acids; Type II, catalpic, α-eleostearic and calendic acids; and Type III, β-eleostearic and all-*trans*-calendic acids.

All oils with conjugated octadecatrienoic acids have electronic-vibration spectra with three apparently visible overlapping bands. The band's maxima wavelength depends upon type of *cis-trans* configuration [18]: exchange of *trans*- by *cis*-configuration causes bathochromic shift. In addition, since chromophore group is far away from carboxylic group in natural octadecatrienoic acid isomers only three different spectra may be registered (Figure 1). Thus, comparison of electronic spectra is a valuable instrument in solute structure differentiation at least into three types. It was proven in present investigation that the differentiation is favored by the absence of solvatochromic effects for the spectra for full range of mobile phase compositions suitable for TGs separation, at least for 34–60% propanol-2 in acetonitrile.

3.2. Retention Analysis of TGs of Plant Sources Seeds with Isomeric Octadecatrienoic Acids

3.2.1. Triglycerides of Oil with (9Z,11E,13E)-octadeca-9,11,13-trienoic Acid Moieties

From the series of plant sources with seed oils enriched with (9Z,11E,13E)-octadeca-9,11,13-trienoic or α-eleostearic acid [17], *Momordica charantia* was chosen as a source with high content of the acid simultaneously with other ordinary acids being suitable for increments calculation. The oil is known [19] to be built with linoleic, oleic, palmitic, stearic and α-eleostearic (α-El) acid.

Mainly 10 types of TGs with different retentions (Figure 2) compose the oil. A search of equal increments of TGs retention alteration together with mass-spectra consideration permits solute composition elucidation, as proposed in Table 1. The results are in a good agreement with published data [20,21].

The main TGs component of the oil is composed of two substituents of α-eleostearic and the third one, stearic acids (α-El_2S). The mole fraction of this component (38.2 mol. %) was calculated taking into account peak areas in Figure 2 and the number of α-eleostearic acid moieties in the TGs (Table 1). Calculated by the same way, fatty acid composition shows the predominant acid to be α-eleostearic (as a sum of all octadecatrienoic acids) (57.7 mol. %), followed by stearic (16.3%), linoleic (14.2%), oleic (7.6%) and palmitic (4.1%) acids.

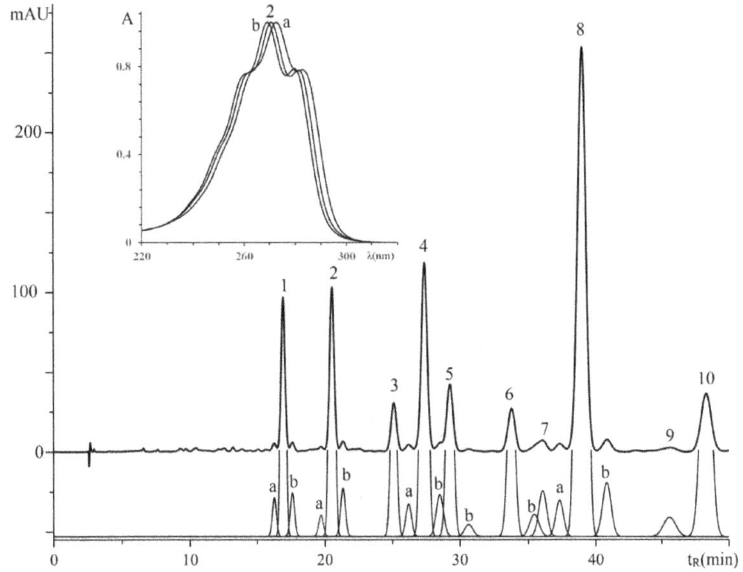

Figure 2. Separation of TGs of *Momordica charantia* seed oil. Column 4.6 mm × 250 mm Kromasil 100-5C18; mobile phase, 35% propanol-2 and 65% acetonitrile, 1 mL/min; column thermostat temperature, 30 °C; detector, 270 nm. For peak number composition, see Table 1.

Table 1. Retention parameters, mass-spectral signals and content of TGs of *Momordica charantia* seed oil for the mobile phase composition 35% propanol-2 and 65% acetonitrile.

No. [a]	TG Composition	t_R (min)	Log*k*	Increment Values				Mole Fraction of TG, %	*M/z* [M + H$^+$]
				X→L	L→O	O→P	P→S		
1	X [b]$_3$	16.94	0.761					4.8	873.8
2	X_2L	20.54	0.858	0.097				9.1	875.8
3	XL_2	25.11	0.956	0.098				5.6	877.7
4	X_2O	27.37	0.998		0.140			13.3	877.7
5	X_2P	29.27	1.030			0.032		4.6	851.8
6	XLO	33.72	1.097	0.099	0.140			6.9	879.8
7	XLP	36.05	1.128	0.098		0.031		2.8	n.d. [c]
8	X_2S	38.95	1.164				0.134	38.2	879.8
9	XO_2	45.65	1.237		0.141			1.5	n.d.
10	XLS(+XOP)	48.27	1.263	0.099			0.135	13.2	881.7 + 855.7
	Middle value			0.098	0.140	0.031	0.134		

[a] numbers of TG are the numbers of peaks in Figure 2; [b] X, moiety of conjugated octadecatrienoic acid (α-eleostearic); [c] n.d., not determined.

The increments calculated in the case of this seed oil were used for the calculation of the compositions of TGs of the other oils discussed below.

3.2.2. Triglycerides of Oil with (9Z,11E,13Z)-octadeca-9,11,13-trienoic Acid Moieties

Punica granatum is a prominent plant source of the oil of punicic acid type, but it has been intensively investigated in a series of papers [19]. Thus, we used a different natural source of the oil enriched with punicic (9Z,11E,13Z)-octadeca-9,11,13-trienoic acid, Pu (*Trichosanthus anguina*) seed oil. According to literature data [22], the seed oil of *Trichosanthus kirilowii* is composed of moieties of linolenic (38.2%), punicic, (38.0%), and oleic (11.8%) acids. In the case of *Trichosanthus anguina* seed oil, we may see five groups of peaks on the chromatogram (Figure 3). The composition of ten main peaks with identical electronic spectra refer to Type I (Figure 1), and hence composed with participation of punicic acid as the only conjugated one, and is calculated by the increment approach (Table 2). The increments for common acids substitutions linoleic by oleic, Δ(L→O), oleic by palmitic, Δ(O→P), and palmitic by stearic, Δ(P→S), are the same as for *Momordica* seed oil, despite different TGs compositions (another isomer of octadecatrienoic acid moieties).

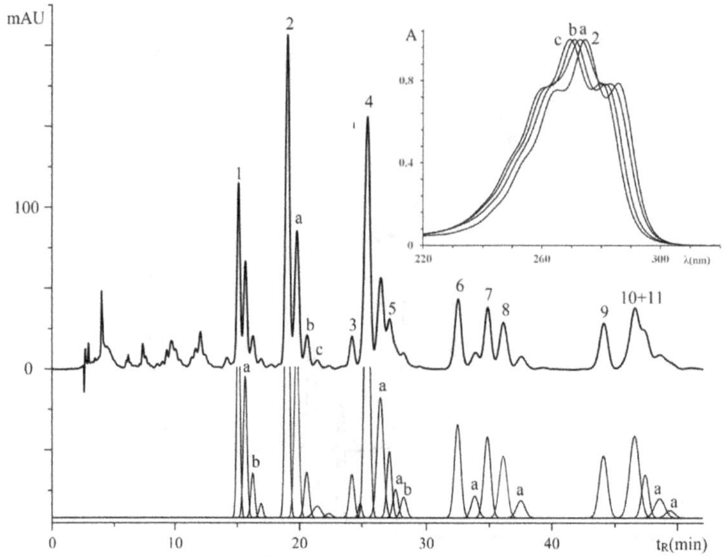

Figure 3. Separation of TGs of *Trichosanthus anguina* seed oil. Column 4.6 mm × 250 mm Kromasil 100-5C18; mobile phase, 35% propanol-2 and 65% acetonitrile, 1 mL/min; column thermostat temperature, 30 °C; detector, 270 nm. For peak number composition, see Table 2.

Table 2. Retention parameters, mass spectral signals and content of TGs of *Trichosanthes anguina* seed oil for the mobile phase composition 35% propanol-2 and 65% acetonitrile.

No. [a]	TG Composition	t_R (min)	Log*k*	Increment Values				Mole Fraction of TG, %	*M*/*z* [M + H$^+$]
				X→L	L→O	O→P	P→S		
1	X [b]$_3$	15.11	0.703					6.18	873.8
2	X_2L	19.04	0.820	0.118				18.43	875.8
3	XL_2	24.20	0.939	0.118				2.96	877.7
4	X_2O	25.39	0.962		0.141			17.18	877.7
5	X_2P	27.10	0.993			0.031		4.33	851.8
6	XLO	32.52	1.080	0.118				10.25	879.8
7	XLP	34.84	1.112			0.032		8.14	853.7
8	X_2S	36.09	1.128				0.135	4.26	879.8
9	XO_2	44.09	1.221		0.142			7.39	881.8
10	XLS(+XOP)	46.59	1.246	0.118			0.135	20.24	881.7 + 855.7
	Mean value:			0.118	0.141	0.032	0.135		

[a] Numbers of TGs are the numbers of peaks on the Figure 3; [b] X, moiety of conjugated octadecatrienoic acid (punicic).

In calculated fatty acid composition, the predominant acid was punicic (as a sum of all octadecatrienoic acids) (51.7 mol. %), followed by oleic (15.1%), linoleic (11.9%), palmitic (9.5%) and stearic (5.8%) acids.

3.2.3. Triglycerides of Oil with (8E,10E,12Z)-octadeca-8,10,12-trienoic Acid Moieties

Calendula officinalis seed represents a unique source of one of the octadecatrienoic isomers: calendic, (8E,10E,12Z)-octadeca-8,10,12-trienoic acid, Cal. According to literature data [23], this acid is predominant in different varieties of the plant (fatty acid composition (40–57%), while the content of other acids are: linoleic (30–37%), oleic (3.6–5.8%), palmitic (3.4–5.4%) and stearic (1.7–4.5%). Chromatogram of our sample of *Calendula officinalis* seed oil indicates the only predominant peak corresponds to Cal_2L composition of TGs (Figure 4). The composition of all TGs types with electronic spectra of the Type II (Figure 1) is presented in Table 3. The increments for common acids substitutions Δ(L→O), Δ(O→P) and Δ(P→S) remain unchanged (see Tables 1 and 2). The proposed types of TGs were supported by TGs' mass spectra (Table 3).

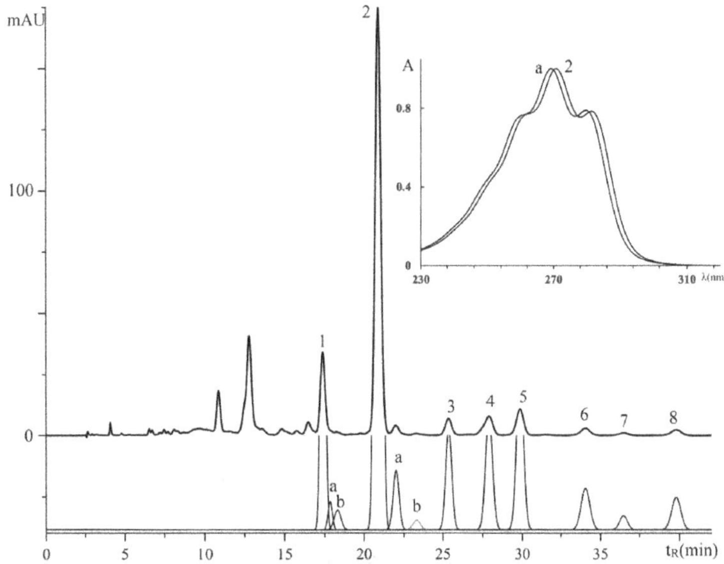

Figure 4. Separation of TGs of *Calendula officinalis* seed oil. Column 4.6 mm× 250 mm Kromasil 100-5C18; mobile phase, 35% propanol-2 and 65% acetonitrile, 1 mL/min; column thermostat temperature, 30 °C; detector, 270 nm. For peak number composition, see Table 3.

Table 3. Retention parameters, mass spectral signals and content of TGs of *Calendula officinalis* seed oil for the mobile phase composition 35% propanol-2 and 65% acetonitrile.

No. [a]	TG Composition	t_R (min)	Log*k*	Increment Values				Mole Fraction of TG, %	*M/z* [M + H⁺]
				X→L	L→O	O→P	P→S		
1	X^{b}_{3}	17.39	0.775					8.25	873.7
2	X_2L	20.90	0.867	0.092				68.91	875.8
3	XL_2	25.36	0.961	0.094				6.06	877.7
4	X_2O	27.89	1.007		0.140			4.66	877.7
5	X_2P	29.86	1.039			0.032		5.42	851.8
6	XLO	34.05	1.101	0.094	0.140			3.75	879.8
7	XLP	36.46	1.133			0.032		1.37	853.8
8	X_2S	39.80	1.174				0.135	1.58	879.8
				0.093	0.140	0.032	0.135		

[a] Numbers of TGs are the numbers of peaks on the Figure 4; [b] X, moiety of conjugated octadecatrienoic acid (calendic).

In calculated by TG area fatty acid composition, the highest content is calendic (as a sum of all octadecatrienoic acids) (65.69 mol. %), followed by linoleic (28.72%), oleic 2.80%) palmitic (2.26%), and stearic (0.53) acids.

3.2.4. Triglycerides of Oil with (9E,11E,13Z)-octadeca-9,11,13-trienoic Acid Moieties

Catalpa species seed are interesting for the present investigation due to the biosynthesis of one octadecatrienoic acid isomer: catalpic, (9E,11E,13Z)-octadeca-9,11,13-trienoic acid, Cat [24]. According to literature data [24], this acid fraction in fatty acids composition is 42.25%, while the contents of other acids are as follows: linoleic (39.95%), oleic (7.71%), palmitic (2.77%), and stearic (2.65%). In addition, it is important to mention that α-linolenic acid was also found (0.56%), along with some other acid. Chromatogram of our sample of *Catalpa ovata* seed oil is somewhat more complicated compared to previously discussed chromatograms (Figure 5). All 15 numbered peaks have the same electronic spectra of Type II (Figure 1).

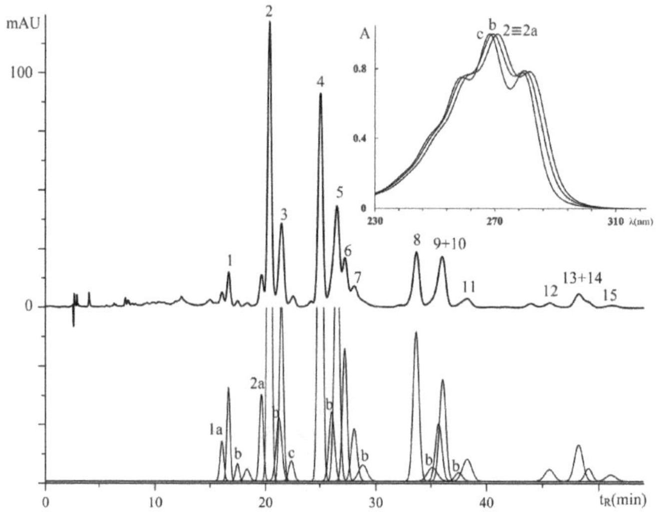

Figure 5. Separation of TGs of *Catalpa ovata* seed oil. Column 4.6 mm × 250 mm Kromasil 100-5C18; mobile phase, 35% propanol-2 and 65% acetonitrile, 1 mL/min; column thermostat temperature, 30 °C; detector, 270 nm. For peak number composition, see Table 4.

Table 4 contains some pairs of TGs known from the previous tables, but, to fit the increment approach as well as mass-spectrometric data, we must include not only one new conjugated octadecatrienoic acid (catalpic) but also another one—an uncommon octadecadienoic acid of unknown structure, non-distorting the electronic spectra of Type II.

Table 4. Retention parameters, mass spectral signals and content of TGs of *Catalpa ovata* seed oil for the mobile phase composition 35% propanol-2 and 65% acetonitrile.

No. [a]	TG Composition	t_R (min)	Log*k*	Increment Values					Mole Fraction of TG, %	*M/z* [M + H⁺]
				X→L	L→Y	L→O	O→P	P→S		
1a	LnX^{b}_2	16.08	0.735						0.56	873.7
1	X_3	16.67	0.753						1.08	
2a	LnXL	19.63	0.836						2.56	
2	X_2L	20.36	0.854	0.100					12.02	875.8
3	X_2Y^{c}	21.50	0.881		0.027				2.87	
4	XL_2	25.04	0.955	0.101					42.24	877.8
5	XYL	26.50	0.982		0.027				9.74	
6	X_2O	27.19	0.995			0.141			2.28	877.8
7	XY_2	28.05	1.009		0.027				4.62	877.8
8	XLO	33.67	1.096	0.101					7.69	879.7
9	XYO	35.69	1.123		0.027				2.84	
10	XLP	36.03	1.127				0.032		5.40	853.7
11	XYP	38.20	1.155		0.027				1.56	
12	XO_2	45.63	1.237			0.141			0.84	881.7
13	XLS	48.11	1.261					0.134	2.31	881.7
14	XOP	48.91	1.269				0.032		0.87	855.8
15	XYS	51.10	1.289		0.028				0.5	881.7
				0.101	0.027	0.141	0.032	0.134		

[a] Numbers of TGs are the numbers of peaks on the Figure 5; [b] X, moiety of conjugated octadecatrienoic acid (catalpic); [c] Y, moiety of uncommon octadecadienoic acid.

3.2.5. Triglycerides of Oil with (8Z,10E,12Z)-octadeca-8,10,12-trienoic Acid Moieties

Fatty acid composition of *Jacaranda* seed oil, determined in Reference [25], includes linoleic (41.4%), jacaric (as (8Z,10E,12Z)-octadeca-8,10,12-trienoic (Jac)) (30.9%), with two isomers ((8Z,10E,12E)-octadeca-8,10,12-trienoic (1.1%) and (8E,10E,12Z)-octadeca-8,10,12-trienoic (1.9%)), oleic (11.1%), stearic (4.3%) and palmitic (3.5%) acids. On the chromatogram of our sample of *Jacaranda mimosifolia* seed oil (Figure 6), there ten main TGs are found (Table 5), with electronic spectra corresponding to Type I (Figure 1). The value of increment Δ(Jac→L) is 0.009 points smaller than that of increment Δ(Pu→L). Thus, the difference of retention (as logarithms of capacity factors) of corresponding TGs with punicic and jacaric acids is similar to the TGs with another pair of conjugated octadecatrienoic acids with the same structure alteration, Δ(Cat→Cal) = 0.008. In this case, this increment may be common for all pairs of TGs for which CH_2-group is shifting form inner to outer part of the acid substituent. It may be proposed that, for substitution of all-*trans* octadeca-9,11,13-trienoic acid moiety by all-trans octadeca-8,10,12-trienoic, one increment must be around 0.009.

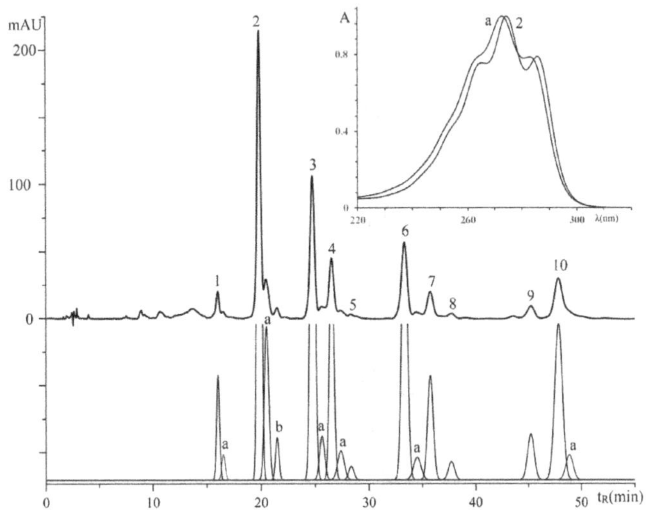

Figure 6. Separation of TGs of *Jacaranda mimosifolia* seed oil. Column 4.6 mm × 250 mm Kromasil 100-5C18; mobile phase, 35% propanol-2 and 65% acetonitrile, 1 mL/min; column thermostat temperature, 30 °C; detector, 270 nm. For peak number composition, see Table 5.

Table 5. Retention parameters, mass spectral signals and content of TGs of *Jacaranda mimosifolia* seed oil for the mobile phase composition 35% propanol-2 and 65% acetonitrile.

No. [a]	TG Composition	t_R (min)	Log*k*	Increment Values				Mole Fraction of TG, %	*M/z* [M + H$^+$]
				X→L	L→O	O→P	P→S		
1	X [b]$_3$	16.02	0.733					1.36	873.8
2	X_2L	19.83	0.841	0.108				22.50	875.8
3	XL_2	24.76	0.950	0.109				24.78	877.7
4	X_2O	26.52	0.983		0.142			6.25	877.7
5	X_2P	28.31	1.014			0.031		0.42	851.8
6	XLO	33.34	1.091	0.108	0.142			18.01	879.8
7	XLP	35.73	1.124	0.110		0.032		6.42	853.7
8	X_2S	37.69	1.148				0.135	0.75	n.d. [c]
9	XO_2	45.33	1.234		0.143			3.83	879.7
10	XLS(+XOP)	47.82	1.258	0.110			0.135	15.67	881.7 + 855.7
	Mean value:			0.109	0.142	0.032	0.135		

[a] Numbers of TGs are the numbers of peaks on the Figure 6; [b] X, moiety of conjugated octadecatrienoic acid (jacaric); [c] n.d., not determined.

Calculated fatty acid composition of the oil is as follows: jacaric (as a sum of all octadecatrienoic acids) (44.30%), followed by stearic (5.69%), linoleic (37.17%), oleic (10.58%) and palmitic (2.27%) acids.

3.2.6. Triglycerides of Oil with (9E,11E,13E)-octadeca-9,11,13-trienoic Acid Moieties

In the first paper with results of investigation of *Centranthus ruber* seed oil, only α-eleostesric acid was found [26,27]. Meanwhile, detection of the mixture of this acid with its all-*E* isomer was reported later [28]. In our case, the seeds were found to be a wonderful source of oil composed by two isomeric octadecatrienoic acid isomers: α-eleostearic and β-elleostearic (9E,11E,13E)-octadeca-9,11,13-trienoic acid (β-El).

The oil is an interesting example with comparable amounts of isomeric octadecatrienoic acids resulting in appearance of thee peaks in region of elution X2L TG species (Figure 7). Opposite to the former cases, peaks have different electronic spectra: Type II, Type III and mixed type. Thus, Table 6 contains more TGs species than the other tables (except of *Catalpa* seed oil). This is just an example to calculate increment for exchange of the two isomeric acids:

$$\Delta(\alpha El \rightarrow \beta\text{-}El) = 0.021 \quad (6)$$

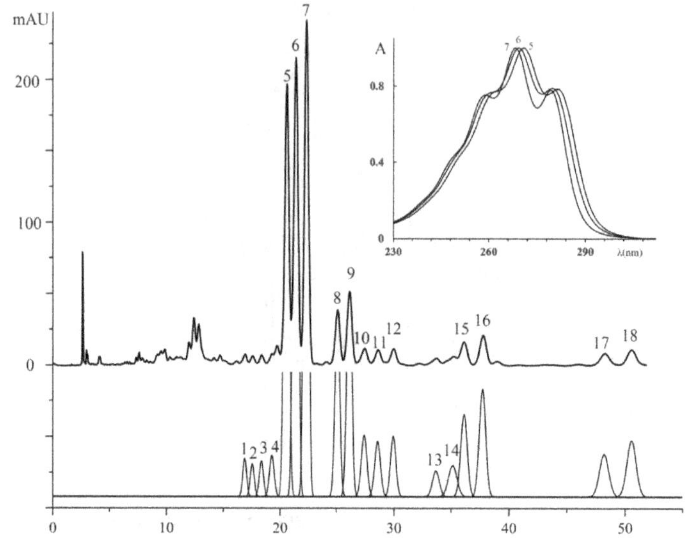

Figure 7. Separation of TGs of *Centranthus ruber* seed oil. Column 4.6 mm × 250 mm Kromasil 100-5C18; mobile phase, 35% propanol-2 and 65% acetonitrile, 1 mL/min; column thermostat temperature, 30 °C; detector, 270 nm. For peak number composition, see Table 6.

Table 6. Retention parameters, mass spectral signals and content of TGs of *Centranthus ruber* seed oil for the mobile phase composition 35% propanol-2 and 65% acetonitrile.

No. [a]	TG Composition	t_R (min)	Log*k*	Increment Values					Mole Fraction of TG, %	*M/z* [M + H⁺]
				αEl→βEl	αEl→L	L→O	O→P	P→S		
1	αEl [b] $_3$	16.94	0.761						0.28	
2	(αEl)$_2$βEl	17.62	0.782	0.020						
3	αEl(βEl)$_2$	18.36	0.802	0.021						873.8
4	βEl [b] $_3$	19.28	0.827	0.022					0.41	
5	αEl$_2$L	20.56	0.859		0.098				15.35	
6	αElβElL	21.39	0.878	0.020					16.66	875.7
7	βEl$_2$L	22.30	0.899	0.020					21.40	
8	αElL$_2$	25.10	0.956		0.097				6.88	877.7
9	βElL$_2$	26.14	0.976	0.020					10.19	
10	αEl$_2$O	27.44	0.999			0.140			2.34	
11	αElβElO	28.60	1.019	0.020		0.140			2.28	877.7
12	βEl$_2$O	29.94	1.040	0.022					3.17	
13	αElLO	33.71	1.096		0.097				1.41	879.8
14	βElLO	35.21	1.117	0.020					2.07	
15	αElLP	36.10	1.128				0.032		4.09	853.7
16	βElLP	37.74	1.149	0.021			0.032		5.52	
17	αElLS	48.24	1.262					0.134	3.01	881.8
18	βElLS	50.53	1.284	0.021				0.134	4.35	
	Mean value:			0.021	0.097	0.140	0.032	0.134		

[a] Numbers of TGs are the numbers of peaks on the Figure 7; [b] αEl and βEl, α-eleostearic and β-eleostearic acids, respectively.

Calculated by the same way, fatty acid composition is as follows: α-eleostearic (23.8%), β-eleostearic (30.8%), linoleic (36%), oleic (3.8%), palmitic (3.2%) and stearic (2.5%) acids. This is close to data published in [28]: α-eleostearic (28%), β-eleostearic (17%), linoleic (36%), oleic (4%), and saturated (18%).

3.2.7. Triglycerides of Three Plant's Seed Oil

For determination of relative retentions parameters of TGs of all basic seed oils, the retention of Pu_3 of *Punica granatum* seed oil [29–31] was chosen as a reference. The reasons for the choice are: (a) the fruit is available all year round in the fruit market and thus the oil may be easily prepared; and (b) there is only one main peak within the seed oil chromatogram (peak of Pu_3), while the others have much smaller peak areas, thus not disturbing the retention of reference solute determination.

The relative retention was determined for three of four different mobile phase compositions for eluent system "propanol-2-acetonitrile". The retentions were registered only for chromatographic isocratic runs performed at chromatographic system equilibrium. The latter was controlled by complete coincidence of two consecutive chromatograms of the sample.

The equations obtained were used to calculate the TGs composition of the three seed oils (*Valeriana officinalis, Vernicia montana* and *Thladiantha dubia*) by minimization of differences of experimental retention parameters and calculated values, while considering the type of electronic spectra (Figures 8–10 and Tables 7–9).

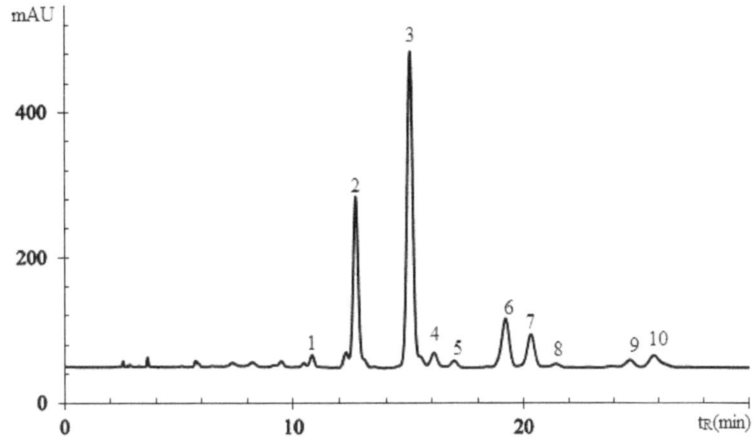

Figure 8. Separation of TGs with α-eleostearic acid moieries of *Valeriana officinalis* seed oil. Column 4.6 mm× 250 mm Kromasil 100-5C18; mobile phase, 42% propanol-2 and 58% acetonitrile, 1 mL/min; column thermostat temperature, 30 °C; detector, 270 nm. For peak number composition, see Table 7.

Table 7. Experimental retention parameters for TGs of *Valeriana officinalis* seed oil versus of relative retention for the mobile phase composition 42% propanol-2 and 58% acetonitrile calculated by equations.

No. [a]	TG Composition	t_R (min)	Log$k$(i)		Content, Mol. %
			Experimental	Calculated	
1	X^{b}_3	10.81	0.522	0.521	0.6
2	X_2L	12.71	0.611	0.609	13.12
3	XL_2	15.04	0.700	0.699	56.12
4	X_2O	16.09	0.735	0.736	1.66
5	X_2P	17.08	0.766	0.767	0.79
6	XLO	19.19	0.825	0.826	12.15
7	XLP	20.42	0.855	0.857	8.19
8	X_2S	21.44	0.879	0.887	0.6
9	XO_2	24.82	0.951	0.953	2.24
10	XLS	26.09	0.975	0.976	4.52

[a] Numbers of TGs are the numbers of peaks on the Figure 8; [b] X, moiety of α-eleostearic acid.

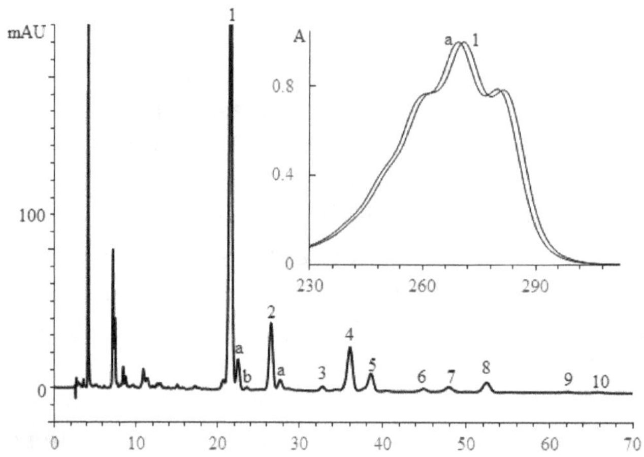

Figure 9. Separation of TGs of *Vernicia montana* seed oil. Column 4.6 × 250 mm Kromasil 100-5C18; mobile phase, 33% propanol-2 and 67% acetonitrile, 1 mL/min; column thermostat temperature, 30 °C; detector, 270 nm. For peak number composition, see Table 8.

Table 8. Experimental retention parameters for TGs of *Vernicia montana* seed oil vs. calculated by equations of relative retention for the mobile phase composition 33% propanol-2 and 67% acetonitrile.

No. [a]	TG Composition	t_R (min)	Log*k*(i)		Content, Mol. %
			Experimental	Calculated	
1	X^{b}_{3}	21.55	0.882	0.880	48.44
2	X_2L	26.51	0.982	0.982	16.89
3	XL_2	32.81	1.084	1.085	3.63
4	X_2O	36.05	1.128	1.129	12.17
5	X_2P	38.56	1.159	1.161	5.28
6	XLO	45.09	1.231	1.233	2.56
7	XLP	48.51	1.265	1.265	4.52
8	X_2S	52.62	1.302	1.303	4.09
9	XO_2	62.29	1.379	1.380	1.17
10	XLS	66.13	1.406	1.408	1.23

[a] Numbers of TGs are the numbers of peaks on the Figure 9; [b] X, moiety of α-eleostearic acid.

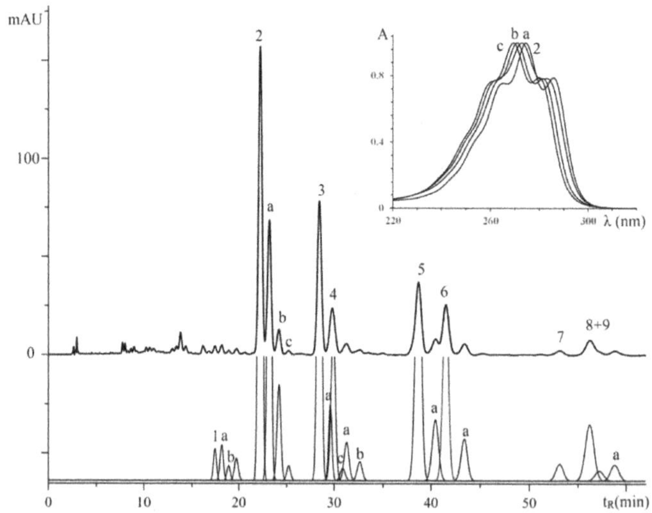

Figure 10. Separation of TGs of *Thladiantha dubia* seed oil. Column 4.6 × 250 mm Kromasil 100-5C18; mobile phase, 34% propanol-2 and 66% acetonitrile, 1 mL/min; column thermostat temperature, 30 °C; detector, 270 nm. For peak number composition, see Table 9.

Table 9. Experimental retention parameters for TGs of *Thladiantha dubia* seed oil vs. calculated by equations of relative retention for the mobile phase composition 34% propanol-2 and 66% acetonitrile.

No. [a]	TG Composition	t_R (min)	Log*k*(i)		Content, Mol. %
			Experimental	Calculated	
1	$X^{b}{}_3$	17.46	0.777	0.777	1.14
2	X_2L	22.18	0.896	0.898	29.14
3	XL_2	28.71	1.021	1.021	27.82
4	X_2O	30.08	1.043	1.045	4.56
5	XLO	38.99	1.164	1.166	17.18
6	XLP	41.88	1.197	1.199	12.34
7	XO_2	53.79	1.312	1.313	1.18
8 + 9	XLS + XOP	56.82	1.337	1.339	6.65

[a] Numbers of TGs are the numbers of peaks on the Figure 10; [b] X, moiety of punicic acid.

α-Eleostearic acid moieties were the main conjugated acid moieties involved in TG structures in the two oils. However, for *Valeriana* seed oil, we did not calculate fatty acid composition because of existence of valuable amounts of TGs without conjugated acid moieties, which was revealed by refractive index detection not included into results of the current paper. *Vernicia montana* seed oil composition: 78.4% of α-eleostearic, 10.8% of linoleic, 5.7% of oleic, 3.3% of palmitic and 1.8% of stearic acid. *Thladiantha* seed oil contains simultaneously α-eleostearic (9.5%), punicic (35.5%), β-eleostearic (0.3%), linoleic (40.1%), oleic (8.3%), palmitic (4.4%) and stearic (2.0%) acids.

4. Discussion

4.1. Some Remarks About Increment Approach

For a given chromatographic system, according to the principle of additivity, the retention (relative to trioleate retention, α, or selectivity) of TG composed by acids A, B and C may be described by Equation [15]:

$$\log\alpha(ABC) = \log\alpha(A) + \log\alpha(B) + \log\alpha(C) \tag{7}$$

For acid C moiety replacement by acid D moiety, we may write Equation (8):

$$\log\alpha(ABD) = \log\alpha(A) + \log\alpha(B) + \log\alpha(D) \tag{8}$$

Subtracting the Equation (7) from Equation (8) we get:

$$\log\alpha(ABD) - \log\alpha(ABC) = \log\alpha(D) - \log\alpha(C) \tag{9}$$

Since, for the given chromatographic system in the state of equilibrium, the retention of trioleate is constant, Equation (9) may be exchanged by Equation (10), where increment is calculated as difference between capacity factors of two TGs:

$$\log k(ABD) - \log k(ABC) = \Delta(C \rightarrow D) \tag{10}$$

Thus, for calculation of retention of all 20 TGs composed by acids A, B, C and D, it is necessary and sufficient to point out retention of a reference solute (e.g., A_3) and values of three increments: $\Delta(A \rightarrow B)$, $\Delta(B \rightarrow C)$ and $\Delta(C \rightarrow D)$.

Utilization of difference values $\Delta(A \rightarrow B)$ instead of partial values of Equation (7) is favorable because they are directly calculated for a set of logarithms of experimental capacity factors responsible for substituents exchange, revealing a dependence between solute structure and retention.

4.2. Comparison of Retention of TGs with Isomeric Conjugated Octadecatrienoic Acid Moieties

Summarizing the results obtained for oil separation in mobile phase containing 35 % propanol-2 and 65% acetonitrile we may built a final table of increments (Table 10).

Table 10. Increments (±0.001) of TGs retention for conjugated octadecatrienoic acid moieties Δ(top→bottom) exchange for mobile phase composition 35% propanol-2 and 65% acetonitrile.

		Pu	Jac	Cat	αEl	Cal	βEl	βCal
Pu	9Z11E13Z							
Jac	8Z10E12Z	0.009						
Cat	9E11E13Z	0.017	0.008					
αEl	9Z11E13E	0.020	0.011	0.003				
Cal	8E10E12Z	0.025	0.016	0.008	0.005			
βEl	9E11E13E	0.041	0.032	0.024	0.021	0.016		
βCal	8E10E12E	0.049	0.040	0.032	0.029	0.024	0.010	

For the same conjugated double bonds position (9,11,13) along the carbon chain, the sequence of elution times is:

$$Pu < Cat < \alpha El < \beta El, \quad (11)$$

that is, retention becomes greater when *cis*-double bond is substituted by *trans*-configuration, the increase being somewhat greater for the substitution of the outer double bond. The difference of TGs retention with catalpic and α-eleostearic acid is not great but enough to differentiate TGs with two or three of these acid moieties. There are some problems for differentiation of TGs with one moiety of the acids (Figure 11).

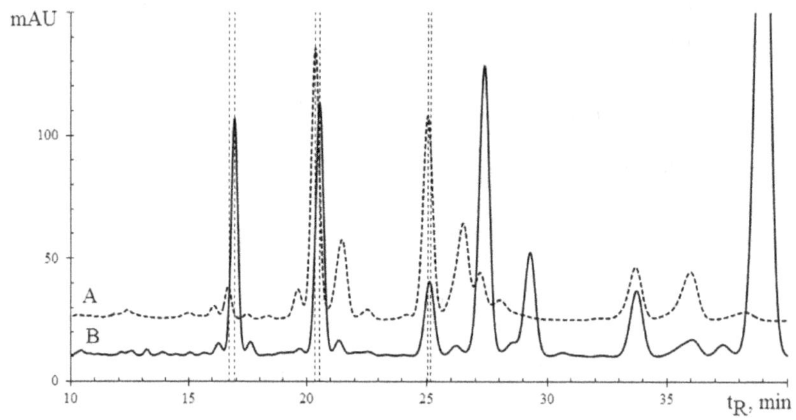

Figure 11. Overlapping of chromatograms of *Catalpa ovata* (A) and *Momordica charantia* (B) seed oil. Column 4.6 × 250 mm Kromasil 100-5C18; mobile phase, 35% propanol-2 and 65% acetonitrile, 1 mL/min; column thermostat temperature, 30 °C; detector, 270 nm.

Exchange of double bonds position from 9,11,13 to 8,10,12 is equal to movement of one CH_2-group from inner part of molecule to the outer one. For mobile phase 35% propanol-2 and 65% acetonitrile, increments for exchange punicic by jacaric acids as well as that for pairs β-eleostearic—β-calendic and catalpic—β-calendic acids are similar:

$$\Delta(Pu \rightarrow Jac) = 0.009 \quad (12)$$

$$\Delta(\beta El \rightarrow \beta Cal) = 0.010 \quad (13)$$

$$\Delta(Cat \rightarrow Cal) = 0.008 \quad (14)$$

Finally, the complete series of acid moieties for the increase of TGs retention time is as follows:

$$\text{Pu} < \text{Jac} < \text{Cat} < \alpha\text{El} < \text{Cal} < \beta\text{El} < \beta\text{Cal}. \quad (15)$$

4.3. *Minor Seed Oil Components Identification*

4.3.1. *Momordica charantia* Seed Oil

Pairs of tiny peaks accomplish the main peaks in Figure 2 with electronic spectra differing from it. The positions of all (including non-separated) peaks in Figure 1 was revealed by utilization of MagicPlot Student 2.7.2 program for chromatogram handling. The peaks that are eluted before the main one (marked with *a*) have bathochromic shifts of electronic spectra maxima, while peaks eluted after the main one (marked with *b*) have hypochromic shifts. By taking into account the previously found increments, it is possible to conclude that the oil contains not only α-eleostearic acid moieties but also earlier eluted TGs with punicic (a) acid moiety:

$$\Delta(\alpha\text{El}\rightarrow\text{Pu}) = \log \boldsymbol{k}(\text{a}) - \log \boldsymbol{k}(2) = -0.020 \quad (16)$$

TGs with β-eleostearic (b) acid moiety have greater retention:

$$\Delta(\alpha\text{El}\rightarrow\beta\text{E}) = \log \boldsymbol{k}(\text{b}) - \log \boldsymbol{k}(2) = 0.021 \quad (17)$$

These results are in a good agreement with the conclusions in Reference [21], although the authors were somewhat uncertain about presence of β-eleostearic acids moieties in the oil.

4.3.2. *Trichosanthus anguina* Seed Oil

The oil contains not only punic acid moieties but also later eluted TGs with one punicic acid moiety substitution with α-eleostearic (a):

$$\Delta(\alpha\text{E}\rightarrow\text{Pu}) = \log \boldsymbol{k}(2) - \log \boldsymbol{k}(\text{a}) = -0.020 \quad (18)$$

and TGs with two consecutive substitutions:

$$2\Delta(\alpha\text{E}\rightarrow\text{Pu}) = \log \boldsymbol{k}(2) - \log \boldsymbol{k}(\text{b}) = -0.039 \quad (19)$$

The validity of increment approach is supported with electronic spectra alteration (Figure 3). Meanwhile, the appearance of tiny peak *c* indicates existence of some other minor acid component, presumably β-eleostearic, but in small quantities.

4.3.3. Calendula Officinalis Seed Oil

The data in Table 3 indicated that calendic acid is distributed between TGs types apparently non-statistically because, for example, the ratio between peak areas (S) of the three TGs composed by moieties of calendic (Cal) and linoleic (L) acids:

$$S(Cal_3):S(Cal_2L):S(CalL_2) \quad (20)$$

is 0.50:1.00:0.08 for experimental data instead of calculated for statistical distribution as 0.97:1.00:0.25.

Meanwhile, the oil also contains isomeric to (α-)calendic acid moiety, leading to later elution of TGs with evident hypochromic shift of spectral maxima with a high value of increment per one new acid permitted to refer the acid as to β-calendic (8E,10E,12E)-octadeca-8,10,12-trienoic (1.56%).

$$\Delta(\alpha\text{Cal}\rightarrow\beta\text{Cal}) = \log \boldsymbol{k}(\text{a}) - \log \boldsymbol{k}(2) = 0.024 \quad (21)$$

$$2\Delta(\alpha\text{Cal}\rightarrow\beta\text{Cal}) = \log k(\text{b}) - \log k(2) = 0.049 \quad (22)$$

4.3.4. *Catalpa ovata* Seed Oil

In Figure 5, for the two main TGs types (No. 1 and No. 2), earlier eluting components are rather noticeable. The increment for exchange of catalpic acid by corresponding acid for this TGs does not fit any increment found for CLnA:

$$\Delta(\text{X}\rightarrow\text{Cat}) = \log k(1) - \log k(1\text{a}) = \log k(2) - \log k(2\text{a}) = 0.018 \quad (23)$$

Electronic spectra of all mentioned peaks are the same, as are their mass-spectra. Thus, the acid X may be non-conjugated α-linolenic, (9Z,12Z,15Z)-octadeca-9,12,15-trienoic acid (Ln). The proposition was confirmed by analysis of fatty acids obtained by alkaline hydrolysis of this oil and flaxseed oil. Moreover, peaks No. 3 and No. 5 have the same electronic and mass spectra as peaks No. 2 and No. 4, respectively. This demands proposing the existence of some uncommon octadecadienoic acid, Y, in the oil. The corresponding exchange is perfectly described by increment model,

$$\Delta(\text{L}\rightarrow\text{Y}) = \log k(3) - \log k(2) = \log k(5) - \log k(4) = \ldots = 0.027 \quad (24)$$

The most probable solution that satisfies the parameters for unknown acid structure may be linoleic acid isomer with double bonds system shifted towards carboxylic group by 3 CH_2-group because for one CH_2-group shift an increment was found above to be around 0.009 (Equations (11)–(13)):

However, the real structure of the acid will be established in upcoming research.

4.3.5. *Jacaranda mimosifolia* Seed Oil

Considerations of electronic spectra and retentions of tiny peaks on the chromatogram of the oil leads to the conclusion that these peaks may appear due to synthesis of calendic and/or isomeric to the calendic—not found in other natural sources (8Z,10E,12E)-octadeca-8,10,12-trienoic acid—as was supposed in Reference [25]. Indeed, the spectrum of peak 2a is non-distinguishable from spectrum of peak 2a of trichosanthus seed oil chromatogram.

4.4. Relative Retention Analysis of TGs with Conjugated Seed Oils

Coefficients of relative retention (Equation (1)) of all TGs of all seed oils under investigation in current paper are listed in Table 11. The data may be explored for seed oil TGs determination by choice of the equations of relative retention fitting experimental data within discrepancy of no more than 0.002 of logarithmic units. The approach was successfully applied to three plant types of seed oils (Tables 8–10).

Table 11. Coefficients of relative retentions of TGs vs. retention of Pu_3 (Equation (1)) for chromatographic system "pronanol-2-acetonitrile" and stationary phase Kromasil 100-5C18.

TG Structure	Coefficients of Equation (1)	X, Conjugated Octadecatrienoic Acid						
		Pu	Jac	Cat	αEl	Cal	βEl	βCal
X_3	a_1	1.000	1.011	1.014	1.010	1.021	1.035	1.054
	a_0	0.000	0.022	0.041	0.052	0.058	0.100	0.113
X_2L	a_1	1.042	1.048	1.052	1.047	1.054	1.062	n.d.
	a_0	0.089	0.104	0.116	0.123	0.127	0.154	n.d.
XL_2	a_1	1.091	1.086	1.090	1.085	1.092	1.087	n.d.
	a_0	0.174	0.187	0.190	0.195	0.195	0.212	n.d.
X_2O	a_1	1.101	1.106	1.112	1.104	1.115	1.120	n.d.
	a_0	0.189	0.206	0.215	0.224	0.225	0.255	n.d.
X_2P	a_1	1.105	1.103	n.d.	1.107	1.118	n.d.	n.d.
	a_0	0.218	0.239	n.d.	0.253	0.256	n.d.	n.d.
XLO	a_1	1.147	1.144	1.151	1.144	1.152	1.135	n.d.
	a_0	0.276	0.281	0.289	0.295	0.294	0.317	n.d.

Table 11. *Cont.*

TG Structure	Coefficients of Equation (1)	X, Conjugated Octadecatrienoic Acid						
		Pu	Jac	Cat	αEl	Cal	βEl	βCal
XLP	a_1	1.148	1.146	1.155	1.146	1.156	1.146	n.d.
	a_0	0.307	0.319	0.318	0.325	0.323	0.346	n.d.
X_2S	a_1	n.d.	1.170	n.d.	1.171	1.183	n.d.	n.d.
	a_0	n.d.	0.327	n.d.	0.343	0.345	n.d.	n.d.
XO_2	a_1	1.209	1.206	1.213	1.200	n.d.	n.d.	n.d.
	a_0	0.374	0.387	0.387	0.396	n.d.	n.d.	n.d.
XLS	a_1	1.216	1.210	1.218	1.214	n.d.	1.216	n.d.
	a_0	0.394	0.408	0.408	0.412	n.d.	0.434	n.d.

The relative retention plot (separation map) for jacaranda seed oil TGs (vs. Jac_3) is presented in Figure 12.

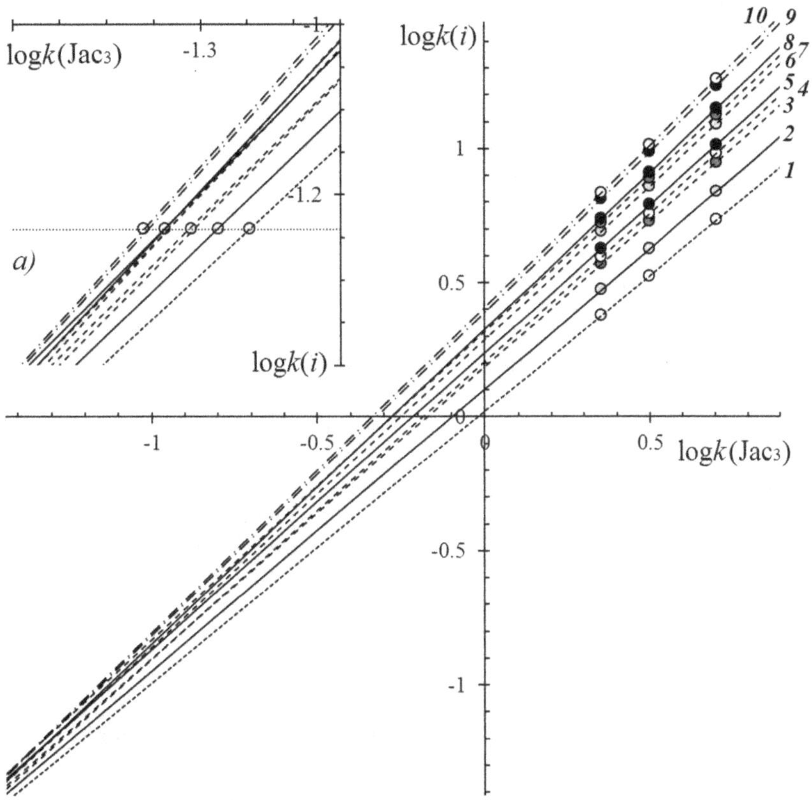

Figure 12. Separation map for *Jacaranda mimosifolia* seed oil. Insert is the left lower part of the map with "zero" points. Column 4.6 × 250 mm Kromasil 100-5C18; mobile phase system "propanol-2-acetonitrile", column thermostat temperature, 30 °C; detector, 270 nm (numbers of lines are numbers of substances in Table 5).

The straight lines of the relationships have a trend to intersect in lower left side corner of the plot (for imaginable mobile phase composition with high elution strength). The first consequence of the plot type is the dependence of increments upon eluent composition: increments decrease with elution strength increase. The second consequence is that selectivity parameter of the pair of solutes separation in the general case is also not a constant for wide range of mobile phase compositions. The valuable property of two-parameter relative retention equations is a non-sensitivity to mobile phase composition.

Another advantage of the relative retention plot in the present case is the possibility of differentiation of TGs by the number of double bonds in a molecule. Indeed, if the mobile phase elution strength is rising, for some mobile phase composition hydrophobic (Van der Waals interactions), solvation properties may become equal for any compound of homologues series [32]. The retention in this case points out "zero" point on the plot. However, because of the existence of another type of interaction, this point moves to the right or left; for example, the more double bonds exist in a molecule, the more to the right the point must be shifted.

For the Jacaranda seed oil, the position of the "zero" point may be determined by intersection of lines for relative retention of the pair of homologous TGs, Jac_2P and Jac_2S (Figure 12). On the plot lines of relative retention of all ten TGs intersect, the horizontal line (for "zero" points positions on the plot) is almost proportional to the number of double bonds in a molecule, despite possible errors at approximation far beyond the compositions of mobile phase used for formation of the plot.

5. Conclusions

The sequence of retention times of the same types of TGs is constant for reasonable compositions of propanol-2-acetonitrile mobile phases and Kromasil 100-5C18 stationary phases: punicic ($C18{:}3^{9Z11E13Z}$) < jacaric ($C18{:}3^{8Z10E12Z}$) < catalpic ($C18{:}3^{9E11E13Z}$) < α-eleostearic ($C18{:}3^{9Z11E13E}$) < calendic ($C18{:}3^{8E10E12Z}$) < β-eleostearic ($C18{:}3^{9E11E13E}$) < all-E calendic ($C18{:}3^{8E10E12E}$) acids. Meanwhile, the retention alteration in series catalpic→α-eleostearic→calendic moieties for TGs of similar structures are quite small but sufficient for solute differentiation when TGs contain at least two moieties of the acids. The migration of CH_2-group from the inner part of the fatty acid moiety to the outer one leads to rise of solute retention by similar values of logarithms of retention capacity (log*k*) for any starting *cis-trans* stereoisomers. Exchange of stereo configurations in direction *cis*→*trans* also results in increase of retention being slightly different for inner and outer *cis*-double bond. Thus, it was proven that it is possible to differentiate TGs of complex composition with moieties of all natural CLnA by retention control accomplished by electronic spectra comparison, even though there are only three types of electronic-vibration spectra for seven isomeric CLnA. Equations of TGs relative retention were found to be useful for preliminary TGs identification.

Author Contributions: Anh Van Nguyen: Experimental investigation and processing of the results. Deineka Victor: Planning of experiments, processing of the results, and writing and preparing manuscript to submission. Deineka Lumila: Sample preparation, and processing of the results. Ngoc Anh Vu Thi: Collection and identification of plant material in Vietnam, and literature survey.

Conflicts of Interest: The authors declare no conflict of interest.

References

1. De Carvalho, E.B.T.; de Melo, I.L.P.; Mancini-Filho, J. Chemical and physiological aspects of isomers of conjugated fatty acids. *Ciênc. Tecnol. Aliment. Campinas* **2010**, *30*, 295–307. [CrossRef]
2. Yuan, G.-F.; Chena, X.-E.; Li, D. Conjugated linolenic acids and their bioactivities: A Review. *Food Funct.* **2014**, *5*, 1360–1368. [CrossRef] [PubMed]
3. Tanaka, T.; Hosokawa, M.; Yasui, Y.; Ishigamori, R.; Miyashita, K. Cancer Chemopreventive Ability of Conjugated Linolenic Acids. *Int. J. Mol. Sci.* **2011**, *12*, 7495–7509. [CrossRef] [PubMed]
4. Tsuzuki, T.; Tokuyama, Y.; Igarashi, M.; Miyazawa, T. Tumor growth suppression by α-eleostearic acid, a linolenic acid isomer with a conjugated triene system, via lipid peroxidation. *Carcinogenesis* **2004**, *25*, 1417–1425. [CrossRef] [PubMed]
5. Aruna, P.; Venkataramanamma, D.; Singh, A.K.; Singh, R.P. Health Benefits of Punicic Acid: A Review. *Compr. Rev. Food Sci. Food Saf.* **2016**, *15*, 16–27. [CrossRef]
6. Suzuki, R.; Yasui, Y.; Kohno, H.; Miyamoto, S.; Hosokawa, M.; Miyashita, K.; Tanaka, T. Catalpa seed oil rich in 9*t*,11*t*,13*c*-conjugated linolenic acid suppresses the development of colonic aberrant crypt foci induced by azoxymethane in rats. *Oncol. Rep.* **2006**, *16*, 989–996. [CrossRef] [PubMed]
7. Shinohara, N.; Ito, J.; Tsuduki, T.; Honma, T.; Kijima, R.; Sugawara, S.; Arai, T.; Yamasaki, M.; Ikezaki, A.; Yokoyama, M.; et al. Jacaric acid, a linolenic acid isomer with a conjugated triene system, reduce stearoyl-CoA desaturase expression in liver of mice. *J. Oleo Sci.* **2012**, *61*, 433–441. [CrossRef] [PubMed]

8. Benjamin, S.; Spener, F. Conjugated linoleic acids as functional food: An insight into their health benefits. *Nutr. Metabol.* **2009**, *6*, 1–13. [CrossRef] [PubMed]
9. Hopkins, C.Y.; Chisholm, M.J. A Survey of the Conjugated Fatty Acids of Seed Oils. *J. Am. Oil Chem. Soc.* **1968**, *45*, 176–182. [CrossRef] [PubMed]
10. Ichihara, K.; Fukubayashi, Y. Preparation of fatty acid methyl esters for gas-liquid chromatography. *J. Lipid Res.* **2010**, *51*, 635–640. [CrossRef] [PubMed]
11. Frede, E. Improved HPLC of Triglycerides by Special Tempering Procedures. *Chromatographia* **1988**, *21*, 29–36. [CrossRef]
12. Barrón, L.J.R.; Santa-Maria, G. Non-Aqueous Reverse-Phase HPLC Analysis of Triglycerides. *Chromatographia* **1987**, *23*, 209–214. [CrossRef]
13. Li, L.; Han, J.; Wang, Z.; Liu, J.; Wei, J.; Xiong, S.; Zhao, Z. Mass Spectrometry Methodology in Lipid Analysis. *Int. J. Mol. Sci.* **2014**, *15*, 10492–10507. [CrossRef] [PubMed]
14. Podlaha, O.; Toregerd, B. A System for Identification of Triglycerides in Reversed Phase HPLC Chromatograms Based on Equivalent Carbon Numbers. *J. Sep. Sci.* **1982**, *5*, 553–558. [CrossRef]
15. Sempoore, G.; Bezard, J. Qualitative and quantitative analysis of peanut oil triacylglycerols by reversed-phase liquid chromatography. *J. Chromatogr.* **1986**, *366*, 261–282. [CrossRef]
16. Deineka, V.I. Experimental substantiation of a method of relative analysis of retention in HPLC. *Russ. J. Phys. Chem.* **2006**, *80*, 425–428. [CrossRef]
17. Hennessy, A.A.; Ross, P.R.; Fitzgerald, G.F.; Stanton, C. Sources and Bioactive Properties of Conjugated Dietary Fatty Acids. *Lipids* **2016**, *51*, 377–397. [CrossRef] [PubMed]
18. Hopkins, C.Y.; Chisholm, M.J.; Orgodnik, J.A. Identity and Configuration of Conjugated Fatty Acids in Certain Seed Oils. *Lipids* **1969**, *4*, 89–92. [CrossRef]
19. Chisholm, M.J.; Hopkin, C.Y. Fatty Acid Composition of Some *Cucurbitaceae* Seed Oils. *Can. J. Chem.* **1964**, *42*, 560–564. [CrossRef]
20. Chang, M.-K.; Conkerton, E.J.; Chapital, D.C.; Wan, P.J.; Vadhwa, O.P.; Spiers, J.M. Chinese Melon (*Momordica charantia* L.) Seed: Composition and Potential Use. *J. Am. Oil Chem. Soc.* **1996**, *73*, 263–265. [CrossRef]
21. Joh, Y.-G.; Kim, S.-J. Analysis of Molecular Species of Triacylglycerols from Vegetable Oils Containing Fatty Acids with Non-Methylene-Interrupted Double Bonds, by HPLC in the Silver-ion Mode. *J. Jpn. Oil Chem. Soc.* **1998**, *47*, 927–936. [CrossRef]
22. Joh, Y.-G.; Kim, S.-L.; Christie, W.W. The Structure of the Triacylglycerols, Containing Punicic Acid, in the Seed Oil of *Trichosanthes kirilowii*. *J. Am. Oil Chem. Soc.* **1995**, *72*, 1037–1042. [CrossRef]
23. Król, B.; Paszko, T.; Król, A. Conjugated linolenic acid content in seeds of some pot marigold (*Calendula officinalis*) cultivars grown in Poland. *Farmacia* **2016**, *64*, 881–886.
24. Takagi, T.; Itabashi, Y. Occurrence of mixtures of geometrical isomers of conjugated octadecatrienoic acids in some seed oils: Analysis by open-tubular gas liquid chromatography and high performance liquid chromatography. *Lipids* **1981**, *16*, 546–551. [CrossRef]
25. Yamasaki, M.; Motonaga, C.; Yokoyama, M.; Ikezaki, A.; Kakihara, T.; Hayasegawa, R.; Yamasaki, K.; Sakono, M.; Sakakibara, Y.; Suiko, M.; et al. Induction of apoptotic call death in HL-60 cells by Jacaranda seed oil derived fatty acids. *J. Oleo Sci.* **2013**, *62*, 925–932. [CrossRef] [PubMed]
26. Hopkins, C.Y.; Chisholm, M.J. Identification of conjugated triene fatty acids in certain seed oils. *Can. J. Chem.* **1962**, *40*, 2078–2082. [CrossRef]
27. Radunz, A.; Heb, P.; Schmid, G.H. Analysis of the Seed Lipids of *Aleurites montana*. *Z. Naturforsch.* **1998**, *53c*, 305–310.
28. Tulloch, A.P. ^{13}C Nuclear Magnetic Resonance Spectroscopic Analysis of Seed Oils Containing Conjugated Unsaturated Acids. *Lipids* **1982**, *17*, 544–550. [CrossRef]
29. Turtygin, A.V.; Deineka, V.I.; Deineka, L.A. Determination of the triglyceride composition of pomegranate seed oil by reversed-phase HPLC and spectrophotometry. *J. Anal. Chem.* **2013**, *68*, 558–563. [CrossRef]
30. Boroushaki, M.T.; Mollazadeh, H.; Afshari, A.R. Pomegranat seed oil: A comprehensive review on its therapeutic effects. *Int. J. Pharm. Sci. Res.* **2016**, *7*, 430–442.

9

HPLC Method Development for Quantification of Doxorubicin in Cell Culture and Placental Perfusion Media

Mansi Shah [1], Luke Bourner [2], Shariq Ali [2,3], Sanaalarab Al-Enazy [2], Menatallah M. Youssef [4], Morgan Fisler [5] and Erik Rytting [1,2,*]

1 Department of Obstetrics & Gynecology, University of Texas Medical Branch, 301 University Boulevard, Galveston, TX 77555-1062, USA; manshah@utmb.edu
2 Department of Pharmacology and Toxicology, University of Texas Medical Branch, Galveston, TX 77555-1062, USA; labourne@utmb.edu (L.B.); sh2ali@utmb.edu (S.A.); saalenaz@utmb.edu (S.A.-E.)
3 School of Medicine, University of Texas Medical Branch, Galveston, TX 77555-1062, USA
4 Department of Pharmaceutical Analytical Chemistry, Ain-Shams University, Cairo 1156, Egypt; mennamagdy@pharma.asu.edu.eg
5 Department of Obstetrics & Gynecology, Medical University of South Carolina, Charleston, SC 29425-2221, USA; fisler@musc.edu
* Correspondence: erik.rytting@utmb.edu

Abstract: Assessment of drug transport across the placenta is important in understanding the effect of drugs on placental and fetal health. These phenomena can be studied in both in vitro cell lines and ex vivo placental perfusions. We have successfully developed a sensitive yet simple high performance liquid chromatography (HPLC) method coupled with fluorescence detection to determine the concentration of doxorubicin (DXR) in cell culture media for transport studies in human trophoblast cells (BeWo, b30 clone) and in fetal media for placental perfusion experiments. The method was developed based on a protein precipitation technique and was validated in both media types for linearity, intra-day, and inter-day precision and accuracy. The relationship of peak area to concentration was linear with R^2 values of 0.99 or greater obtained over the concentration range of 1.5 to 15,000 ng/mL. Despite the high concentrations of albumin in fetal perfusion media (30 mg/mL), the lower limits of detection and quantification for DXR were found to be 1.5 and 5 ng/mL, respectively. This analytical method may be used to study the transport of DXR across BeWo cells and human placenta during placental perfusion studies.

Keywords: doxorubicin; BeWo cell culture media; placenta; high performance liquid chromatography; fluorescence; anthracycline; analytical method; perfusion

---

1. Introduction

The occurrence of breast cancer in pregnant women has increased over the past few years, complicating approximately 1 in 3000 pregnancies. Moreover, treatment strategies involving chemotherapy can be toxic to the fetus [1]. Pregnancy is a dynamic state in which the body undergoes several anatomical and physiological changes. These changes can alter the pharmacokinetics and pharmacodynamics of the drug compared to non-pregnant subjects [2]. However, pharmacokinetic studies have rarely been performed on pregnant women receiving chemotherapy. Doxorubicin is commonly prescribed during pregnancy for the treatment of malignancies such as breast cancer, ovarian cancer, and solid tumors [3], despite the fact that the FDA has defined it as pregnancy category D (positive evidence of risk). Therefore, it is imperative to study the transplacental transfer of the drug itself and its liposomal formulation (Doxil®) and understand its implications on fetal development.

Doxorubicin (DXR) is an anthracycline antibiotic isolated from the bacterium *Streptomyces peucetius* var. *caesius* and is available as a hydrochloride salt. Doxorubicin has multiple mechanisms of action, including DNA intercalation and prevention of transcription, inhibition of topoisomerase II, and induction of cytotoxicity by the formation of oxygen free radicals [4]. DXR is known to be rapidly and extensively distributed to various organs such as the heart, liver, kidney, and spleen, but little is known about its transport and accumulation in the placenta [4–6]. Therefore, to investigate the mechanisms involved in the transplacental transfer of doxorubicin, various in vitro and ex vivo models can be studied [7]. The BeWo (b30 clone) cell line, which is a model of the human placental trophoblast, and the dually perfused human placental lobule are both widely used in vitro and ex vivo models to study the transplacental transfer of various drugs and their metabolites [7,8]. In order to study the transplacental transport of DXR across BeWo cells and placenta during placental perfusion studies, it is necessary to have simple, selective, and validated high performance liquid chromatography (HPLC) methods for determining DXR concentrations in media.

A handful of analytical methods for the determination of doxorubicin in biological fluids are available; however, to the best of our knowledge, none have been published for quantification in DMEM/F12 cell culture media and/or placental perfusion media. Some of the methods involve sample preparations such as liquid-liquid extraction [9–12] or solid phase extraction [13], which can be tedious and require the use of an internal standard. Other methods with protein precipitation involved the use of perchloric acid [14], which can cause deterioration of both the analyte and typical silica-based reversed-phase HPLC columns. While screening materials to use as a protein precipitant, we observed instability in DXR in the presence of perchloric acid. Other methods involve mass spectrometry, which may be highly sensitive but cost-prohibitive for many laboratories [15]. If the drug of interest is inherently fluorescent, fluorescence detection is often preferable because it is simple, inexpensive, and more sensitive in comparison to ultraviolet absorption. A recent method based upon protein precipitation was used to determine doxorubicin in arthritic joints, but it has a lower limit of quantification of only 0.8 μg/mL [16]. Therefore, even though several methods have been used to detect DXR in biological fluids, there is great benefit to developing a simple method with a low limit of quantification (5 ng/mL) for the assessment of DXR in cell culture and placental perfusion media.

For these reasons, we have developed a sensitive reversed-phase HPLC method using fluorescence detection for the quantification of DXR in DMEM/F12 cell culture medium and M199 placental perfusion medium containing 30 mg/mL human serum albumin. Since DXR exhibits high protein binding (45–85%), it is very important to use the physiological concentration of albumin, which can affect the transplacental transport of DXR [17,18]. Method development in placental perfusion medium is particularly challenging due to this high concentration of albumin. The method developed was reliable and highly sensitive. We believe that the simplicity of the sample preparation method and the portability between the two different types of media makes this an accessible and valuable method for laboratories studying doxorubicin's interactions with the placenta.

2. Materials and Methods

2.1. Chemicals and Reagents

Doxorubicin hydrochloride was obtained from Oakwood Chemical (Estill, SC, USA). Cell culture medium for the BeWo b30 cell line contains Dulbecco's Modified Eagle's Medium/Ham's F-12 50/50 mixture without phenol red (Mediatech, Manassas, VA, USA) with 10% fetal bovine serum (Atlanta Biologicals, Flowery Branch, GA, USA), antibiotic/antimycotic containing 10,000 units/mL penicillin, 10,000 μg/mL streptomycin, and 25 μg/mL amphotericin B (Gibco®, Carlsbad, CA, USA), 100X MEM nonessential amino acid solution (Sigma, St. Louis, MO, USA), and 200 mM L-glutamine (Mediatech, Manassas, VA, USA). Placental perfusion medium was prepared using dextran (Sigma

Aldrich, St. Louis, MO, USA), human serum albumin fraction V (Calbiochem, La Jolla, CA, USA), and medium 199 (Sigma Aldrich). Acetonitrile of HPLC grade, acetone, monobasic potassium phosphate, and phosphoric acid were obtained from Fisher Scientific (Fair Lawn, NJ, USA). Deionized water was obtained from a Millipore Synergy UV Ultrapure Water System (Millipore, Billerica, MA, USA).

2.2. Chromatographic Instruments and Conditions

The HPLC system (Waters, Milford, MA, USA) was comprised of a Waters 1525 Binary HPLC pump, a Waters 2707 Autosampler, and a Waters 2475 multi λ fluorescence detector. For both methods, a C-18 column (150 mm × 4.6 mm, 5 μm particle size, and 100 Å pore size, AkzoNobel/Kromasil, Brewster, NY, USA) was used at ambient temperature.

The mobile phase consisted of an aqueous phase (A) containing 20 mM KH_2PO_4 and 0.1% (*v*/*v*) phosphoric acid and an organic phase (B), which was acetonitrile. Flow was set at 1.0 mL/min. A gradient method was used in which the mobile phase started as 75% A and 25% B and changed in a linear manner to 60% A and 40% B by 7 min. Between 7 and 10 min, the mobile phase ratio was held constant. The mobile phase then reverted to 75% A and 25% B by 12 min and was held constant to baseline until 15 min. A wash vial containing initial mobile phase was injected between samples to avoid sample carryover. Peaks were monitored using fluorescence detection at an excitation wavelength of 480 nm and an emission wavelength of 550 nm using the fluorescence detector.

2.3. Sample Preparation

Cell culture medium for the analytical method validation of DXR was prepared as per the protocol in which BeWo b30 cells are typically grown [7]. In brief, cell culture media containing Dulbecco's Modified Eagle's Medium/Ham's F-12 50/50 mixture without phenol red (Mediatech, Manassas, VA, USA) containing 10% fetal bovine serum (Atlanta Biologicals, Flowery Branch, GA, USA), antibiotic/antimycotic containing 10,000 units/mL penicillin, 10,000 μg/mL streptomycin, and 25 μg/mL amphotericin B (Gibco®, Life Technologies, Carlsbad, CA, USA), 100X MEM nonessential amino acid solution (Sigma, St. Louis, MO, USA) and 200 mM L-glutamine (Mediatech, Manassas, VA, USA) was prepared and stored at 4 °C until further use.

For the analytical method validation of DXR in placental perfusion medium, fetal perfusion media was used. Here, we have selected only fetal perfusion medium as the only difference between maternal and fetal perfusion media is the level of dextran. Maternal perfusion medium contains 3.0 mg/mL, whereas fetal perfusion medium contains 12.0 mg/mL dextran. Fetal perfusion medium was prepared using deionized water in which solid medium 199 with Earle's salts (11 mg/mL), dextran (12 mg/mL), heparin (29 μg/mL), gentamicin (33.3 μg/mL), sulfamethoxazole (80 μg/mL), trimethoprim (16 μg/mL), and human serum albumin (30 mg/mL) were dissolved and the pH of the media was adjusted to 7.4 using sodium bicarbonate [18]. All of the above mentioned chemicals used in fetal perfusion media were obtained from Sigma Aldrich (St. Louis, MO, USA). The prepared perfusion medium was stored at 4 °C and was used within a week.

A stock solution of 300 μg/mL of doxorubicin hydrochloride was prepared in purified water. Using this stock solution, 15 μg/mL DXR hydrochloride in cell culture media (DMEM/F12, phenol red free) was prepared. The series of dilutions were prepared in cell culture media. Thereafter, in order to remove the proteins, samples for HPLC analysis were prepared by a protein precipitation technique. Briefly, in a microcentrifuge tube, 250 μL of media samples with known concentrations of doxorubicin were added to 1.0 mL of acetone (1:4 ratio). Samples were then centrifuged at 16,100× *g* for 20 min to pellet the precipitated proteins. At the end of the centrifugation, 1.0 mL of supernatant was collected in a glass vial and allowed to evaporate in a vacuum oven (Napco Model No. 5831, Napco, Tualatin, OR, USA) using a chemical resistant pump (Labnet, Model No. D4000, Labnet, Edison, NJ, USA) for 30 min, followed by drying under a nitrogen stream (ZipVap 24, Glas-Col, Terre Haute, IN, USA) until complete dryness. Once dried, 100 μL of initial mobile phase (75% aqueous phase and 25% organic phase) was used to reconstitute the residual content, and the sample was transferred to an HPLC vial containing an insert, and 50 μL was injected into the HPLC system using partial loop with needle overfill.

2.4. Transplacental Transfer of DXR across BeWo Cells and the Dually Perfused Human Placental Lobule

To further confirm the validity and suitability of our validated HPLC method, a short experiment in the presence of BeWo cells and a placental perfusion of DXR was performed. For transport across the BeWo cells, a method was followed as per a previous report [19], and sample was collected from the basolateral chamber at 30 min. In the case of placental perfusion, a published method was followed with slight modification [20]. In brief, a placenta from an uncomplicated pregnancy at the John Sealy Hospital Labor and Delivery unit was immediately collected. After thorough evaluation, one placental lobule was selected, and a fetal artery and vein were catheterized. The maternal side of the placenta was perfused at a flow rate of 12.0 mL/min, whereas the fetal side was perfused at 3 mL/min. After addition of doxorubicin into the material reservoir, 1.0 mL of sample was withdrawn at 30 min from the fetal reservoir and analyzed by HPLC for doxorubicin determination to check the suitability of the HPLC method. Care was taken to supply 95% nitrogen and 5% carbon dioxide to the fetal reservoir and 95% oxygen and 5% carbon dioxide to the maternal reservoir (250 mL) to ensure acceptable oxygen transfer and placental viability.

2.5. Assay Validation

2.5.1. Linearity, Precision, and Accuracy

The calibration curve was prepared using the peak areas obtained from the chromatograms. Linear regressions were made using SigmaPlot (version 13.0) based on $1/y$ weighting to minimize the sum of squares of relative errors across the concentration range (1.5–15,000 ng/mL). In order to calculate various parameters including intra-day precision, inter-day precision, and accuracy, ICH guidance for method validation of analytical procedures was followed [21]. Intra-day precision was calculated based upon relative standard deviation (%RSD) of the peak area from three concentrations (6 determinations per concentration) prepared on the same day and inter-day precision was calculated as the %RSD of peak areas of three concentrations ($n = 6$ determinations per concentration per day) over three days ($n = 18$). Accuracy was determined as the average peak area for three concentrations ($n = 6$) as a percent of the value for their corresponding concentrations based on the linear regression calculation. The matrix effect of doxorubicin was investigated in six batches of each type of media with low and high concentrations undergoing the same centrifugation, drying, and reconstitution steps. The matrix factor (MF) was calculated as MF = (analyte peak area of the matrix-matched calibration standard)/(analyte peak area of a pure standard sample) [22]. The variation of the matrix factor (relative standard deviation) was <15%.

2.5.2. Limits of Detection and Quantification

The peak area from the blank samples at the same retention time of the DXR was analyzed to determine the noise. The lower limit of detection (LLOQ) and lower limit of quantification (LLOQ) were based on the concentration at which the peak area of the drug was at least three times and ten times higher than the noise (blank samples) at that retention time, respectively.

2.5.3. Stability of Samples

Stability of the samples in each of the media were studied for 1 week at 4 °C. These conditions were chosen based on the temperature and maximum storage time to which typical experimental samples are subjected. Results were calculated as a percent change. Percent change was calculated by the difference obtained in the peak area of the same sample on Day 1 and on Day 8, divided by the peak area of that sample on Day 1.

3. Results and Discussion

3.1. HPLC Method Development

Due to the intrinsic fluorescence of doxorubicin, concentrations were analyzed using a fluorescence detector (λ_{ex}: 480 nm; λ_{em}: 550 nm). Samples were analyzed in both DMEM/F12 media and fetal perfusion media. Irrespective of the media used, the DXR peak was at a retention time of 6.0 ± 0.1 min with no interfering peaks (Figures 1 and 2). In addition, low and consistent noise obtained following the sample treatment confirmed the reproducibility and absence of contamination from samples even after the solvent evaporation step.

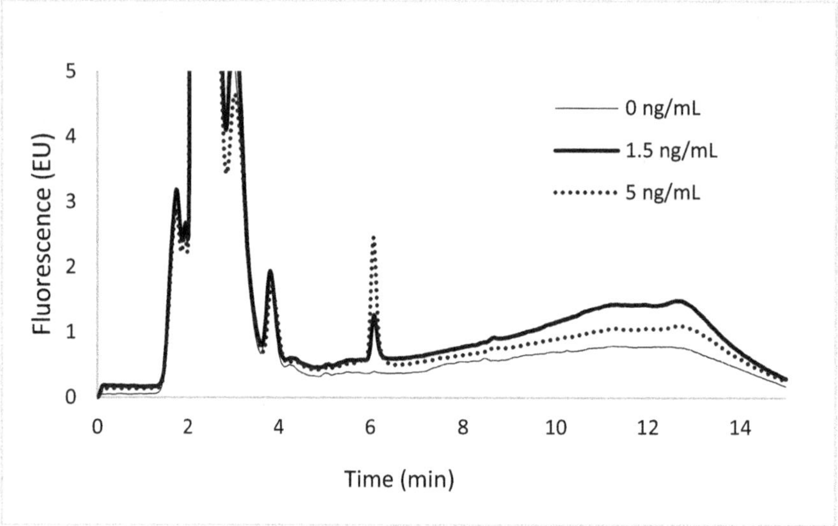

Figure 1. Chromatograms showing a peak for doxorubicin in DMEM/F12 medium analyzed with a fluorescence detector for a blank (0 ng/mL), a lower limit of detection (1.5 ng/mL), and a lower limit of quantification (5 ng/mL) at a retention time of 6.0 ± 0.1 min.

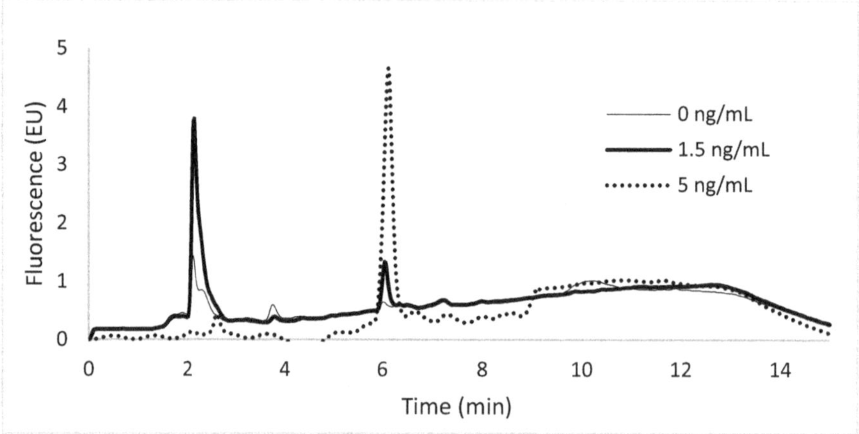

Figure 2. Chromatograms showing a peak for doxorubicin in fetal perfusion medium analyzed with a fluorescence detector for blank (0 ng/mL), lower limit of detection (1.5 ng/mL), and lower limit of quantification (5 ng/mL) samples at a retention time of 6.0 ± 0.1 min.

3.2. Linearity, Precision, and Accuracy

The relationship of peak area to concentration was found to be linear throughout the concentration range studied for both media. Using SigmaPlot, linear regressions were generated based on $1/y$

weighting and the results of slope, intercept, and correlation coefficient are shown in Table 1. A correlation coefficient of 0.99 or greater was obtained for the concentration range (1.5–15,000 ng/mL) for all three days with both types of media. The data calculated for intra-day precision, inter-day precision, and accuracy are shown in Table 2. The intra-day precision and inter-day precision were found to be less than 15% for all three studied concentrations in DMEM/F12 and placental perfusion media, which was within acceptable limits. Accuracy was also found to be between 90% and 110% for all the concentrations studied. As shown in Table 3, the matrix factor of doxorubicin is in the range of 97–113% with a variation <11%. These results indicate that no significant matrix effect was observed in either type of media.

Table 1. Linear regression characteristics.

Regression Parameters	Day 1	Day 2	Day 3	Average	Standard Deviation	RSD (%)
			DMEM/F-12			
Intercept	4191.9	2513.3	3749.7	3485.0	870.1	24.9
Slope	3.761×10^7	3.940×10^7	3.753×10^7	3.818×10^7	1.06×10^6	2.7
R^2	0.9996	0.9988	0.9954	0.9979	0.002	0.2
			Fetal M199			
Intercept	10253.5	9697.6	10453.9	10135.0	391.9	3.9%
Slope	4.592×10^7	4.341×10^7	4.228×10^7	4.387×10^7	1.86×10^6	4.2%
R^2	0.9974	0.9966	0.9993	0.9978	0.001	0.1%

Table 2. Precision and accuracy.

Concentration (μg/mL)	Intra-Day Precision (%)			Inter-Day Precision (%)	Accuracy (%)
	Day 1	Day 2	Day 3		
		DMEM/F-12			
0.05	5.7	1.2	1.9	7	96.4
5	2.6	5.3	9.1	3.3	101.6
10	2.7	10.3	6.4	1.2	100.1
		Fetal M199			
0.05	4.1	3.5	3.6	11.3	103.1
5	2.8	2.8	3.4	0.9	100.2
10	0.5	1.6	3.2	6.3	102.8

Table 3. Matrix factor of doxorubicin in DMEM/F12 and fetal M199 media (n = 6).

Measured Concentration (μg/mL)	Matrix Factor (%)	
	DMEM/F12	Fetal M199
0.01	113.1 ± 6.4	101.1 ± 10.8
10.0	109.7 ± 7.0	96.6 ± 9.4

3.3. Limits of Detection and Quantification

Using the protein precipitation technique, the lower limit of detection (LLOD) was found to be 1.5 ng/mL in both types of media. This was of particular concern because the albumin concentration in perfusion media (30 mg/mL) is much higher than that in cell culture media (22 mg/mL), which could have led to differences in LLOD at low concentrations. Even though the precision at the LLOD was found to be poor due to background noise interference at this very low concentration, this analytical method still had a limit of quantification of 5 ng/mL in DMEM/F12 and placental perfusion media with acceptable accuracy and precision. This LLOQ is comparable to that of methods requiring liquid-liquid extraction [23].

3.4. Sample Stability

Doxorubicin was found to be stable for 1 week at 4 °C in DMEM/F12 as well as in placental perfusion media. Peak areas of doxorubicin in both types of media at all concentrations tested were comparable with minimal change following storage (Table 4). No additional peaks or alterations in retention time were observed. These results indicate that the storage conditions chosen are suitable for experimental samples.

Table 4. Stability of samples at 4 °C for 1 week.

Nominal Concentration (µg/mL)	Measured Concentration (µg/mL)		
	Day 1	**Day 8**	**% Change**
	In DMEM/F12 cell culture media		
0.05	0.050	0.049	1.3%
5	5.02	4.44	11.5%
10	10.19	9.57	6.1%
	In fetal placental perfusion media		
0.05	0.052	0.048	8.5%
5	4.80	4.45	7.2%
10	10.50	10.53	0.3%

3.5. Transplacental Transfer of DXR across BeWo Cells and the Dually Perfused Human Placental Lobule

The samples collected after the transport study and placental perfusion were analyzed using the developed HPLC method. Representative chromatograms are shown in Figure 3. It appears that the obtained chromatograms show DXR eluting at the same retention time, irrespective of whether the drug was studied using a BeWo cell monolayer or a placental lobule. Despite the fact that the placental lobule contains blood and other substances, the protein precipitation technique applied herein has successfully allowed the detection of DXR without any interference.

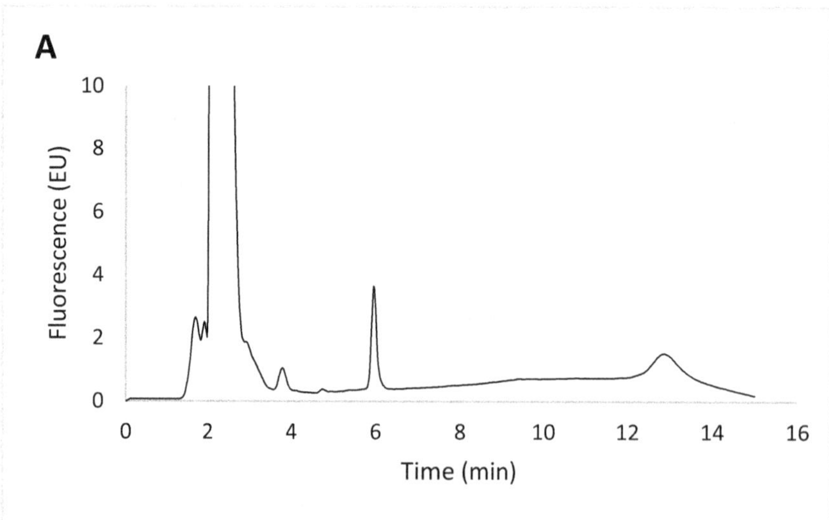

Figure 3. *Cont.*

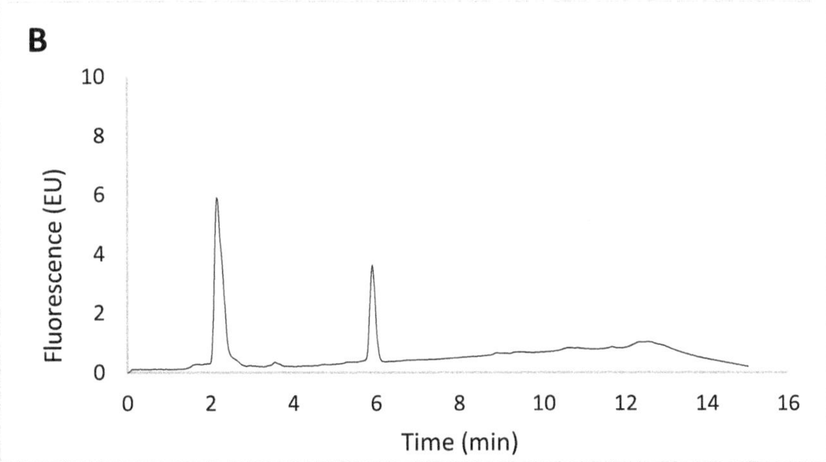

Figure 3. Chromatograms showing a peak for doxorubicin (retention time of 6 min) from samples collected from a transport study in BeWo cells (**A**) and placental perfusion (**B**) analyzed with the developed HPLC method.

4. Conclusions

The analytical method developed in this study was found to be simple and sensitive in both types of media tested. Despite the high albumin concentration used in fetal perfusion media along with other components, the method was able to resolve, identify, and quantify DXR without substantial interference from matrix components. Excellent linearity, precision, and accuracy across all the concentrations studied were obtained using this technique. This method involving protein precipitation and fluorescence detection had lower limits of detection and quantification of 1.5 ng/mL and 5.0 ng/mL, respectively, while bypassing the need for more expensive instrumentation such as LC/MS. Therefore, this method can be used further with great confidence to understand the transplacental transfer of doxorubicin across BeWo cells and in placental perfusion experiments.

Acknowledgments: This work was supported by the *Eunice Kennedy Shriver* National Institute of Child Health and Human Development (U54HD047891). The authors express gratitude to Xiaoming Wang for technical advice.

Author Contributions: M.S. designed the study, collected data, performed statistical analysis, and wrote the first draft of the manuscript. L.B., S.A., S.A.-E., M.M.Y., and M.F. collected data. E.R. assisted with statistical analysis and writing.

Conflicts of Interest: The authors declare no conflict of interest.

References

1. Daw, J.R.; Hanley, G.E.; Greyson, D.L.; Morgan, S.G. Prescription drug use during pregnancy in developed countries: A systematic review. *Pharmacoepidemiol. Drug Saf.* **2011**, *20*, 895–902. [CrossRef] [PubMed]
2. Loebstein, R.; Lalkin, A.; Koren, G. Pharmacokinetic changes during pregnancy and their clinical relevance. *Clin. Pharmacokinet.* **1997**, *33*, 328–343. [CrossRef] [PubMed]
3. Soininen, S.K.; Repo, J.K.; Karttunen, V.; Auriola, S.; Vähäkangas, K.H.; Ruponen, M. Human placental cell and tissue uptake of doxorubicin and its liposomal formulations. *Toxicol. Lett.* **2015**, *239*, 108–114. [CrossRef] [PubMed]
4. Speth, P.A.J.; van Hoesel, Q.G.C.M.; Haanen, C. Clinical Pharmacokinetics of Doxorubicin. *Clin. Pharmacokinet.* **1988**, *15*, 15–31. [CrossRef] [PubMed]
5. Gabizon, A.; Shmeeda, H.; Barenholz, Y. Pharmacokinetics of Pegylated Liposomal Doxorubicin. *Clin. Pharmacokinet.* **2003**, *42*, 419–436. [CrossRef] [PubMed]
6. Chiannilkulchai, N.; Ammoury, N.; Caillou, B.; Devissaguet, J.P.; Couvreur, P. Hepatic tissue distribution of doxorubicin-loaded nanoparticles after i.v. administration in reticulosarcoma M 5076 metastasis-bearing mice. *Cancer Chemother. Pharmacol.* **1990**, *26*, 122–126. [CrossRef] [PubMed]

7. Cartwright, L.; Poulsen, M.; Nielsen, H.; Pojana, G.; Knudsen, L.E.; Saunders, M.; Rytting, E. In vitro placental model optimization for nanoparticle transport studies. *Int. J. Nanomed.* **2012**, *7*, 497–510.
8. Saunders, M. Transplacental transport of nanomaterials. *Wiley Interdiscip. Rev. Nanomed. Nanobiotechnol.* **2009**, *1*, 671–684. [CrossRef] [PubMed]
9. Alhareth, K.; Vauthier, C.; Gueutin, C.; Ponchel, G.; Moussa, F. HPLC quantification of doxorubicin in plasma and tissues of rats treated with doxorubicin loaded poly(alkylcyanoacrylate) nanoparticles. *J. Chromatogr. B Anal. Technol. Biomed. Life Sci.* **2012**, *887–888*, 128–132. [CrossRef] [PubMed]
10. Daeihamed, M.; Haeri, A.; Dadashzadeh, S. A Simple and Sensitive HPLC Method for Fluorescence Quantitation of Doxorubicin in Micro-volume Plasma: Applications to Pharmacokinetic Studies in Rats. *Iran. J. Pharm. Res. IJPR* **2015**, *14*, 33–42. [PubMed]
11. Beijnen, J.H.; Meenhorst, P.L.; Van Gijn, R.; Fromme, M.; Rosing, H.; Underberg, W.J.M. HPLC determination of doxorubicin, doxorubicinol and four aglycone metabolites in plasma of AIDS patients. *J. Pharm. Biomed. Anal.* **1991**, *9*, 995–1002. [CrossRef]
12. Lucas, A.T.; O'Neal, S.K.; Santos, C.M.; White, T.F.; Zamboni, W.C. A sensitive high performance liquid chromatography assay for the quantification of doxorubicin associated with DNA in tumor and tissues. *J. Pharm. Biomed. Anal.* **2016**, *119*, 122–129. [CrossRef] [PubMed]
13. Camaggi, C.M.; Comparsi, R.; Strocchi, E.; Testoni, F.; Pannuti, F. HPLC analysis of doxorubicin, epirubicin and fluorescent metabolites in biological fluids. *Cancer Chemother. Pharmacol.* **1988**, *21*, 216–220. [CrossRef] [PubMed]
14. Urva, S.R.; Shin, B.S.; Yang, V.C.; Balthasar, J.P. Sensitive high performance liquid chromatographic assay for assessment of doxorubicin pharmacokinetics in mouse plasma and tissues. *J. Chromatogr. B* **2009**, *877*, 837–841. [CrossRef] [PubMed]
15. Arnold, R.; Slack, J.; Straubinger, R. Quantification of doxorubicin and metabolites in rat plasma and small volume tissue samples by liquid chromatography/electrospray tandem mass spectroscopy. *J. Chromatogr. B* **2004**, *808*, 141–152. [CrossRef] [PubMed]
16. Niu, H.; Xu, M.; Li, S.; Chen, J.; Luo, J.; Zhao, X.; Gao, C.; Li, X. High-Performance Liquid Chromatography (HPLC) Quantification of Liposome-Delivered Doxorubicin in Arthritic Joints of Collagen-Induced Arthritis Rats. *Med. Sci. Monit. Basic Res.* **2017**, *23*, 150–158. [CrossRef] [PubMed]
17. Berveiller, P.; Vinot, C.; Mir, O.; Broutin, S.; Deroussent, A.; Seck, A.; Camps, S.; Paci, A.; Gil, S.; Tréluyer, J.M. Comparative transplacental transfer of taxanes using the human perfused cotyledon placental model. *Am. J. Obstet. Gynecol.* **2012**, *207*, 514.e1–514.e7. [CrossRef] [PubMed]
18. Nekhayeva, I.A.; Nanovskaya, T.N.; Pentel, P.R.; Keyler, D.E.; Hankins, G.D.; Ahmed, M.S. Effects of nicotine-specific antibodies, Nic311 and Nic-IgG, on the transfer of nicotine across the human placenta. *Biochem. Pharmacol.* **2005**, *70*, 1664–1672. [CrossRef] [PubMed]
19. Albekairi, N.A.; Al-Enazy, S.; Ali, S.; Rytting, E. Transport of digoxin-loaded polymeric nanoparticles across BeWo cells, an in vitro model of human placental trophoblast. *Ther. Deliv.* **2015**, *6*, 1325–1334. [CrossRef] [PubMed]
20. Earhart, A.; Patrikeeva, S.; Wang, X.; Abdelrahman, D.R.; Hankins, G.; Ahmed, M.; Nanovskaya, T. Transplacental transfer and metabolism of bupropion. *J. Matern. Neonatal Med.* **2009**, *23*, 1–10. [CrossRef] [PubMed]
21. ICH Guidlines. Guidances (Drugs)—Q2 (R1) Validation of Analytical Procedures: Text and Methodology. 2005. Available online: https://www.fda.gov/Drugs/GuidanceComplianceRegulatoryInformation/Guidances/ucm265700.htm (accessed on 27 February 2017).
22. Clark, C.R.; Dowling, P.M.; Boison, J.O. Development and validation of a method for determination of tilmicosin residues in equine plasma and tissues using HPLC. *J. Liq. Chromatogr. Relat. Technol.* **2009**, *32*, 2839–2856. [CrossRef]
23. Van Calsteren, K.; Verbesselt, R.; Beijnen, J.; Devlieger, R.; De Catte, L.; Chai, D.C.; Van Bree, R.; Heyns, L.; de Hoon, J.; Amant, F. Transplacental transfer of anthracyclines, vinblastine, and 4-hydroxy-cyclophosphamide in a baboon model. *Gynecol. Oncol.* **2010**, *119*, 594–600. [CrossRef] [PubMed]

10

Development and Validation of an HPLC-DAD Method for the Simultaneous Extraction and Quantification of Bisphenol-A, 4-Hydroxybenzoic Acid, 4-Hydroxyacetophenone and Hydroquinone in Bacterial Cultures of *Lactococcus lactis*

Angelos T. Rigopoulos [1], Victoria F. Samanidou [2] and Maria Touraki [1,*]

[1] Laboratory of General Biology, Division of Genetics, Development and Molecular Biology, Department of Biology, School of Sciences, Aristotle University of Thessaloniki (A.U.TH.), 54 124 Thessaloniki, Greece; angelos.rigopoulos@embl.de

[2] Laboratory of Analytical Chemistry, Department of Chemistry, School of Sciences, Aristotle University of Thessaloniki (A.U.TH.), 54 124 Thessaloniki, Greece; samanidu@chem.auth.gr

* Correspondence: touraki@bio.auth.gr

Abstract: Bisphenol-A, a synthetic organic compound with estrogen mimicking properties, may enter bloodstream through either dermal contact or ingestion. Probiotic bacterial uptake of bisphenol can play a major protective role against its adverse health effects. In this paper, a method for the quantification of BPA in bacterial cells of *L. lactis* and of BPA and its potential metabolites 4-hydroxybenzoic Acid, 4-hydroxyacetophenone and hydroquinone in the culture medium is described. Extraction of BPA from the cells was performed using methanol–H_2O/TFA (0.08%) (5:1 v/v) followed by SPE. Culture medium was centrifuged and filtered through a 0.45 μm syringe filter. Analysis was conducted in a Nucleosil column, using a gradient of A (95:5 v/v H_2O: ACN) and B (5:95 v/v H_2O: ACN, containing TFA, pH 2), with a flow rate of 0.5 mL/min. Calibration curves (0.5–600 μg/mL) were constructed using 4-*n*-Octylphenol as internal standard ($1 > R^2 > 0.994$). Limit of Detection (LOD) and Limit of Quantification (LOQ) values ranged between 0.23 to 4.99 μg/mL and 0.69 to 15.1 μg/mL respectively. A 24 h administration experiment revealed a decline in BPA concentration in the culture media up to 90.27% while the BPA photodegradation levels were low. Our results demonstrate that uptake and possible metabolism of BPA in *L. lactis* cells facilitates its removal.

Keywords: HPLC-DAD; bisphenol A; 4-Hydroxybenzoic Acid; 4-Hydroxyacetophenone hydroquinone; *Lactococcus lactis*

---

1. Introduction

Bisphenol-A is the organic synthetic compound 4, 4′-Isopropylidenediphenol, which is widely used in thermal paper industry and as a component of synthetic plastic (vinyl-chloride) due to its mechanical and extreme-temperature resistance [1]. BPA, was found to exhibit estrogenic properties, arising from its structural resemblance to the human 17β-estradiol [2], thus it is also referred to as xenoestrogen or endocrine disruptor. The main routes of human exposure to BPA are the consumption of food in BPA containing packages [3] and through dermal contact with BPA rich materials [4]. Following exposure, BPA is mainly metabolized in humans to BPA-glucuronide through the hepatic glucuronide transferase and excreted from the body [5,6]. However, a percentage of BPA, referred to as "free" or "active" BPA, remains in the blood circulation for up to a week and can interfere with endogenous biological processes [4]. Due to its estrogenic and genotoxic effects and its wide occurrence, BPA presents a risk to humans and animals and hence it is the object of great environmental concern [7–9].

Although part of bisphenol is degraded via photo-oxidation [10–12], monitoring of its levels is deemed necessary in various environmental sources [13]. To this end, biodegradation has been proposed as an advanced technique for BPA elimination [13–15], since a variety of microorganisms have been identified to decompose BPA [15,16]. Probiotics are bacteria beneficial to the host, which have been recognized as safe for human consumption and are used in the production of food products [17,18]. Recently they have been reported to be ideal against heavy metal toxicity since they bind and sequester metals [19]. Moreover, the probiotic bacteria *Bifidobacterium breve* and *Lactobacillus casei* when administered to rats showed a prophylactic effect against the detrimental effects of BPA, by reducing its intestinal absorption and facilitating its excretion, [18]. Likewise, the ability of *Bacillus* strains [20–22], certain *Lactococcus* strains [23] and *Shingomonas paucimobilis* [24] to degrade bisphenol has been documented. Recently a simplified method for the extraction of bisphenol from bacterial culture suspensions has been reported [25]. Moreover, the compounds 4-hydroxy acetophenone (HAP), 4-hydroxy benzoic acid (HBA) and Hydroquinone (HQ) have been proposed as major bacterial metabolites of BPA bacterial degradation pathways [14,26,27]. However, all these studies employed the determination of bisphenol in the cell free culture supernatant and not in the bacterial cells even though they examine the kinetics of bisphenol.

Analytical methods for the determination of bisphenol in various food samples [3], as well as in biological fluids and environmental samples [28], have been extensively reviewed. The determination of BPA has also been investigated in more complex matrices such as rat tissues [29], human breast milk [30] resin-based dental restorative materials [31] and saliva [32]. However, to the best of our knowledge, no method has been reported for the determination of bisphenol in bacterial cells.

In the present study, the development and validation of an analytical method for the quantification of BPA in bacterial cells is reported for the first time using the widespread probiotic bacterium *Lactococcus lactis* (ATC 11454). The method is also validated in bacterial culture supernatants for the simultaneous determination of the bacterial metabolites reported to be produced during the bacterial major metabolic pathway [14], namely 4-hydroxy-acetophenone (HAP), 4-hydroxy benzoic acid (HBA) and Hydroquinone (HQ). The application of this method demonstrates the uptake of bisphenol by *Lactococcus lactis* cells and allows its quantification.

2. Materials and Methods

2.1. Chemicals and Reagents

Acetonitrile was purchased from VWR Chemicals (Paris, France). Methanol and Absolute ethanol were purchased from Fisher Scientific (Loughborough, UK). Trifluoroacetic acid (TFA), KH_2PO_4 and $FeCl_3$ were purchased from AppliChem GmbH (Darmstadt, Germany), while NaCl, $CaCl_2$, $(NH_4)_2SO_4$ and $MgSO_4$ were purchased from Merck KGaA (Darmstadt, Germany). All the chemicals used were HPLC Grade, unless specified otherwise.

Bisphenol-A (BPA) and the internal standard 4-*n*-Octylphenol (OP) were purchased from Alfa Aesar (Karlsruhe, Germany). 4-Hydroxyacetophenone (HAP), 4-Hydroxybenzoic Acid (HBA) and Hydroquinone (HQ) were purchased from Acros Organics (Ceel, Belgium).

The bacterial strain from which the liquid culture supernatant and the pellet originated and used for the spike-recovery studies, was *Lactococcus lactis* (ATCC 11454). The liquid culture medium that was used for the growth of *L. lactis* was Brain Heart Infusion broth (BHI broth) and was purchased from AppliChem GmbH (Darmstadt, Germany). An artificial bacterial culture medium was prepared, namely Basal Salts Medium (BSM), composed of KH_2PO_4 and $(NH_4)_2SO_4$ 1 g/L each, $MgSO_4$ 0.1 g/L, NaCl and $CaCl_2$ 0.05 g/L each and $FeCl_3$ 0.01 g/L. BPA was used as a sole carbon source in BSM in varying concentrations for the feeding experiments. For the retrieval of the bacterial culture supernatant and pellet 100 μL of an *L. lactis* glycerol stock solution were inoculated in a 250 mL flash with BHI broth. Following a 24-h incubation at 28 °C under constant stirring, the culture was centrifuged at 5000 rpm and the precipitate was transferred to a 250 mL flask containing BSM. At the end of the experiment, the culture was centrifuged, the precipitate was placed on a filtering paper and its weight was recorded following removal of excess water. The recorded weight is referred as 'wet cell weight'. The culture supernatant and the wet cell weight were used for the spike-recovery assays.

Solid Phase Extraction columns, frits and C_{18} sorbent material, 35–75 U were purchased from Grace Davison Discovery Sciences (Bannockburn, IL, USA). Membrane filters with 0.45 μm diameter pores were purchased from Merck KGaA (Darmstadt, Germany).

2.2. *Standard Preparation*

2.2.1. Stock Solution Preparation

Stock solutions 500 μg/mL of BPA, HAP, HBA and HQ were prepared by dissolving 25 mg of the analyte in 50 mL methanol. The stock solution of BPA (50 mg/mL) was prepared by dissolving 2.5 g of BPA in 50 mL ethanol. The internal standard stock solution of 500 μg/mL was prepared by dissolving 25 mg of OP in 50 mL methanol.

2.2.2. Working Standard Solution Preparation

Seven-point and six-point working standard solutions were prepared for analytes at concentrations of 0.5, 10, 50, 100, 200, 400, and 600 μg/mL for BPA, at 0.5, 2, 4, 6, 8 and 10 μg/mL for HAP and HQ and at 1, 2, 4, 6, 8 and 10 μg/mL for HBA and the internal standard 4-*n*-octylphenol (75 μg/mL).

2.3. *Sample Preparation*

For the bacterial pellet method development, 0.1 g of wet cell weight *L. lactis* was added in a test tube with 8 mL methanol, 2 mL of double distilled water containing 0.08% (*v*/*v*) TFA and 150 μL internal standard from the 500 μg/mL stock. The concentration of the internal standard in the final 1 mL volume was 75 μg/mL. The test tube was transferred to an ice-filled container and was placed under a homogenizer for 5 min. The homogenate was separated in ten Eppendorf centrifuge tubes and was centrifuged for 10 min in 8000 rpm. The supernatant was collected and mixed with an equal volume of ddH_2O containing 0.08% TFA. BPA extraction was performed using Solid Phase Extraction (SPE). An SPE cartridge was prepared using 0.5 g of C_{18} sorbent in an SPE column. The SPE cartridge was conditioned with 10 mL methanol, 5 mL ddH_2O and 5 mL ddH_2O containing 0.08% TFA. The sample was loaded on the cartridge and a manual flow of 1 mL/min was employed. The cartridge was, then, washed with 5 mL ddH_2O and elution was performed using 4 mL methanol. The eluate was collected and placed under a rotary evaporator in a 37 °C water bath, to dryness. Finally the sample was reconstituted with 1 mL acetonitrile and stored in an aluminum foil-covered Eppendorf tube at −20 °C. All experiments were conducted in dark or covered tubes to avoid photodegradation of BPA.

For the bacterial supernatant, an 8.5 mL sample of the culture was added in a test tube along with 1.5 mL internal standard from the stock solution. The sample was placed in 10 Eppendorf centrifuge tubes (1 mL per tube), and was centrifuged for 10 min in 8000 rpm. The supernatant was collected and filtered through a 0.45 μm diameter pore filter. The filtrate was stored in an aluminum foil-covered Eppendorf tube at −20 °C.

2.4. *BPA Administration (Feeding) Assay*

For the feeding assays 6 falcon tubes of 50 mL volume were filled with BHI broth and were inoculated with 100 μL of a *L. lactis* glycerol stock solution. The cultures were incubated for 24 h at 28 °C under constant stirring. The cultures were, then, centrifuged and the medium was discarded. A total of 50 mL of fresh BSM was added to each culture along with varying quantities from the BPA stock solution (50 mg/mL) so that the final BPA concentration in each culture would be 50, 100, 150, 200, 400 και 500 μg/mL respectively. The cultures were incubated for another 24 h at 28 °C under constant stirring. The remaining BPA in the cultures was quantified using the method and the calibration curves that were described.

2.5. Photodegradation Assay

To investigate potential photodegradation of BPA, a falcon tube of 50 mL containing BSM and BPA at a concentration of 50 μg/mL was prepared and was incubated under the same conditions as the feeding assays. BPA was quantified at the beginning and the end of the 24 h incubation period. The presence of photodegradation products was examined in the resulting chromatograms.

2.6. HPLC-DAD Conditions and Methods

The instrument used consisted of an $LC20_{AD}$ pump and an SPD-20A photodiode array detector (DAD) purchased from Shimadzu (Kyoto, Japan). Separation of BPA was achieved on a Nucleosil 100 C_{18} column (250 × 4.6 mm, 5 μm) purchased by Macherey-Nagel GmbH & Co. (Duren, Germany) with a binary mobile phase system at room temperature. The injection was performed using a 100 μL Hamilton syringe, with an injection volume of 80 μL. Mobile Phase A consisted of a 0.1% TFA in ddH_2O solution and Acetonitrile at a ratio of 95:5, *v*/*v*. Mobile Phase B consisted of a 0.08% TFA in ddH_2O solution and Acetonitrile at a ratio of 5:95, *v*/*v*. The gradient started at 100% Mobile Phase A which was brought down to 80% over 1 min, 70% over 4 min, 55% at 15 min, 40% at 17 min, 20% at 20 min and finally to 10% over 2 min and then was kept at this level until the end on the analysis. The flow rate was constant at 0.5 mL/min. BPA was detected at 220 nm with a band width of 2.0 nm. HAP, HBA and HQ were detected at 280 nm with a band width of 2.0 nm. Identification was conducted by matching the peak retention times (*Rt*) with those of the pure standards. Internal standard calibration was used by plotting the peak area ratio of the analyte to the peak area of the internal standard against the concentration of the analyte.

3. Results and Discussion

3.1. Optimization of Sample Preparation

The homogenization of the sample was performed to lyse the cells and release the adsorbed BPA for the feeding assay. Initially, ethyl acetate was used instead of methanol and a manual homogenization was attempted with a glass homogenizer, but cell lysis was deemed inadequate. Hence methanol was selected as the extraction solvent. The addition of an equal volume ddH_2O containing 0.08% (*v*/*v*) TFA after centrifugation aimed to increase the solubility of BPA due to further dilution.

An elaborate and efficient vortex assisted liquid-liquid microextraction with octanol was used for the trace analysis of bisphenol in water and wastewater samples [33]. However, an attempt to use liquid-liquid extraction with ethyl acetate in the samples of the present study resulted to the formation of an emulsion zone between the two phases, a fact that led to greatly varying percentages of BPA recovery due to the loss of the analyte during the procedure. Thus, solid phase extraction was implemented, since SPE is a common step in the extraction procedures of bisphenol from complex matrices, such as rat tissues [29] or human serum [5] and its benefits in the biomonitoring of bisphenol have been demonstrated [28,33]. In the present study a flow rate of 1 mL/min was selected throughout the SPE procedure, as effective and timely favorable, while higher flow rates did not provide full retention of BPA by the C_{18} material, especially at high BPA concentrations employed and resulted in lower recovery values. A 37 °C water bath was selected to facilitate the evaporation of methanol after SPE. The slightly higher evaporation temperatures of 45 °C that have been reported for the analysis of bisphenol and 4-nonylphenol [29] resulted in heat-induced decomposition of the internal standard when used in the present study. Our results for BPA extraction are in accordance with a recently published report on a methanol extraction of bisphenols from bacterial cultures and solid matrices [25] in which, however, SPE was not employed and lower concentrations of BPA were used [25], resulting in recoveries up to 95%. The SPE step used in the present study greatly improved recovery which amounted to 99.38–100.18% in culture supernatants and to 90.8–102.46% in bacterial cells, in the wide range of concentrations employed (Table 1).

Table 1. Analyses of the calibration curves for analytes in the cell culture supernatant and the cell culture precipitate.

Analyte in Media/Std. Curve Equation/R^2	Nominal Conc. (μg/mL)	Mean Calculated Concentration (μg/mL) (Mean ± SD)	Relative Bias (%)	Precision		Recovery Rate (Mean ± SD)
				Intra-Day (RSD %)	Inter-Day (RSD %)	
BPA in BSM Supernatant y = 0.0295x + 0.0077 1	0.5	0.48 ± 0.05	−3.6	8.7	10.4	96 ± 13
	10	9.94 ± 0.08	−0.6	0.1	0.8	99.4 ± 0.7
	50	50.2 ± 0.7	0.4	0.8	1.4	100 ± 1
	100	100 ± 2	0.05	0.8	1.7	100 ± 1
	200	200 ± 2	−0.2	0.7	1.2	100 ± 1
	400	401 ± 3	0.2	0.2	0.8	100.2 ± 0.6
	600	600 ± 3	−0.01	0.3	0.6	100.0 ± 0.5
BPA in *L. lactis* precipitate y = 0.0232x + 0.0092 0.9999	0.5	0.41 ± 0.03	−18.5	3.2	7.4	82 ± 5
	10	9.1 ± 0.4	−9.2	2.3	4.7	91 ± 4
	50	48± 1	−3.7	2.7	2.7	96 ± 2
	100	98 ± 3	−2.0	2.0	2.9	98 ± 2
	200	205 ± 3	2.5	0.4	1.4	103 ± 1
	400	400.6 ± 0.9	0.2	0.2	0.2	100 ± 0.2
	600	597 ± 5	−0.6	0.1	0.8	99.4 ± 0.7
HAP in BSM y = 0.202x + 0.0134 0.9994	0.5	0.52 ± 0.02	3.4	2.8	3.8	103 ± 3
	2	2.06 ± 0.05	2.9	0.2	2.4	103 ± 2
	4	3.84 ± 0.04	−4.1	0.7	1.0	95.9 ± 0.8
	6	6.1 ± 0.1	1.1	1.0	2.3	101 ± 2
	8	8.0 ± 0.2	0.6	2.1	2.2	101 ± 2
	10	10.0 ± 0.1	−0.3	0.3	1.0	99.7 ± 0.8
HBA in BSM y = 0.0407x + 0.0073 0.994	1	0.91 ± 0.03	−9.5	2.7	3.3	90 ± 2
	2	1.86 ± 0.04	−6.9	0.3	2.2	93 ± 3
	4	4.1 ± 0.2	2.3	3.2	4.2	102 ± 2
	6	6.2 ± 0.1	3.0	2.1	2.3	103 ± 2
	8	8.3 ± 0.3	4..2	3.0	3.1	104 ± 3
	10	9.62 ± 0.04	−3.8	0.3	0.4	96.2 ± 0.3
HQ in BSM y = 0.0351x − 0.0082 0.999	0.5	0.48 ± 0.01	−3.8	1.4	2.1	96 ± 2
	2	2.2 ± 0.1	7.5	3.4	6.5	108 ± 5
	4	3.91 ± 0.03	−2.2	0.03	0.8	97.8 ± 0.6
	6	5.91 ± 0.09	−1.5	1.6	1.5	98 ± 1
	8	7.9 ± 0.2	−1.4	0.5	2.0	99 ± 2
	10	10.1 ± 0.2	1.3	1.4	2.2	101 ± 2

In an effort to perform the analysis of BPA in the BHI medium, which is recommended for growth of *L. lactis*, the nutrients in the broth resulted in the presence of peaks in the chromatogram that appeared to interfere with the tested bacterial metabolites peaks. The artificial bacterial medium (BSM) was prepared to tackle the problems that arose in the analysis using the reference BHI medium. Moreover, in this manner BPA was the sole carbon source for the bacterium, a fact that contributes in its biotransformation.

3.2. Method Validation

The results of the linear regression analysis performed using the "Linear Regression" package of Microsoft Excel 2013 software are presented in Table 1, while detection limits and Retention time variance are presented in Table 2. The coefficient of determination values (R^2) for the standard curves in BSM supernatant and *L. lactis* cells were 1 and 0.999 respectively, while for HAP, HBA and HQ in BSM supernatant the values exceeded 0.9994, 0.994 and 0.999 respectively, indicating excellent linearity. The Relative Bias was calculated by computing the percentage of the difference of the mean calculated concentration to the nominal concentration divided by the nominal concentration of the analyte. All results are within the accepted range of ±10%. Intra-day precision was calculated by performing a total of three injections of the nominal concentration on the same day, while inter-day precision was calculated by performing six injections of the nominal concentration over three days. Both were computed as % R.S.D. = (S.D./mean) × 100. None of the values exceeded 6.0% for all analytes, with the exception of the values obtained for the lowest nominal value of 0.5 μg/mL of BPA in bacterial cells and in the medium, which might be due to the fact that this nominal value is below LOD (Table 2). Since an elaborate extraction protocol was required for the bacterial cells, the use of an internal standard was deemed necessary, to compensate for sample losses during sample preparation. The internal standard 4-*n*-octylphenol was selected, due to the fact that it is well resolved and it has a structure similar to bisphenol. Moreover, it is not expected to be present in samples, since only 4-*t*-octyphenol has been reported to be found at very low concentrations in environmental samples and human milk [34,35]. Our results, as presented in Table 1, indicate that the method is highly precise and repeatable. Recovery rates were well within range of 10% error, reaching close to 100% recovery, thus further supporting the precision of this method. The recovery results presented in Table 1 are in good correlation with those reported for rat serum (95%) and higher than those reported for rat testis (78.6%) and liver (84.0%) [29].

Table 2. Analyses LOD, LOQ and *Rt* values for all analytes.

Std. Curve	Retention Time (Min) Mean ± SD (RSD %)	LOD (μg/mL)	LOQ (μg/mL)
BPA in *L. lactis*	26.68 ± 0.13 (0.41)	4.99	15.01
BPA in BSM	26.71 ± 0.09 (0.38)	0.64	1.93
HAP in BSM	15.66 ± 0.05 (0.29)	0.24	0.73
HBA in BSM	12.77 ± 0.12 (0.23)	0.73	2.21
HQ in BSM	10.80 ± 0.08 (0.90)	0.32	0.98

The nominal concentrations of 0.5 μg/mL BPA in BSM and cell precipitate were below the calculated LOD values (Table 2) and were thus excluded from the estimation of recovery results described above. The values for the Limit of Detection (LOD) and Limit of Quantification (LOQ) were calculated from the calibration curves. The LOD was calculated to 3.3 times the ratio of the Standard Error (intercept) to the Coefficients x, while the LOQ was 10 times this ratio. The retention time means, standard deviations and RSDs were calculated using the data from one chromatogram of each of the nominal concentrations used. The value of an LOD of 4.99 μg/mL estimated in the present study for BPA in bacterial cells at the range of 10–600 μg/mL, is higher than the value of 2.8 ng/mL reported for rat tissues, but over a narrower range of concentrations of 0.15–150 μg BPA/mL [29]. A lower value of LOD was found for BPA in the less complex sample of BSM, amounting to 0.64 μg/mL (Table 2), although the same range of concentrations was employed (Table 1), possibly due to lower matrix complexity. Similar low LOD values were estimated for all possible metabolites tested, as shown in Table 2. Although these compounds have been proposed as possible bisphenol bacterial metabolites, they were identified using MS and NMR analysis of the HPLC isolated fractions [27]. In fact, in the studies on the determination of bisphenol degradation using HPLC-DAD [20,23], no validation was included. The method presented in this work allows the simultaneous quantification of bisphenol A and three of its major bacterial metabolites.

3.3. Feeding Results

A typical chromatogram of HQ, HBA and HAP and BPA is presented in Figure 1. Analyte peaks of HQ, HBA, HAP, BPA and internal standard OP are shown with a retention time (*Rt*) of 10.671, 12.728, 15.579, 26.241 and 40.771 min respectively. The peak that appears at 25.6 (Figure 1) might be attributed to a BPA photodegradation product, since it was also identified during the photodegradation experiment and is presented in Figure 2. Photodegradation of BPA via photooxidation has been confirmed [10,12], and the products of this process presented higher toxicity than the parent compound [11]. Since the feeding experiments were conducted under normal light, it was deemed appropriate to test if any BPA loss was due to photodegradation. In these experiments the initial BPA concentration was calculated at 48.65 µg/mL and the concentration following the 24 h incubation amounted to 44.18 µg/mL. Hence, the BPA loss due to photodegradation was 9.19%. These values are in accordance with those reported for bisphenol A degradation by UV irradiation in the absence of hydrogen peroxide [36].

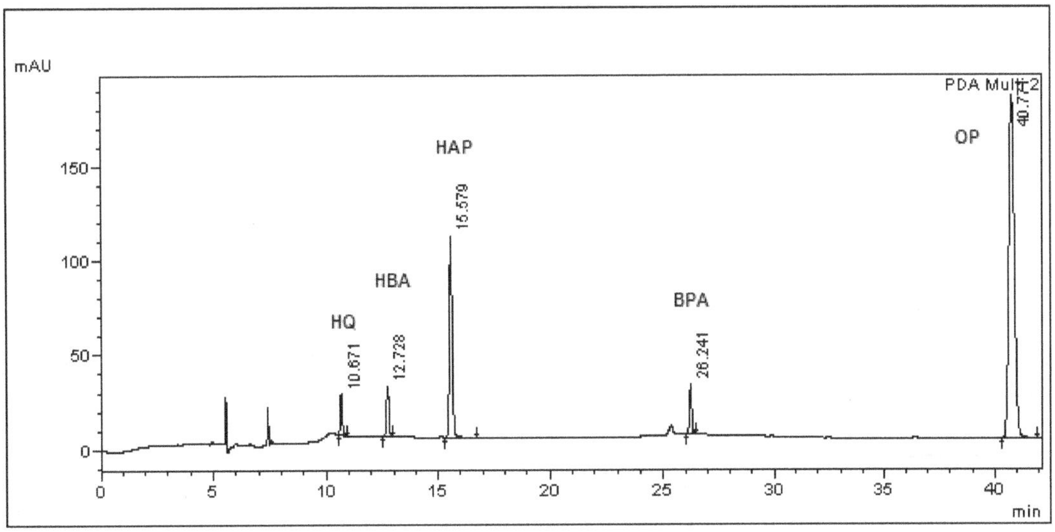

Figure 1. A chromatogram containing all analytes at a concentration of 10 µg/mL and internal standard at 75 µg/mL in fresh BSM without *L. lactis* inoculation. The first two peaks with *Rt* of 5.7 min and 7.5 min are background noise from acetonitrile solvent and TFA.

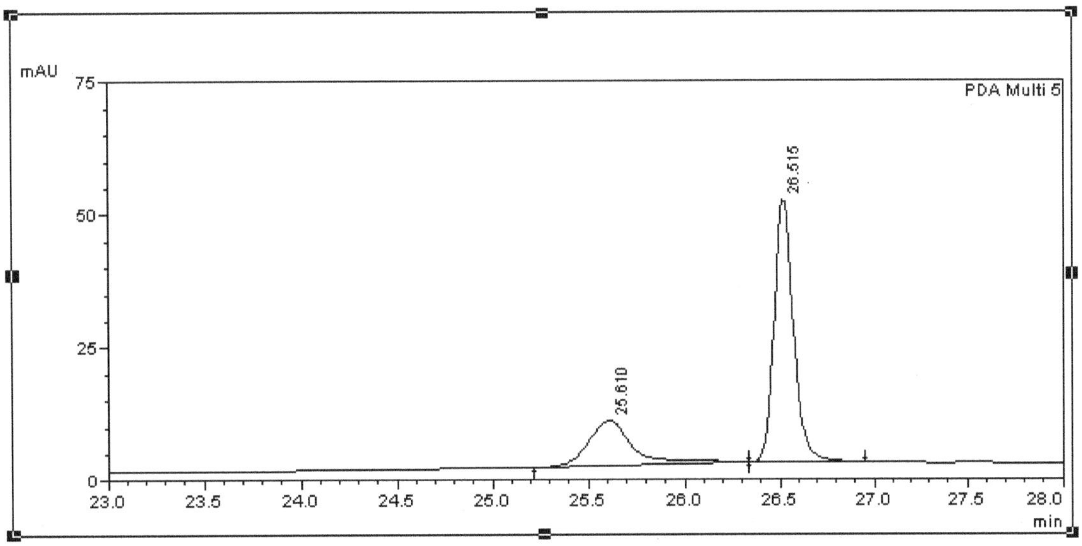

Figure 2. Chromatogram displaying BPA and its identified photodegradation product with a *Rt* of 26.515 min and 25.610 min respectively. The compounds were detected at a wavelength of 300 nm.

Typical chromatograms of the feeding assays are presented in Figure 3 (bacterial cells) and Figure 4 (bacterial culture medium). The relative blank chromatograms resulting from bacterial cells and from their culture medium any additions are shown in the Supplemental section. The peaks appearing at 10.2 min in both chromatograms does not represent one of the metabolites described in this paper, as confirmed by comparing the absorbance spectra of the peak to those of the pure metabolites analyzed.

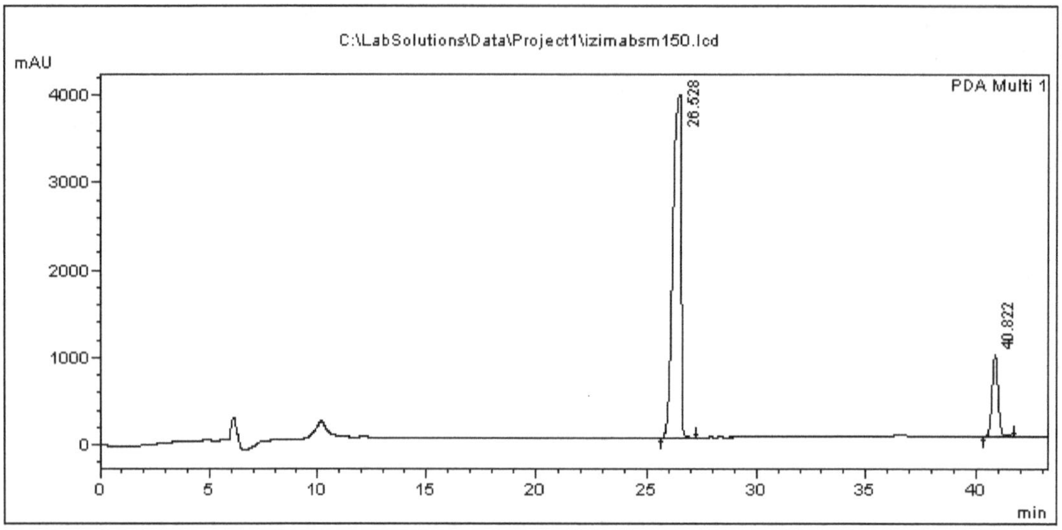

Figure 3. Chromatogram of cell precipitate originating from the administration of 150 μg/mL BPA in the bacterial cells. The peaks at 26.528 min and 40.822 min correspond to BPA and the internal standard.

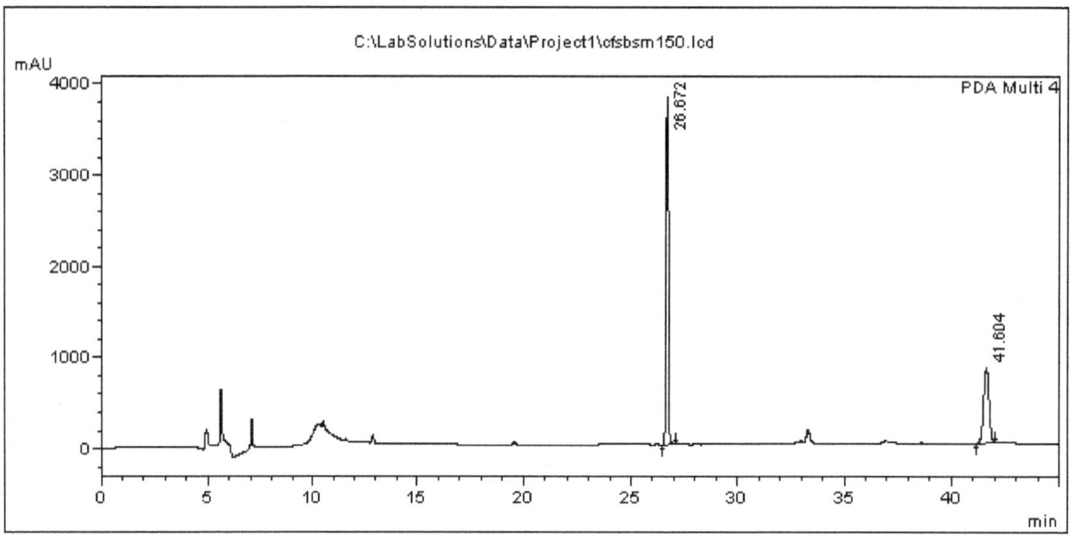

Figure 4. Chromatogram of culture supernatant originating from the administration of 150 μg/mL BPA in the bacterial cells. The peaks at 26.672 min and 41.604 min correspond to BPA and the internal standard.

Two major pathways have been proposed for the metabolism of bisphenol A by bacteria, the first leading to mineralization, with HBA and HAP as intermediates and the second to hydroxylation [6,16]. However, it has been pointed out that the metabolic pathways might be strain specific [14], since certain metabolites produced by one strain are not detected when another bacterial strain is used. In a similar manner, the analytes HQ, HBA and HAP were not detected in the chromatograms of the feeding assays. Absence of bacterial metabolites has also been reported for *Bacillus pumilus* [20] and certain *Lactococcus* strains [23]. In all these reports, including the present work, the presence of metabolites was examined at 24 h of culture. However, when various *Bacilli* and *Sphingomonas* strains were tested for bisphenol degrading capacity, metabolites were detected for three of them and only from 8–12 h, in a declining manner [37]. These results indicate that the time period of 24 h used in the present paper might be too long, since the metabolites under investigation might have formed and then further transformed. On the other hand, the bacterial metabolites examined in the present paper were reported to be produced by *Bacillus* sp., yet only after 3 days of incubation in the presence of 20 mM BPA [21]. Moreover, the photodegradation of BPA metabolites has been reported for hydroxybenzoic acid as well as for hydroquinone [38,39]. It is evident that more experiments are needed in the course of bacterial incubation, since metabolism is an ongoing process and each strain presents particular metabolic characteristics.

A depletion of up to 58% in total BPA (μg) was observed in our experiments, in all bacterial cultures as shown in Table 3. Complete degradation of BPA has been reported following a 48 h incubation period [37] or after 9 days of incubation [21]. In our experiments BPA was reduced by 90.27%, 64.16%, 58.60%, 67.74%, 79.83% and 85.05% in the cultures corresponding to initial BPA concentration of 50, 100, 150, 200, 400 and 500 μg/mL. Our results suggest that the initial BPA concentration added has a profound effect on its degradation. Many bacterial strains that can degrade BPA have been reported, and this degrading ability varies among the species [14]. It should also be noted that different initial concentrations of BPA are used in the reports employing bacteria, ranging from 0.1 to 60 mg/L [22,37,40] and for varying incubation periods. Bacteria can clearly employ a variety of metabolic pathways to exploit pollutants as energy sources, under the prerequisite that the pollutants are bioavailable and at levels that could trigger such mechanisms [16].

Table 3. Comparative data of BPA administered (μg) and found in feeding assays.

BPA Administered in 50 mL Culture (μg)	BPA Found in Precipitate	BPA Found in 50 mL Supernatant	Total BPA Found	Concentration Decline in Cultures (%)
		(μg)		
2500 (50 μg/mL)	46.77	196.46	243.23	90.27
5000 (100 μg/mL)	110.48	1681.74	1792.22	64.16
7500 (150 μg/mL)	255.75	2849.48	3105.23	58.60
10,000 (200 μg/mL)	262.20	2964.01	3226.22	67.74
20,000 (400 μg/mL)	437.24	3595.83	4033.07	79.83
25,000 (500 μg/mL)	421.16	3316.41	3737.57	85.05

4. Conclusions

The developed analytical method provides a reliable, robust and sensitive approach for the simultaneous detection and quantification of bisphenol A and its potential metabolites in the complex matrix sample of bacterial cells. Good chromatography separation was achieved on a Nucleosil 100 C_{18} column with a longer run time, which however allows the detection and separation of metabolites. LC-DAD combined with solid phase extraction and the use of 4-*n*-octylphenol as the internal standard, is capable of operating with a detection limit of 4.99 μg/mL for bisphenol and lower limits for metabolites in bacterial cells. The method was also successfully applied in the analysis of bacterial culture medium with a detection limit of 0.64 μg/mL for bisphenol. The feeding experiments revealed that *L. lactis* is capable to uptake of bisphenol and contribute to its elimination from the culture medium. Further experiments are needed for the detection and quantification of metabolites as a function of culture time and administrated bisphenol concentration.

Acknowledgments: No additional funding was received in support of the present research work.

Author Contributions: M.T. conceived the experiments; M.T. and V.F.S. designed the experiments; A.T.R. performed the experiments; A.T.R. analyzed the data, A.T.R., M.T. and V.F.S. wrote the paper.

Conflicts of Interest: The authors declare no conflict of interest.

References

1. López-Cervantes, J.; Sánchez-Machado, D.; Paseiro-Losada, P.; Simal-Lozano, J. Effects Of Compression, Stacking, Vacuum Packing and Temperature on the Migration of Bisphenol A from Polyvinyl Chloride Packaging Sheeting into Food Simulants. *Chromatographia* **2003**, *58*, 327–330.
2. Kang, J.-H.; Kondo, F.; Katayama, Y. Human exposure to bisphenol A. *Toxicology* **2006**, *226*, 79–89. [CrossRef] [PubMed]
3. Ballesteros-Gómez, A.; Rubio, S.; Pérez-Bendito, D. Analytical methods for the determination of bisphenol A in food. *J. Chromatogr. A* **2009**, *16*, 449–469. [CrossRef] [PubMed]
4. Liu, J.; Martin, J.W. Prolonged exposure to bisphenol A from single dermal contact events. *Environ. Sci. Technol.* **2017**, *51*, 9940–9949. [CrossRef] [PubMed]
5. Völkel, W.; Colnot, T.; Csanady, G.; Filser, J.G.; Dekant, W. Metabolism and kinetics of bisphenol A in humans at low doses following oral administration. *Chem. Res. Toxicol.* **2002**, *15*, 1281–1287. [CrossRef] [PubMed]
6. Chouhan, S.; Yadav, S.; Prakash, K.; Swati, J.; Singh, S.P. Singh Effect of bisphenol A on human health and its degradation by microorganisms: A review. *Ann. Microbiol.* **2014**, *64*, 13–21. [CrossRef]
7. Kang, J.H.; Kondo, F. Bisphenol A Degradation by Bacteria Isolated from River Water. *Arch. Environ. Contam. Toxicol.* **2002**, *43*, 265–269. [CrossRef] [PubMed]
8. Tsai, W.T. Human Health Risk on Environmental Exposure to Bisphenol-A: A Review. *J. Environ. Sci. Health* **2006**, *24*, 225–255. [CrossRef] [PubMed]
9. Oehlmann, J.; Oetken, M.; Schulte-Oehlmann, U. A critical evaluation of the environmental risk assessment for plasticizers in the freshwater environment in Europe, with special emphasis on bisphenol A and endocrine disruption. *Environ. Res.* **2008**, *108*, 140–149. [CrossRef] [PubMed]
10. Staples, C.A.; Dorn, P.B.; Klecka, G.M.; O'Block, S.T.; Harris, L.R. A review of the environmental fate, effects, and exposures of bisphenol-A. *Chemosphere* **1998**, *36*, 2149–2173. [CrossRef]
11. Da Silva, J.C.C.; Reis Teodoro, J.A.; Afonso, R.J.; Aquino, S.F.; Augusti, R. Photodegradation of bisphenol A in aqueous medium: Monitoring and identification of by-products by liquid chromatography coupled to high-resolution mass spectrometry. *Rapid Commun. Mass Spectrom.* **2014**, *28*, 987–994. [CrossRef] [PubMed]
12. Diepens, M.; Gijsman, P. Photodegradation of bisphenol A polycarbonate. *Polym. Degrad. Stab.* **2007**, *92*, 397–406. [CrossRef]
13. Suzuki, T.; Nakagawa, Y.; Takano, I.; Yaguchi, K.; Yasuda, K. Environmental Fate of Bisphenol A and Its Biological Metabolites in River Water and Their Xeno-estrogenic Activity. *Environ. Sci. Technol.* **2004**, *38*, 2389–2396. [CrossRef] [PubMed]
14. Zhang, W.; Yin, K.; Chen, L. Bacteria-mediated bisphenol A degradation. *Appl. Microbiol. Biotechnol.* **2013**, *97*, 5681–5689. [CrossRef] [PubMed]
15. Kang, J.H.; Katayama, Y.; Kondo, F. Biodegradation or metabolism of bisphenol A: From microorganisms to mammals. *Toxicology* **2006**, *217*, 81–90. [CrossRef] [PubMed]
16. Kolvenbach, B.A.; Helbling, D.E.; Kohler, H.P.E.; Corvini, P.F.X. Emerging chemicals and the evolution of biodegradation capacities and pathways in bacteria. *Curr. Opin. Biotechnol.* **2014**, *27*, 8–14. [CrossRef] [PubMed]
17. Giarma, E.; Amanetidou, E.; Toufexi, A.; Touraki, M. Defense systems in developing *Artemia franciscana* nauplii and their modulation by probiotic bacteria offer protection against a *Vibrio anguillarum* challenge. *Fish Shellfish Immunol.* **2017**, *66*, 163–172. [CrossRef] [PubMed]
18. Oishi, K.; Sato, T.; Yokoi, W.; Yoshida, Y.; Ito, M.; Sawada, H. Effect of probiotics, *Bifidobacterium breve* and *Lactobacillus casei*, on bisphenol A exposure in rats. *Biosci. Biotechnol. Biochem.* **2008**, *72*, 1409–1415. [CrossRef] [PubMed]
19. Monachese, M.; Burton, J.P.; Reid, G. Bioremediation and tolerance of humans to heavy metals through microbial processes: A potential role for probiotics? *Appl. Environ. Microbiol.* **2012**, *78*, 6397–6404. [CrossRef] [PubMed]
20. Yamanaka, H.; Moriyoshi, K.; Ohmoto, T.; Ohe, T.; Sakai, K. Degradation of bisphenol A by *Bacillus pumilus* isolated from kimchi, a traditionally fermented food. *Appl. Biochem. Biotechnol.* **2007**, *136*, 39–51. [CrossRef] [PubMed]
21. Vijayalakshmi, G.; Ramadas, V.; Nellaiah, H. Isolation, indentification and degradation of Biphenol A by *Bacillus* sp. from effluents of thermal paper industry. *Int. J. Sci. Eng. Res.* **2013**, *4*, 366–374.

22. Zühlke, M.K.; Schlüter, R.; Henning, A.K.; Lipka, M.; Mikolasch, A.; Schumann, P.; Giersberg, M.; Kunze, G.; Schauer, F. A novel mechanism of conjugate formation of bisphenol A and its analogues by *Bacillus amyloliquefaciens*: Detoxification and reduction of estrogenicity of bisphenols. *Int. Biodeterior. Biodegrad.* **2016**, *109*, 165–173. [CrossRef]
23. Endo, Y.; Kimura, N.; Ikeda, I.; Fujimoto, K.; Kimoto, H. Adsorption of bisphenol A by lactic acid bacteria, *Lactococcus*, strains. *Appl. Microbiol. Biotechnol.* **2007**, *74*, 202–207. [CrossRef] [PubMed]
24. Fujiwara, H.; Soda, S.; Fujita, M.; Ike, M. Kinetics of bisphenol A degradation by *Sphingomonas paucimobilis* FJ-4. *J. Biosci. Bioeng.* **2016**, *122*, 341–344. [CrossRef] [PubMed]
25. Im, J.; Yip, D.; Lee, J.; Löffler, F.E. Simplified extraction of bisphenols from bacterial culture suspensions and solid matrices. *J. Microbiol. Methods* **2016**, *126*, 35–37. [CrossRef] [PubMed]
26. Spivack, J.; Leib, T.K.; Lobos, J.H. Novel Pathway for Bacterial Metabolism of Bisphenol A. *J. Biol. Chem.* **1994**, *269*, 7323–7329. [PubMed]
27. Lobos, J.H.; Leib, T.K.; Su, T.M. Biodegradation of bisphenol A and other bisphenols by a gram-negative aerobic bacterium. *Appl. Environ. Microbiol.* **1992**, *58*, 1823–1831. [PubMed]
28. Rykowska, I.; Wasiak, W. Properties, threats, and methods of analysis of bisphenol A and its derivatives. *Acta Chromatogr.* **2006**, *16*, 7–27.
29. Xiao, Q.; Li, Y.; Ouyang, H.; Xu, P.; Wu, D. High-performance liquid chromatographic analysis of bisphenol A and 4-nonylphenol in serum, liver and testis tissues after oral administration to rats and its application to toxicokinetic study. *J. Chromatogr. B* **2006**, *830*, 322–329. [CrossRef] [PubMed]
30. Samanidou, V.F.; Frysali, M.A.; Papadoyannis, I.N. Matrix solid phase dispersion for the extraction of bisphenol A from human breast milk prior to HPLC analysis. *J. Liquid Chromatogr. Relat. Technol.* **2014**, *37*, 247–258. [CrossRef]
31. Samanidou, V.F.; Kerezoudi, C.; Tolika, E.; Palaghias, G. A Simple Isocratic HPLC Method for the Simultaneous Determination of the Five Most Common Residual Monomers Released from Resin-Based Dental Restorative Materials. *J. Liquid Chromatogr. Relat. Technol.* **2015**, *38*, 740–749. [CrossRef]
32. Samanidou, V.; Hadjicharalampous, M.; Palaghias, G.; Papadoyannis, I. Development and validation of an isocratic HPLC method for the simultaneous determination of residual monomers released from dental polymeric materials in artificial saliva. *J. Liquid Chromatogr. Relat. Technol.* **2012**, *35*, 511–523. [CrossRef]
33. Asimakopoulos, A.G.; Thomaidis, N.S.; Koupparis, M.A. Recent trends in biomonitoring of bisphenol A, 4-*t*-octylphenol, and 4-nonylphenol. *Toxicol. Lett.* **2012**, *210*, 141–154. [CrossRef] [PubMed]
34. Chen, G.W.; Ding, W.H.; Ku, H.Y.; Chao, H.R.; Chen, H.Y.; Huang, M.C.; Wang, S.L. Alkylphenols in human milk and their relations to dietary habits in central Taiwan. *Food Chem. Toxicol.* **2010**, *48*, 1939–1944. [CrossRef] [PubMed]
35. Luo, L.; Yang, Y.; Wang, Q.; Li, H.P.; Luo, Z.F.; Qu, Z.P.; Yang, Z.G. Determination of 4-*n*-octylphenol, 4-*n*-nonylphenol and bisphenol A in fish samples from lake and rivers within Hunan Province China. *Microchem. J.* **2017**, *132*, 100–106. [CrossRef]
36. Neamtu, M.; Frimmel, F.H. Degradation of endocrine disrupting bisphenol A by 254 nm irradiation in different water matrices and effect on yeast cells. *Water Res.* **2006**, *40*, 3745–3750. [CrossRef] [PubMed]
37. Matsumura, Y.; Hosokawa, C.; Sasaki-Mori, M.; Akahira, A.; Fukunaga, K.; Ikeuchi, T.; Oshiman, K.; Tsuchido, T. Isolation and characterization of novel bisphenol-A-degrading bacteria from soils. *Biocontrol Sci.* **2009**, *14*, 161–169. [CrossRef] [PubMed]
38. Beltran-Heredia, J.; Torregrosa, J.; Dominguez, J.R.; Peres, J.A. Comparison of the degradation of p-hydroxybenzoic acid in aqueous solution by several oxidation processes. *Chemosphere* **2001**, *42*, 351–359. [CrossRef]
39. Abbas, S.S.; Elghobashy, M.R.; Bebawy, L.I.; Shokry, R.F. Stability-indicating chromatographic determination of hydroquinone in combination with tretinoin and fluocinolone acetonide in pharmaceutical formulations with a photodegradation kinetic study. *RSC Adv.* **2015**, *5*, 43178–43194. [CrossRef]
40. Eio, E.J.; Kawai, M.; Tsuchiya, K.; Yamamoto, S.; Toda, T. Biodegradation of bisphenol A by bacterial consortia. *Int. Biodeterior. Biodegrad.* **2014**, *96*, 166–173. [CrossRef]

11

Enhancement of Thermal Stability and Selectivity by Introducing Aminotriazine Comonomer to Poly(Octadecyl Acrylate)-Grafted Silica as Chromatography Matrix

Abul K. Mallik [1,*], Hiroki Noguchi [2], Yige Han [2], Yutaka Kuwahara [2], Makoto Takafuji [2,3] and Hirotaka Ihara [2,3,*]

1 Department of Applied Chemistry and Chemical Engineering, University of Dhaka, Dhaka 1000, Bangladesh
2 Department of Applied Chemistry and Biochemistry, Kumamoto University, 2-39-1 Kurokami, Chuo-ku, Kumamoto 860-8555, Japan; Hiroki_Noguchi@kumadai.jp (H.N.); 122d9104@st.kumamoto-u.ac.jp (Y.H.); kuwahara@kumamoto-u.ac.jp (Y.K.); takafuji@kumamoto-u.ac.jp (M.T.)
3 Kumamoto Institute for Photo-Electro Organics (Phoenics), 3-11-38 Higashimachi, Higashi-ku, Kumamoto 862-0901, Japan
* Correspondence: abulkmallik@du.ac.bd (A.K.M.); ihara@kumamoto-u.ac.jp (H.I.);

Abstract: This paper introduces a poly(octadecyl acrylate) (pODA)-based organic phase on silica, which is assisted by 2-vinyl-4,6-diamino-1,3,5-triazine (AT), for a chromatography stationary phase. The ODA-AT copolymer grafting onto silica surface was characterized by elemental analysis, FT-IR spectroscopy, scanning electron microscopy, thermo gravimetric analysis and differential scanning calorimeter (DSC). An endothermic peak top of the copolymer-grafted silica was increased to 46 °C from 38 °C, which was a peak top of pODA homopolymer. For high performance liquid chromatography (HPLC) application, the molecular selectivity increased with an increase in the AT contents of the ODA-AT copolymer as organic phase. The co-existence of an aminotriazine moiety in the copolymer promoted side-chain ordering of the poly(octadecyl) moiety, thus enhancing molecular planarity selectivity for PAHs in reversed-phase liquid chromatography.

Keywords: organic-inorganic hybrid; molecular shape selectivity; crystalline-to-isotropic phase transition; HPLC

1. Introduction

Surface modification of materials by polymers is an effective technique to improve the material properties and incorporate the functions of the polymer [1–3]. Surface modification with a copolymer system is another important issue for the preparation of materials with tailor-made properties that are composed of two different monomers having diverse chemical and/or physical properties [4–6]. Inorganic carrier particles with polymeric/copolymeric phases on their surface have long been used as the stationary phase for HPLC [7–11] because such tailor-made surface properties directly influence the retention and separation efficiency. However, molecular ordering of copolymeric phases results in supramolecular functionality. A simple example of poly(octadecyl acrylate) and its copolymer systems has been reported in this regard [12]. Poly(octadecyl acrylate) copolymers grafted onto a porous silica surface exhibit an ordered-to-disordered phase transition in the octadecyl side chains, similar to lipid bilayer membrane systems [13,14], as well as extremely high selectivity for the shape recognition of geometrical isomers.

In this paper, we introduce the versatility of 2-vinyl-4,6-diamino-1,3,5-triazine (AT) as a co-monomer component in a polymeric phase on silica (Figure 1). AT is known to be effective for improving the thermal stability of polymers [15] and increasing the strength of hydrogels [16,17]; in addition, it can adsorb DNA through hydrogen bonding interactions [18–21]. This is the first report demonstrating an increase in the thermal stability of copolymer systems composed of octadecyl acrylate (ODA) and AT due to the co-existence of AT, and the induction of selectivity enhancement for molecular planarity against polyaromatic hydrocarbons in HPLC applications.

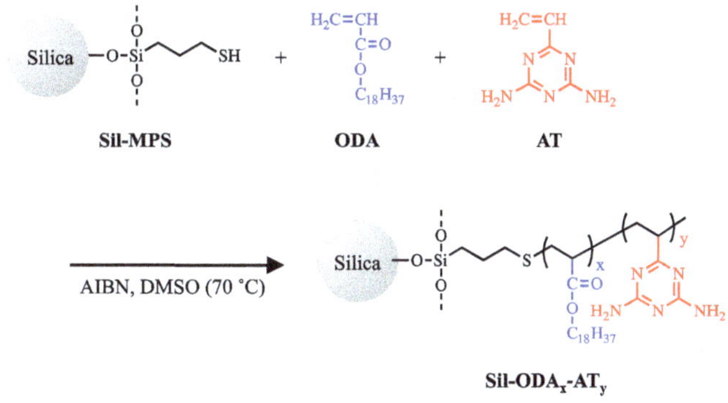

Figure 1. Reaction scheme for poly(octadecyl acrylate-co-2-vinyl-4,6-diamino-1,3,5-triazine)-grafted silica Sil-ODA$_x$-AT$_y$ from 3-mercaptopropylated silica (Sil-MPS). ODA: octadecyl acrylate; AT: 2-vinyl-4,6-diamino-1,3,5-triazine.

2. Materials and Methods

ODA, AT, and 3-mercaptopropyltrimethoxysilane (MPS) were purchased from TCI (Tokyo, Japan). ODA and AT were used after removing the polymerization inhibitors. 2,2′-Azobis(isobutyronitrile) (AIBN) was purchased from Nacalai tesque (Kyoto, Japan) and recrystallized from methanol. HPLC-grade solvents were used in the chromatographic separations.

Copolymers (ODA$_x$-AT$_y$) were introduced onto porous silica via a *"grafted-from"* method [22] based on 3-mercaptopropylated silica (Sil-MPS). The typical procedure is as follows. Sil-MPS was prepared by using porous silica particles diameter 5 μm, pore size 12 nm, and surface area 300 m^2 g^{-1}, (YMC CO., LTD., Kyoto, Japan) and MPS according to a previously reported method [23]. Sil-MPS was mixed with given amounts of ODA, AT, and AIBN as a radical initiator in DMSO, and then, the mixture was gently stirred at 70 °C for 6 h. The reaction mixture was filtered and sufficiently washed with DMSO, chloroform, and methanol. The prepared stationary phases characterized using diffuse reflectance infrared Fourier transform (DRIFT). The measurements were conducted with JASCO FT/IR-4100 (Tokyo, Japan) and DRIFT measurement accessory DR PRO410-M (JASCO CO., Tokyo, Japan). Elemental analyses were carried out on a Yanaco CHN Corder MT-6 Apparatus (Yanaco Analytical Science Inc., Kyoto, Japan). The amount of copolymer loaded on silica was determined by thermogravimetric analysis (TGA) and elemental analysis (EA). TGA was performed on a Seiko Exstar 6000 TG/DTA 6200 thermal analyzer (Seiko Instruments Inc., Chiba, Japan) in air from 100 to 800 °C and a heating rate of 10 °C/min.

For comparison of the chromatographic results, ODA polymer-grafted silica (Sil-ODA$_{10}$) without AT [13,24] was also prepared. The orientation of copolymer side chain on silica surface was evaluated by differential scanning calorimetry (DSC) in methanol suspension. DSC analysis of Sil-ODA$_{10}$-AT$_1$ and Sil-ODA$_5$-AT$_5$ were carried out using DSC6200 and EXTRA6000 (Seiko Instruments Inc., Chiba, Japan). The sample (about 20 mg) was added in 50 μL methanol taken in a silver pan and another silver pan was used as a reference and DSC was measured between 10 and 70 °C at heating and cooling rate of 2 °C/min.

The obtained surface-modified silicas were packed into stainless-steel columns for use in HPLC. The chromatographic system consisted of a Gulliver PU-1580 intelligent HPLC pump with a Rheodyne sample injector (JASCO CO., Tokyo, Japan). A JASCO multi-wavelength UV detector MD 1510 plus was used. The column temperature was maintained by using a column jacket having a heating and cooling system. Data were acquired using a JASCO ChromNAV (Ver. 1.17) system. Chromatography-grade methanol was used for the mobile phase. The retention factor (k) was determined by the relationship $(t_R - t_0)/t_0$, where t_R and t_0 are the retention times of the samples and methanol, respectively. The separation factor (α) was expressed as the ratio of the retention factors. The flow rate of the mobile phase was 1.0 mL/min for all analyses.

3. Results and Discussion

3.1. Characterization of Grafted Copolymers on Silica Surface

To estimate the amount of organic loading onto the silica particles, the resultant composites were subjected to EA and TGA measurements. Table 1 summarizes the results. An increase in the nitrogen content (with the C/N ratio) was observed upon increasing the initial molar ratio of the AT component in the copolymerization, indicating the introduction of the AT component. However, as shown in Figure 2, the silica surface was thoroughly covered with the polymer with an increase in the AT ratio, which decreased retentivity in the reversed-phase HPLC (RP-HPLC) mode, especially for Sil-AT_{10} produced without the ODA component. Therefore, further investigation was carried out using Sil-ODA_{10}-AT_1 and Sil-ODA_5-AT_5.

Table 1 includes the results of weight loss between 100 and 800 °C, as observed by TGA (Figure S1). The loaded amounts for Sil-ODA_{10}-AT_1 and Sil-ODA_5-AT_5 were determined to 21.4 and 23.4 wt %, respectively, which are sufficient for HPLC applications.

Table 1. Fundamental characterization of the prepared silica composites.

Silica Composites	Molar Ratio		EA			TGA	
	x	y	C (%)	N (%)	C/N	Loss (%)	Polymer (%)
Sil-MPS	-	-	03.81	0	-	11.3	-
Sil-ODA_{10}-AT_1	10	1	21.75	00.64	33.98	32.7	21.4
Sil-ODA_5-AT_5	5	5	19.55	04.09	04.78	34.7	23.4
Sil-AT_{10}	0	10	20.68	23.11	00.89	28.4	17.1

The values of x and y correspond to the initial molar ratios of ODA and AT in the copolymerization process, respectively, and the obtained polymer-grafted silicas are abbreviated as Sil-ODA_x-AT_y.

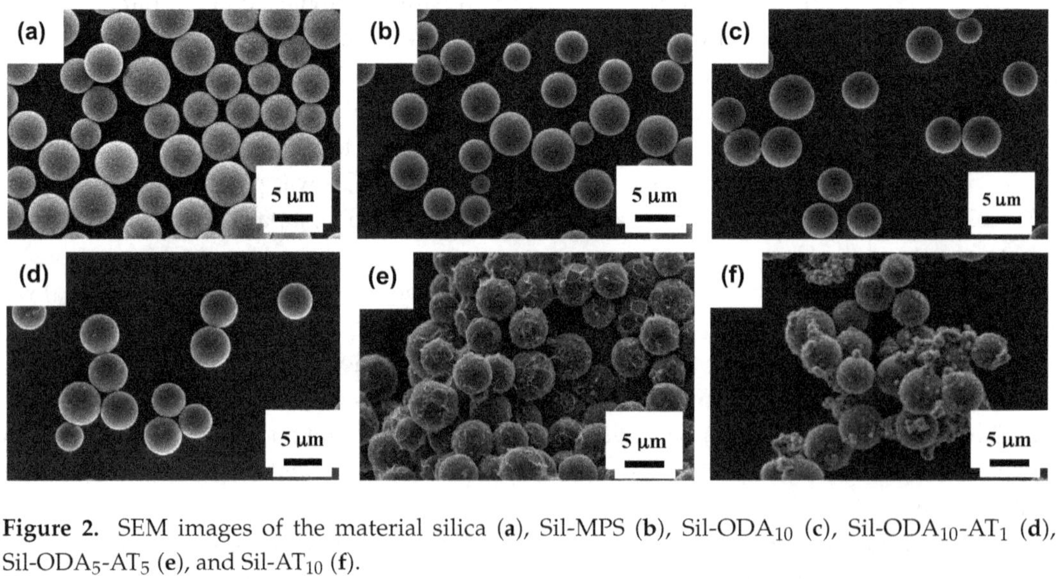

Figure 2. SEM images of the material silica (**a**), Sil-MPS (**b**), Sil-ODA_{10} (**c**), Sil-ODA_{10}-AT_1 (**d**), Sil-ODA_5-AT_5 (**e**), and Sil-AT_{10} (**f**).

3.2. Evaluation of Side-Chain Orientation in Grafted Polymeric Phase

The DSC thermogram of poly(octadecyl acrylate) (pODA) exhibited a sharp endothermic peak having a peak top at 48 °C, at the heating rate of 2 °C min^{-1}. This peak was attributed to a crystalline-to-isotropic phase transition based on the octadecyl moiety [25]. Figure 3 indicates that the peak-top temperature decreases from 48 °C to 38 °C, with peak shape broadening. This was due to perturbation from the silanol moiety of the base silica.

However, when a AT unit as a copolymer component was introduced into pODA, the peak-top temperatures increased from 44 °C to 46 °C with an increase in the molar ratio of AT for Sil-ODA_{10}-AT_1 to Sil-ODA_5-AT_5. These results indicated that the molecular ordering (crystallinity) of the ODA side chains is promoted by the co-existence of AT. It is presumed that pODA forms a homopolymeric block and that a AT unit interacts with ODA through hydrogen bonding interactions, in addition to interacting with a silanol moiety of the base silica. As a result, the pODA moiety becomes free from perturbations due to the silanol group. As a confirmation of this speculation, the diffuse reflectance infrared fourier transform (DRIFT) spectra exhibited distinct shifts for the carbonyl group: 1739 and 1733 in Sil-ODA_5-AT_5 and Sil-ODA_{10}-AT_1, respectively (Figure S2).

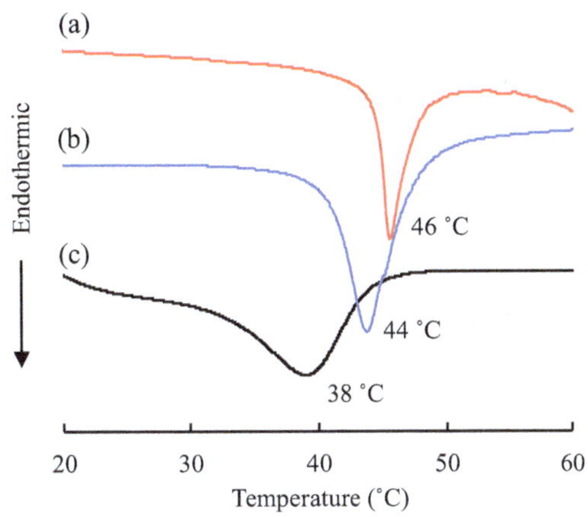

Figure 3. DSC thermograms of (a) Sil-ODA_5-AT_5, (b) Sil-ODA_{10}-AT_1, and (c) Sil-ODA_{10}.

3.3. HPLC

The liquid chromatographic performance was evaluated using polycyclic aromatic hydrocarbons (PAHs; terphenyl isomers, stilbene isomers, triphenylene, benz[*a*]anthracene, naphthacene, Figure S3) as the elutes in RP-HPLC mode. Table 2 summarizes the retention factor (k) and selectivity (α) for various PAHs. The PAHs used were broadly separated into two categories: planar compounds having different molecular lengths; and non-planar, twisted compounds [26–28]. The retention order of planar PAHs such as triphenylene, benzo[*a*]naphthalene, and naphthalene agreed with the log P order of the PAHs in all the columns. This trend is similar to that observed simple hydrophobized silicas such as ODS, thus indicating that the RP-HPLC mode was included in these columns. However, there is no significant increase in the α value among these planar PAHs, while the k value depended on the AT content. For example, the k value of naphthacene increased from 6.34 in Sil-ODA_{10} to 12.5 in Sil-ODA_{10}-AT_1, but decreased to 4.38 in Sil-ODA_5-AT_5. This was probably due to the ODA contents, which were calculated to be 22.4 wt %, and 15.7 wt % in Sil-ODA_{10}-AT_1 and Sil-ODA_5-AT_5, respectively.

However, a significant increase in selectivity was observed between *o*-terphenyl and triphenylene (Figure S4). As shown in Table 2, the selectivity increased from 2.84 to 3.83 and then to 5.01 with an increase in the AT contents. A similar increase in selectivity was also observed between *cis*- and *trans*-stilbenes. *o*-Terphenyl and *cis*-stilbene can be categorized as non-planar compounds, but triphenylene and *trans*-stilbene are planar. Therefore, the molecular planarity selectivity increases with increasing AT content.

Table 2. Retention factors (k) and separation factors (α) for various polycyclic aromatic hydrocarbons (PAHs) with pure ODA and AT-containing ODA phases.

PAHs as Elutes		Sil-ODA_5-AT_5		Sil-ODA_{10}-AT_1		Sil-ODA_{10}	
		k	α	k	α	k	α
o-Terphenyl	Non-planar	0.39	1	1.30	1	0.97	1
m-Terphenyl	Non-planar	0.88	2.27	2.47	1.90	1.48	1.52
p-Terphenyl	Non-planar	1.34	3.48	3.76	2.89	1.77	1.83
Triphenylene	Planar	1.93	5.01	4.98	3.83	2.76	2.84
cis-Stilbene	Non-planar	0.11	1	1.01	1	0.80	1
trans-Stilbene	Planar	0.24	2.26	1.64	1.62	1.15	1.43
Triphenylene	Planar	1.93	1	4.98	1	2.76	1
Benzo[α]anthracene	Planar	2.42	1.25	6.16	1.24	3.39	1.23
Naphthacene	Planar	4.38	2.27	12.5	2.51	6.34	2.30

Temperature: 30 °C, Mobile phase: MeOH-H_2O (90:10). The separation factors are calculated based on the retention factors of *o*-terphenyl, *cis*-stilbene, and triphenylene.

The selectivity increase for molecular planarity can be confirmed by investigating the temperature dependence. Figure 4 shows the temperature dependence of the selectivity for *o*-terphenyl and triphenylene. At all measured temperatures, the higher the AT content, the better was the selectivity. The selectivity of all phase was increased with decrease of temperature. The selectivity of Sil-ODA_{10} was enhanced below 40 °C, which was closed to the phase transition temperature, 38 °C (Figure 3). On the other hand, for Sil-ODA_{10}-AT_1 and Sil-ODA_5-AT_5, the enhancement of their selectivities was observed in the higher range than 40 °C. This suggests due to higher phase transition temperature (44 and 46 °C, respectively, Figure 3) than Sil-ODA_{10}. Therefore, it was concluded that the co-existence of AT promotes the molecular ordering of the long-chain alkyl groups in ODA; consequently, the molecular planarity selectivity can be enhanced by an increase in the molecular ordering.

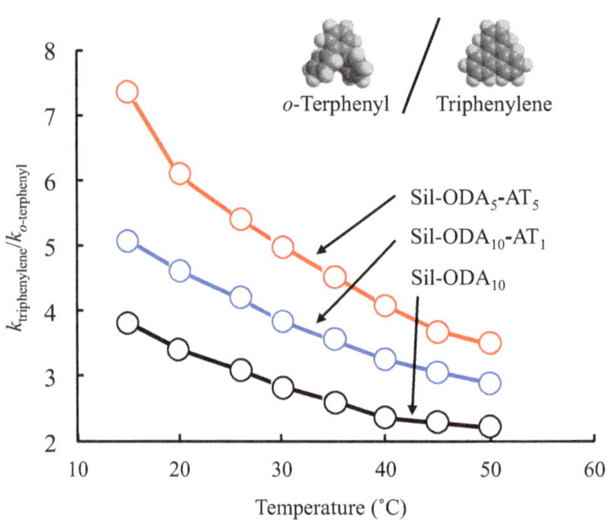

Figure 4. Temperature-dependent selectivity (α, $k_{triphenylene}/k_{o\text{-terphenyl}}$) with Sil-$ODA_x$-$AT_y$. Mobile phase: methanol-water (90:10), flow rate: 1.0 mL min^{-1}, UV detection: 275 nm.

4. Conclusions

We have demonstrated the positive effects of AT in a copolymer component for ODA. The most attractive effect is the increased thermal stability of the molecular ordering of ODA, which indicates an elevation of the phase transition temperature. Therefore, the concept of using AT as a co-monomer with ODA is a powerful way to enhance the crystalline properties of the ODA polymer. The obtained ODA-AT-copolymer-grafted porous silica can be used as a chromatographic stationary phase, as well as an adsorbent and in applications where the thermal stability of the organic phase is important. This effective approach will encourage analysts and material scientists to synthesize new materials with better performance for research and industrial applications.

Acknowledgments: This work was partially supported by a Grant-in-Aid for Scientific Research from the Ministry of Education, Culture, Sports, Science and Technology, Japan.

Author Contributions: H.I. and A.K.M. conceived and designed the experiments; A.K.M. and Y.H. performed the experiments; A.K.M., H.N. and M.T. analyzed the data; A.K.M., Y.K. and H.I. wrote the paper.

Conflicts of Interest: The authors declare no conflict of interest.

References

1. Kato, K.; Uchida, E.; Kang, E.-T.; Uyama, Y.; Ikada, Y. Polymer surface with graft chains. *Prog. Polym. Sci.* **2003**, *28*, 209–259. [CrossRef]
2. Goddard, J.M.; Hotchkiss, J.H. Polymer surface modification for the attachment of bioactive compounds. *Prog. Polym. Sci.* **2007**, *32*, 698–725. [CrossRef]
3. Kango, S.; Kalia, S.; Celli, A.; Njuguna, J.; Habibi, Y.; Kumar, R. Surface modification of inorganic nanoparticles for development of organic-inorganic nanocomposites—A review. *Prog. Polym. Sci.* **2013**, *38*, 1232–1261. [CrossRef]
4. Sumerlin, B.S.; Vogt, A.P. Macromolecular engineering through click chemistry and other efficient transformations. *Macromolecules* **2009**, *43*, 1–13. [CrossRef]
5. Li, X.; Wang, M.; Wang, L.; Shi, X.; Xu, Y.; Song, B.; Chen, H. Block copolymer modified surfaces for conjugation of biomacromolecules with control of quantity and activity. *Langmuir* **2013**, *29*, 1122–1128. [CrossRef] [PubMed]
6. Demir, B.; Broughton, R.M.; Huang, T.S.; Bozack, M.; Worley, S.D. Polymeric antimicrobial *N*-halamine-surface modification of stainless steel. *Ind. Eng. Chem. Res.* **2017**, *56*, 11773–11781. [CrossRef]
7. Mallik, A.K.; Takafuji, M.; Ihara, H. Molecular-shape selectivity tuned by donor-acceptor type copolymers as organic phase in reversed-phase high-performance liquid chromatography. *J. Chromatogr. A* **2009**, *1216*, 7433–7439. [CrossRef] [PubMed]
8. Nagase, K.; Kumazaki, M.; Kanazawa, H.; Kobayashi, J.; Kikuci, A.; Akiyama, Y.; Annaka, M.; Okano, T. Thermoresponsive polymer brush surfaces with hydrophobic groups for all-aqueous chromatography. *ACS Appl. Mater. Interfaces* **2010**, *2*, 1247–1253. [CrossRef] [PubMed]
9. Nagase, K.; Kobayashi, J.; Kikuchi, A.; Akiyama, Y.; Kanazawa, H.; Okano, T. High stability of thermoresponsive polymer-brush-grafted silica beads as chromatography matrices. *ACS Appl. Mater. Interfaces* **2012**, *4*, 1998–2008. [CrossRef] [PubMed]
10. Hiruta, Y.; Kanazashi, R.; Ayano, E.; Okano, T.; Kanazawa, H. Temperature-responsive molecular recognition chromatography using phenylalanine and tryptophan derived polymer modified silica beads. *Analyst* **2016**, *141*, 910–917. [CrossRef] [PubMed]
11. Bo, C.M.; Wang, C.; Wei, Y.M. Preparation and evaluation of diblock copolymer-grafted silica by sequential surface initiated-atom transfer radical polymerization for reverse-phase/ion-exchange mixed-mode chromatography. *J. Sep. Sci.* **2017**, *40*, 4700–4708. [CrossRef] [PubMed]
12. Ihara, H.; Nagaoka, S.; Tanaka, H.; Sakaki, S.; Hirayama, C. Lipid membrane analogue-immobilized silica gels for separation with molecular recognition. *J. Liq. Chromatogr. Relat. Technol.* **1996**, *19*, 2967–2984. [CrossRef]
13. Hirayama, C.; Ihara, H.; Mukai, T. Lipid membrane analogs. Specific retention behavior in comb-shaped telomer-immobilized porous silica gels. *Macromolecules* **1992**, *25*, 6375–6376. [CrossRef]
14. Takafuji, M.; Fukui, M.; Ansarian, H.R.; Derakhshan, M.; Shundo, A.; Ihara, H. Conformational effect of silica-supported poly(octadecyl acrylate) on molecular-shape selectivity of polycyclic aromatic hydrocarbons in RP-HPLC. *Anal. Sci.* **2004**, *20*, 1681–1685. [CrossRef] [PubMed]

15. Nair, K.P.; Breedveld, V.; Weck, M. Complementary hydrogen-bonded thermoreversible polymer networks with tunable properties. *Macromolecules* **2008**, *41*, 3429–3438. [CrossRef]
16. Gao, H.; Wang, N.; Hu, X.; Nan, W.; Han, Y.; Liu, W. Double hydrogen-bonding pH-sensitive hydrogels retaining high-strengths over a wide pH range. *Macromol. Rapid Commun.* **2013**, *34*, 63–68. [CrossRef] [PubMed]
17. Xu, B.; Li, Y.; Gao, F.; Zhai, X.; Sun, M.; Lu, W.; Cao, Z.; Liu, W. High strength multifunctional multiwalled hydrogel tubes: Ion-triggered shape memory, antibacterial, and anti-inflammatory efficacies. *ACS Appl. Mater. Interfaces* **2015**, *7*, 16865–16872. [CrossRef] [PubMed]
18. Asanuma, H.; Ban, T.; Gotoh, S.; Hishiya, T.; Komiyama, M. Hydrogen bonding in water by poly(vinyldiaminotriazine) for the molecular recognition of nucleic acid bases and their derivatives. *Macromolecules* **1998**, *31*, 371–377. [CrossRef]
19. Asanuma, H.; Ban, T.; Gotoh, S.; Hishiya, T.; Komiyama, M. Precise recognition of nucleotides and their derivatives through hydrogen bonding in water by poly(vinyldiaminotriazine). *Supramol. Sci.* **1998**, *5*, 405–410. [CrossRef]
20. Kuo, S.-W.; Tsai, H.-T. Complementary multiple hydrogen-bonding interactions increase the glass transition temperatures to PMMA copolymer mixtures. *Macromolecules* **2009**, *42*, 4701–4711. [CrossRef]
21. Tang, L.; Yang, Y.; Bai, T.; Liu, W. Robust MeO_2MA/vinyl-4,6-diamino-1,3,5-triazine copolymer hydrogels-mediated reverse gene transfection and thermo-induced cell detachment. *Biomaterials* **2011**, *32*, 1943–1949. [CrossRef] [PubMed]
22. Mallik, A.K.; Rahman, M.M.; Czaun, M.; Takafuji, M.; Ihara, H. A new route for preparation of high-density organic phase to high selective HPLC for polycyclic aromatic hydrocarbons by atom-transfer radical polymerization of octadecyl acrylate on silica. *Chem. Lett.* **2007**, *36*, 1460–1461. [CrossRef]
23. Mallik, A.K.; Qiu, H.; Takafuji, M.; Ihara, H. Selectivity enhancement for the separation of tocopherols and steroids by integration of highly ordered weak interaction sites along the polymer main chain. *Anal. Bioanal. Chem.* **2012**, *404*, 229–238. [CrossRef] [PubMed]
24. Fukumoto, T.; Ihara, H.; Sakaki, S.; Shosenji, H.; Hirayama, C. Chromatographic separation of geometrical isomers using highly oriented polymer-immobilized silica gels. *J. Chromatogr. A* **1994**, *672*, 237–241. [CrossRef]
25. Ihara, H.; Fukumoto, T.; Hirayama, C. Molecular length recognition of polyaromatics using lipid membrane analogue-immobilized silica gels. *Anal. Sci.* **1993**, *9*, 711–713. [CrossRef]
26. Jinno, K.; Kawasaki, K. Correlation between the tetention data of polycyclic aromatic hydrocarbons and several description in reversed-phase HPLC. *Chromatgraphia* **1983**, *17*, 445–449. [CrossRef]
27. Wise, S.A.; Bonnett, W.J.; Guenther, F.R.; May, W.E. A relationship between reversed-phase C_{18} liquid chromatographic retention and the shape of polycyclic aromatic hydrocarbons. *J. Chromatogr. Sci.* **1981**, *19*, 457–465. [CrossRef]
28. Tanaka, N.; Tokuda, Y.; Iwaguchi, K.; Araki, M. Effect of stationary phase structure on retention and selectivity in reversed-phase liquid chromatography. *J. Chromatogr. A* **1982**, *239*, 761–772. [CrossRef]

12

Assessment, Validation and Application to Real Samples of an RP-HPLC Method for the Determination of Guayulins A, B, C and D in Guayule Shrub

Nadia Spano [1,*], Paola Meloni [1], Ilenia Idda [1], Alberto Mariani [1], Maria Itria Pilo [1], Valeria Marina Nurchi [2], Joanna Izabela Lachowicz [2], Ernesto Rivera [3], Ancelmo Orona-Espino [4] and Gavino Sanna [1,*]

1 Dipartimento di Chimica e Farmacia, Università degli Studi di Sassari, via Vienna 2, 07100 Sassari, Italy; paola94meloni@gmail.com (P.M.); ilenia.idda@gmail.com (I.I.); mariani@uniss.it (A.M.); mpilo@uniss.it (M.I.P.)

2 Dipartimento di Scienze Chimiche e Geologiche, Cittadella Universitaria di Monserrato, 09042 Monserrato, Italy; nurchi@unica.it (V.M.N.); lachowicz@unica.it (J.I.L.)

3 Instituto de Investigaciones en Materiales UNAM, Circuito Exterior Ciudad Universitaria, CP 04510 Mexico DF, Mexico; riverage@unam.mx

4 Facultad de Ciencias Biológicas, Universidad Juárez del Estado de Durango, 35010 Gómez Palacio, Mexico; chemo74@hotmail.com

* Correspondence: nspano@uniss.it (N.S.); sanna@uniss.it (G.S.)

Abstract: Guayule (*Parthenium argentatum* Gray) is a shrub native to the arid regions of Mexico. In the last decades, significant attention to its cultivation has arisen because it is the raw material for the production of hypoallergenic natural rubber. Guayule biomass also contains high amounts of resin, which is not normally exploited in any way. Among other sesquiterpenic esters, guayulins (i.e., the parteniol esters of cinnamic acid, guayulin A, or of anisic acid, guayulin B) are contained in resin. In addition, minor amounts of guayulin C and guayulin D are formed by degradation/oxidation of guayulins A and B, respectively. Guayulins likely act as cinnamate and *p*-anisate reservoirs for the Guayule shrub; in addition, it has been postulated that they might have a key role in the chemical defense system of Guayule. Furthermore, it seems reasonable that guayulins may possess significant biological properties (e.g., antibacterial and anticancer activities), in close analogy with those shown by sesquiterpene lactones contained in many other species of *Parthenum* genus. As a matter of fact, guayulins A and B play an important role in the synthesis of antineoplastics used in breast cancer treatment. In this contribution we propose an original and validated RP-HPLC approach to the simultaneous quantification of guayulins A, B, C and D. The procedure of resin extraction from Guayule biomass has been optimized in terms of both extraction method and solvent. RP-HPLC separation has been accomplished by an Ascentis® C18 column under isocratic elution with a 80:20 (*v:v*) acetonitrile:water mixture. Validation was carried out in terms of limits of detection and quantification, linearity, precision, and trueness. Finally, the method was tested with a number of fresh and seasoned samples of spontaneous Guayule shrub from Mexico.

Keywords: guayule; *Parthenium argentatum* Gray; resin; guayulins; RP-HPLC

1. Introduction

Guayule (*Parthenium argentatum* Gray) is a perennial and low-growing shrub belonging to the Asteraceae family, native to arid regions of Mexico and the southwestern United States. Since the early years of last century it has gained considerable industrial and scientific interest for the production of a kind of natural rubber exhibiting properties similar to that obtained from the *Hevea brasiliensis* tree [1,2]. In recent decades, the discovery that guayule latex is free from the allergenic proteins contained in the *Hevea* latex led to a reemphasis and expansion of guayule research, mainly aimed at the production of hygiene and medical products compliant with the strictest standards of quality for hypoallergenic rubber [3]. Together with rubber, Guayule plant also contains a resinous material (the so-called 'resin'), which has to be separated from rubber in order not to compromise its physical and technological properties. Rubber (8–26% of dry weight) and resin (5–25% of dry weight) are present in amounts roughly equivalent in Guayule shrub [1]. Since the resin fraction is soluble in polar solvents (like acetone), whereas the rubber one can be dissolved only in non-polar solvents (like hexane), sequential or simultaneous extractions have to be made in order to separate these fractions from plant tissue and, eventually, the resin from the latex [1]. The high manufacturing costs of the process of rubber extraction from Guayule make the success of its exploitation dependent to a large extent on the possibility of using as much as possible the residual plant fractions, like resin or bagasse, as raw materials to obtain high-value coproducts [4].

Guayule resin is usually obtained from the extraction by the ground whole shrub or the coagulated latex with polar solvents, usually acetone [5]. Despite the fact that it is often not valorized and hence discarded, resin is a very interesting fraction from both an academic and an industrial point of view. Indeed, it is not only rich in sesquiterpenoids and triterpenes of potential commercial value [6], it also contains a variety of interesting secondary metabolites of the shrub, like guayulins [6–8]. Guayulins A ((1R,2S,4Z,8Z,10S)-4,8,11,11-tetramethylbicyclo[8.1.0]undeca-4,8-dien-2-yl(2E)-3-phenylprop-2-enoate) and B ((4Z,8Z)-4,8,11,11-tetramethylbicyclo[8.1.0]undeca-4,8-dien-2-yl 4-methoxybenzoate) are respectively the cinnamic and anisic esters of the partheniol ((1S,2E,6E,9S,10R)-3,7,11,11-tetramethylbicyclo[8.1.0]undeca-2,6-dien-9-ol) [7], whereas the guayulin C ((1AR,1BR,2R, 4AR,7S,7AR)-2-hydroxy-1,1,2-trimethyl-5-methylidene-decahydro-1H- cyclopropa [E]azulen-7-yl (2E)-3-phenylprop-2-enoate) and guayulin D ((1AR,1BR,2R,4AR,7S,7AR)-2-hydroxy-1,1,2-trimethyl-5-methylidene-decahydro-1H-cyclopropa[E]azulen-7-yl 4-methoxybenzoate) are likely formed by the oxidation of the guayulines A and B, respectively [6,8]. Scheme 1 reports the structures of these molecules.

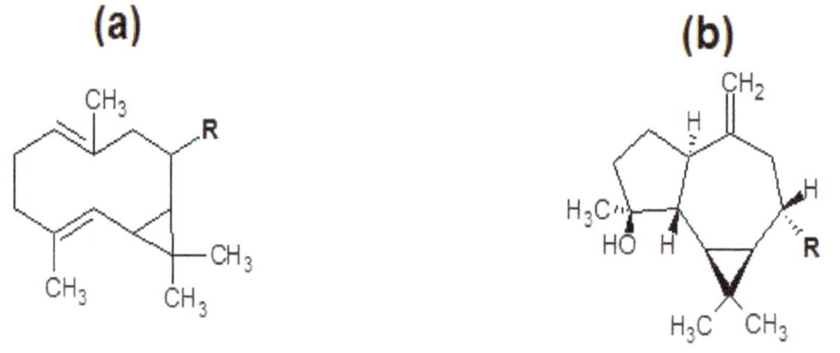

R = *t*-cinnamate, guayuline A
R = *p*-anisate, guayuline B

R = *t*-cinnamate, guayuline C
R = *p*-anisate, guayuline D

Scheme 1. Structures of guayulins A and B (**a**) and guayulins C and D (**b**).

Guayulins A and B are abundant in the resin. Namely, up to 10% of the resin mass is constituted of guayulin A [9–11], while the relevant amount of guayulin B in the resin from India is between 0.1% and 3% [10]. The wide variability in the amounts of guayulins A and B in the resin depends on a number of factors, such as: soil nature, climate, phenological phases, cultivar, age and, mainly, the nature of the different parts of the shrub [10,11]. On the other hand, the concentrations of guayulins C and D in resin seems to be much lower than those of guayulins A and B. Schloman et al. estimated that the amounts of guayulins C and D in resin are between 50% and less than 10% of the amount of the guayulin A and B, respectively [6]. Guayulins likely act as cinnamate and *p*-anisate reservoirs for the Guayule shrub: when it is required by the plant, a metabolic turnover releases the corresponding free acid [4]. The hypothesis that guayulins and rubber may have a common biochemical link [12] suggested Teetor et al. [11] to use the amounts of guayulins A and B as possible predictive tools in order to evaluate the rubber content in Guayule shrub, but the results obtained were absolutely unsatisfactory. It is noteworthy that, in *Heliantheae* tribe and *Parthenium* genus, only *Parthenium argentatum* contains both guayulins and rubber, whereas it does not contain any sesquiterpene lactone (i.e., one of the principal families of compounds easily found in almost all the other species of *Parthenum* genus) [13]. Since the sesquiterpene lactones are synthesized in the *Parthenum* genus plants as chemical defense agents against animal attacks (these molecules are toxic for livestock, a deterrent for insects and cause severe contact allergies in mammalians) [14], it seems reasonable that guayulins might play a similar physiological role also in *Parthenium argentatum*. Results of pioneering studies conducted by Rodriguez et al. supported this insight [15]. As a matter of fact, guayulin A has been found to be a powerful contact allergen for guinea pigs [8] while the same behavior has not been observed for guayulin B. For humans, the sensitizing power of guayulin A is much smaller than for laboratory animals. Hence, the possibility that traces of guayulin A may be present in the guayule rubber at a sensitizing level seems to be remote at the moment [16]. In analogy to what was observed for sesquiterpene lactones in most of the species of *Parthenum* genus, it is likely that guayulins may also exhibit other significant biological properties (i.e., antibacterial and anticancer activities) [15]. Indeed, guayulines A and B act as biological triggers in the synthesis of lychnostatine and paclitaxel, which are antineoplastic agents used in breast cancer treatment.

It is evident that the continuous rise of interest in the properties of guayulins needs the development of reliable, accurate and sensitive analytical methods for their identification and quantification. Until now, the analytical methods reported in literature for the measurement of the amount of guayulins in Guayule resin [3,4,8,11,13,17–19] or latex [16] are scarce and all lacking of any validation protocol. Obviously, chromatographic methods are dominant in the analytical characterization of such analytes and, among them, HPLC is the preferred approach. While few papers reported the presence of guayulins C and D in extracts from resin [6,8,11], the quantification of these analytes has been never simultaneously carried out to that of guayulins A and B [6,8]. Hence, it is evident that at present a validated HPLC procedure aimed to simultaneously quantify the amounts of all known guayulins is not reported in literature. For this reason, the principal aim of this contribution is to develop, validate and test with real samples an original RP-HPLC method devoted to the quantification of guayulins A, B, C and D in different parts of Guayule shrub from Mexico. As a side result of this study, different approaches of extraction of analytes from the matrix were also compared in order to increase efficiency and make the procedure more eco-friendly.

2. Materials and Methods

2.1. Samples and Sample Preparation

Spontaneous Guayule was collected in the province of San Pedro del Gallo, Durango, Mexico, along the highways where it naturally grows. All the harvested Guayule shrubs were about 60 cm high, thus estimated to be between 15 and 18 year old [20]. These plants were harvested in 2015 in two sites located in the Durango desert. The first one (sampling of January 2015) is in the Mexican Federal Highway 30 (road section Bermejillo-El Palmito, GPS coordinates: 25°52′20.68″ N, 104°06′26.07″ W), whereas the second one (sampling of September 2015) is along the highway Paso Nacional-San Pedro del Gallo (GPS coordinates: 25°40′50.41″ N, 104°16′33.53″ W). A total of twenty plants (ten for each sampling) were harvested following the recommendations specified by the Official Mexican Rule [21], which establishes the procedures for the sampling, transportation and storage. Shrubs collected in the harvest of January 2015 were air-dried at room temperature in open air for eight months, whereas shrubs collected in the harvest of September 2015 did not undergo any drying process. Only a few millimeters of rain fell during the first harvest. Just arrived in the laboratory, all shrubs were cut up to constitute three fractions: leaves (L), fine stems (less than 10 mm in diameter, FS) and coarse stems (more than 10 mm of diameter, CS). All fractions were stored in a freezer at −18 °C until sample preparation. Before extraction, samples were allowed to reach the room temperature. Later, all samples were dried at 50 °C until reaching constant weight, roughly pieced and then ground.

2.2. *Chemicals and Reagents*

All reagents were at least of analytical grade. Ethanol (assay ≥99.8%), acetone (≥99.8%) and acetonitrile (≥99.9%) were purchased from Sigma-Aldrich (Milan, Italy), whereas ultrapure (Type 1) water (specific resistance 18 MΩ) was used throughout the analyses. Guayulins A, B, C and D were from 1717 CheMall (Mundelein, IL, USA).

2.3. *Instrumentation*

Ultrapure water was from a New Human Power II Scholar UV apparatus (Human Corporation, Seoul, Korea). A Retsch Grindomix GM 200 knife mill (Verder Scientific, Torre Boldone, Italy) was used to grind the samples before extraction. ASE extractions were performed by an Accelerated Solvent Extractor Dionex ASE 200, equipped with a Dionex Solvent Controller System (Dionex Corporation, Sunnyvale, CA, USA), whereas ultrasonic assisted extractions were performed by a Bandelin-Sonorex model RK 255 H ultrasonic bath. Solvent was evaporated by extracted samples by a Büchi Rotovapor model EL130 (Büchi, Cornaredo, Italy). The HPLC equipment consisted of a Series 200 binary pump, a sampling valve, a 20 μL sample loop, a Series 200 LC column oven, and a Series 200 UV-Vis variable wavelength detector, all from PerkinElmer, Milan, Italy. Data were processed using a Turbochrom Workstation Software (PerkinElmer, Milan, Italy). Before use, the mobile phase was filtered through a 0.45 μm membrane from Millipore (Bedford, MA, USA) to remove any particulates.

2.4. *Validation*

Validation of the proposed method was accomplished on the basis of limit of detection (LoD), limit of quantification (LoQ), linearity, precision (measured as both repeatability and intermediate precision) and trueness (measured by recovery tests of each analyte on real Guayule samples). LoD was calculated according to the Upper Limit Approach (ULA1) approved by IUPAC [22]. For each analyte, four different solutions at increasing concentrations not far from the expected LoD (i.e., between 0.02 and 0.1 mg L^{-1} for each analyte) were prepared and analyzed. Each measurement was performed in triplicate. In addition, the ULA1 approach recommends that the LoQ value is three times the relevant LoD value. Linearity was checked on at least three orders of magnitude of concentration, as a function of the relative abundance of each guayulin in the different parts of the Guayule shrub. Precision was evaluated in terms of both repeatability (i.e., the CV measured in five consecutive replicates of the same sample in the same analytical session) and intermediate precision (i.e., the CV obtained in five consecutive replicates of the same sample in different analytical sessions across two weeks). The acceptability of these precision scores was checked in terms of HorRat ratio values (i.e., the ratio between experimental and theoretical CV measured on the basis of Horwitz's theory) [23]. Due to the lack of any certified reference materials or of reliable independent analytical methods, trueness was estimated through recovery tests. Four aliquots of the acetone (or ethanol) extract of the stems were submitted to the overall analytical procedure after the addition of increasing amounts of each guayulins to three of them. For each analyte, a plot of analytical concentration versus the added amounts of mass was obtained: the recovery is represented by the percent slope value of the regression line. The estimation of bias was made in duplicate. Acceptability of the trueness values has been established according to the guidelines described in the AOAC manual for peer-verified methods [24].

3. Results

3.1. Choice of the Extraction Procedure

The choice of extraction technique of resin by Guayule biomass (i.e., leaves, fine stems, coarse stems) was accomplished by means of a preliminary comparison among three well-known procedures: the traditional extraction with warm solvents (40 °C) (WSE), the ultrasonic-assisted extraction (UAE) and the accelerated solvent extraction (ASE). Acetone (i.e., the most used solvent for the extraction of the resin from Guayule biomass) and ethanol, one of the most important green polar solvent, were the pure solvents used in this phase. Quantification of resin in organic extracts was performed by means of UV-Vis spectroscopy according to literature methods [25].

3.1.1. Procedures of Extraction

Extraction with Warm Solvents (WSE)

Firstly, 2.5 g of selected Guayule biomass were suspended in 15 mL of acetone (or ethanol). The mixture was stirred and heated up to 40 °C in a thermostatic bath for 20 min. After the separation of the solvent from the biomass and its replacement with a new 15 mL aliquot, the extraction procedure was repeated for an additional two times. The three extracts were collected and made up to the final volume of 50 mL with pure solvent.

Ultrasonic Assisted Extraction (UAE)

Firstly, 2.5 g of selected Guayule biomass were suspended in 15 mL of acetone (or ethanol). The mixture was stirred at room temperature in an ultrasonic bath for 20 min. After the separation of the solvent from the biomass and its replacement with a new 15 mL aliquot, the extraction procedure was repeated for an additional two times. The three extracts were collected and made up to the final volume of 50 mL with pure solvent.

Accelerated Solvent Extraction (ASE)

Firstly, 5.0 g of selected Guayule biomass were transferred in a 22 mL thimble. Extraction was performed using 20 mL of the chosen solvent (acetone or ethanol). The extraction temperature was set at 40 °C. Three consecutive extraction cycles have been performed for each biomass aliquot. The organic extracts were joined and made up to 100 mL with pure solvent.

Figure 1 reports a comparison of resin yields (% on the initial biomass amount) measured for each extraction method and each biomass analyzed.

Data obtained substantiate that the UAE method is characterized by the least recovery in resin, whereas the performances of WSE and ASE methods are roughly comparable as a function of the nature of Guayule biomass and the extraction solvent. However, it is possible to note that the net amount of resin extracted from the fine stems by means of ASE method is statistically higher than that obtained by WSE, whereas the yields in resin measured in the extraction from coarse stems are not statistically different among them from both WSE and ASE methods. Keeping into account the better intrinsic reproducibility of the ASE approach as compared with the WSE method, the ASE extraction method with acetone and ethanol was chosen as the technique used of the present analytical method.

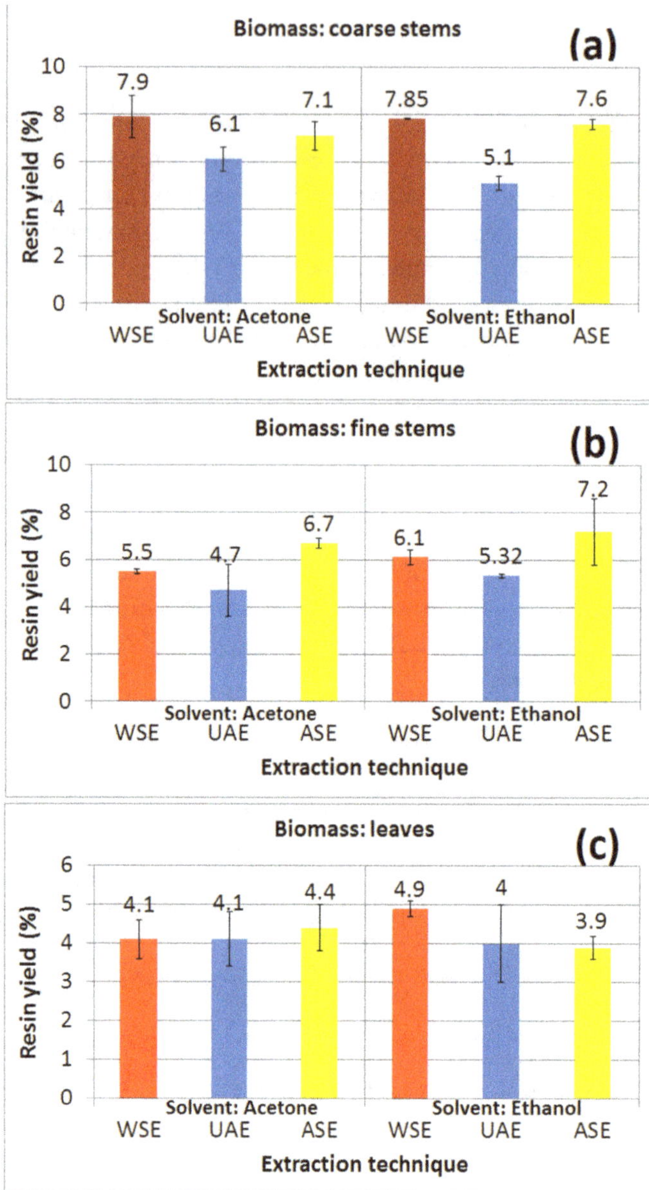

Figure 1. Comparison of average solvent extraction yields in resin (% on the biomass amount), $n = 3$. Solvents: acetone or ethanol. (**a**) biomass: coarse stems; (**b**) biomass: fine stems; (**c**) biomass: leaves. WSE: warm solvent extraction; UAE: ultrasonic-assisted extraction; ASE: accelerated solvent extraction.

3.2. Assessment of the Chromatographic Method

The first RP-HPLC methods aimed to quantify guayulins in Guayule resin were developed more than thirty-five years ago, and were only aimed at the determination of guayulins A and B. On the other hand, reliable RP-HPLC methods devoted to measure the concentration of guayulins C and D are still absent in the literature. For the reader's convenience, Table 1 summarizes the key features of different literature RP-HPLC methods for the determination of guayulins.

Furthermore, it is of utmost significance to underline that no analytical method proposed until now has been validated, and this poses serious doubts about the overall reliability of the data reported. The results of preliminary tests performed on selected literature methods suggested the assessment of an original procedure, based on a more recent C18 stationary phase than those reported in Table 1. Ascentis® C18 is a monomeric-type, fourth-generation, octadecylsiloxane-bonded silica stationary phase, introduced in the market in the early 2000s, and characterized by unusually high surface area

(450 m^2 g^{-1}) if compared with other C18 phases having similar porosity (10 μm) and surface coverage (3.7 μmol m^{-2}). Since Ascentis® stationary phases have already been successfully used for the HPLC determination of terpenoids in vegetal matrices [26], we considered useful to check the performances of an Ascentis® C18 (25 cm × 4.6 mm × 5 μm) column in the separation of guayulins A, B, C and D. Further steps of the method assessment have been devoted to optimize the composition of the mobile phase and to the choice of the best wavelength of UV detection. Starting from the literature methods, different methanol:water and acetonitrile:water mixtures have been tested, working both in isocratic and in gradient elution. An acetonitrile:water 80:20 (*v*:*v*) solution provided the best results. In addition, the acquisition of chromatograms in the wavelength range between 250 nm and 280 nm demonstrates that the best compromise among different UV maximum absorptions of the analytes is the λ at 262 nm, as previously reported in the literature [4,6,10,11]. Table 2 summarizes the operating conditions of the chromatographic method proposed.

Table 1. Key features of selected literature RP-HPLC methods of analysis for guayulins in resin extracts from Guayule.

Stationary Phase (mm × mm × μm)	Mobile Phase (*v*:*v*)	Flow Rate (mL min^{-1})	λ of Quantification (nm)	Guayulins Quantified	Reference
MicroPak MCH-10 (300 × 4.6 × 10)	from CH_3CN:H_2O 50:50 to CH_3CN:H_2O 75:25	n.r.	262	A and B [1]	[6]
C18 column [2]	MeOH:H_2O gradient elution [2]	n.r.	215	C and D	[8]
RP-C18 column [2]	MeOH:H_2O 89:11	1	254	A and B	[10]
Microsorb-MV (250 × 4.6 × 5)	from CH_3CN:H_2O 80:20 to pure CH_3CN	1.5	262	A and B [3]	[11,16,17,19]
Altex Ultrasphere ODS type (150 × 4.6 × 5)	MeOH:H_2O 93:7	1	254	A and B	[13]
Lichrosorb RP-18 (250 × 4.6 × 10)	CH_3CN:H_2O 75:25	2.5	254	A and B	[18]

[1] Guayulins C and D were quantified using a GC method; [2] no further details were provided in the paper; [3] in ref. [11], guayulins C and D were only identified, but not quantified, n.r. not reported in the original study.

Table 2. Chromatographic conditions proposed for the contemporary RP-HPLC separation and quantification of guayulins A, B, C and D in resin extracts from Guayule biomass.

Stationary Phase (Length, mm × Diameter, mm × Particle Size, μm)	Ascentis® C18 (250 × 4.6 × 5)
Mobile phase (*v*:*v*)	CH_3CN:H_2O 80:20
Elution mode	Isocratic
Volume (μL)	20
Flux of mobile phase (mL min^{-1})	1.5
UV wavelength (nm)	262
Column temperature (°C)	25
Length of the chromatographic run (min)	25

The adoption of an Ascentis® C18 RP-HPLC column, more recent in comparison to those reported in literature (see Table 1) have allowed us to develop an isocratic method achieving the complete resolution at baseline level of all analytes in less than 20 min, thus with large time savings of the chromatographic run as compared with the most recent literature method [11].

3.3. *Analysis of the Resin Extracts*

First, 2.5 mL of acetone (or ethanol) ASE extract of the resin from guayule biomass were evaporated, at room temperature and under reduced pressure, to dryness. The resin residue was dissolved in five milliliters of acetonitrile, and then filtered through a 0.45 μm nylon filter before HPLC analysis. Figure 2 shows typical chromatograms of ASE resin extracts from Guayule leaves (Figure 2a) and coarse stems (Figure 2b).

Guayulins' peaks were identified by comparing retention times with those of standard solutions, and the attribution of each analyte was confirmed by spiking each peak in the real sample with a standard solution containing known amounts of pure guayulins. In order to determine the retention times, the reference standards were injected both individually and as a mixture. Quantification was accomplished by external linear calibration on three different concentration levels in the relevant linearity interval of each analyte. Each sample was analyzed three times, and each analytical datum is reported as the average value ± the relevant standard deviation.

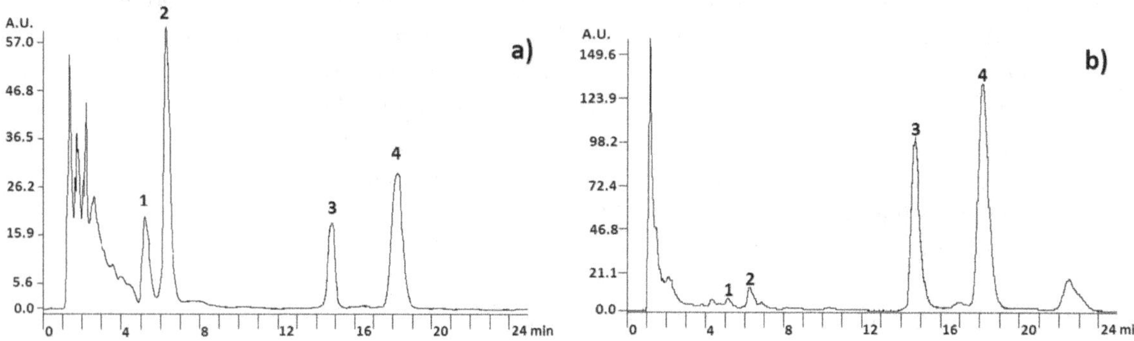

Figure 2. Guayulins A, B, C and D in resin extracts from (**a**) Guayule leaves (sample L1, extraction solvent: acetone) and (**b**) Guayule coarse stems (sample CS2, extraction solvent: ethanol). Peak attribution: 1—guayulin D; 2—guayulin C; 3—guayulin B; 4—guayulin A. A.U. = arbitrary units.

3.4. Validation

Table 3 lists the validation parameters for the method proposed. Low LoD values (always below 0.1 mg L^{-1}) supported a good sensitivity of the method proposed, which is able to quantify analytes at concentration levels between 0.1 and 0.2 mg L^{-1}, as a function of the analyte. Also, linearity is more than satisfactory. It has been checked only within the three orders of magnitude of concentration (i.e., inside the range of variability of the guayulins in the different parts of Guayule shrub), obtaining values of the determination coefficients R^2 ranging between 0.9994 and 0.9984, but it is likely that the linear dynamic interval for each analyte may largely overcome these ranges. In addition, the analysis of the residuals of the regression line has excluded any possible deviation from linearity of each calibration plot. A preliminary evaluation of the samples has substantiated a very wide scattering of the concentration of the analytes in them. Due the fact that the meaning of the precision measurements crucially depends on analyte concentration [23], repeatability and intermediate precision were evaluated on the Guayule extracts characterized, for each analyte, by a concentration that was as close as possible to its average value in the relevant calibration interval. For this reason, the ethanol extract of sample FS1, the acetone extract of sample FS2, the acetone extract of sample FS1 and the ethanol extracts of sample CS1 were chosen for the precision measurements for guayulins A, B, C and D, respectively. The acceptability of the levels of the precision data, ranging between 1.3% and 3.2% (guayulin A) and 3.6% and 10% (guayulin D) for repeatability and intermediate precision, respectively, were evaluated by a fitness-for-purpose methodology based on the Horwitz's theory [23]. Operatively, acceptable values of the HorRat ratio (i.e., the ratio between the experimental CV and the theoretical CV calculated on the basis of Horwitz's theory) should be less than 1.5. Since the HorRat ratios for the procedure subject of this work never exceeded 0.6, the precision levels obtained from all analytes can be considered acceptable. Trueness evaluations have been accomplished with the same Guayule samples involved in the precision measurements. Recoveries obtained ranged between 73 ± 5% (guayulin D) and 91 ± 2% (guayulin B) for acetone extracts, whereas ethanol extracts gave recoveries between 77 ± 8% (guayulin D) and 89 ± 6% (guayulin A). The evaluation of the recovery values obtained on the basis of the AOAC guidelines [24] have allowed to substantiate, for all analytes, a slight underestimation bias. Unfortunately, the complete absence in the literature of validated methods for the determination of guayulins prevented us to make any comparison with results from previous studies.

Table 3. Validation data.

Guayulin	Sensitivity (n = 3)		Linearity		Repeatability (n = 5)		Intermediate Precision (n = 5)		Bias (n = 2)	
	LoD (mg L^{-1})	LoQ (mg L^{-1})	Concentration Range (mg L^{-1})	R^2	$CV_{exp,r}$ [a]	$HorRat_r$ [b]	$CV_{exp,IP}$ [c]	$HorRat_{exp,IP}$ [d]	Recovery (% ± SD [e]) on Acetone	Recovery (% ± SD [e]) on Ethanol
A	0.032 ± 0.004	0.10 ± 0.01	0.2–500	0.9994	1.3	0.26	3.2	0.43	87 ± 4	89 ± 6
B	0.051 ± 0.004	0.15 ± 0.01	0.2–250	0.9991	1.7	0.29	3.5	0.41	91 ± 2	88 ± 8
C	0.034 ± 0.009	0.10 ± 0.03	0.1–200	0.9989	1.5	0.58	4.0	0.39	81 ± 9	80 ± 10
D	0.055 ± 0.005	0.16 ± 0.02	0.16–200	0.9984	3.6	0.39	10	0.26	73 ± 5	77 ± 8

[a] $CV_{exp,r}$ is the experimental coefficient of variation of repeatability, measured in the ethanol extract of FS1 for guayulin A, in the acetone extract of FS2 for guayulin B, in the acetone extract of FS1 for guayulin C and in the ethanol extract of CS1 for guayulin D, respectively; [b] $HorRat_r$ is the ratio between $CV_{exp,r}$ and the theoretical repeatability data ($CV_{H,r}$) according to Horwitz's theory; [c] $CV_{exp,IP}$ is the experimental coefficient of variation of intermediate precision, measured in the ethanol extract of FS1 for guayulin A, in the acetone extract of FS2 for guayulin B, in the acetone extract of FS1 for guayulin C and in the ethanol extract of CS1 for guayulin D, respectively; [d] $HorRat_{IP}$ is the ratio between $CV_{exp,IP}$ and the theoretical intermediate precision data ($CV_{H,IP}$) according to Horwitz's theory; [e] standard deviation.

3.5. *Application on Real Samples*

Table 4 reports the analytical data obtained for the RP-HPLC determination of guayulins A, B, C and D in acetone and ethanol ASE extracts from different Guayule biomass (leaves (L samples), fine stems (FS samples) and coarse stems (CS samples)) harvested in January 2015 (dried samples, 1) and in September 2015 (fresh samples, 2).

Table 4. Concentration of guayulins A, B, C and D in acetone or ethanol extracts of Guayule biomass (leaves (L), fine stems (FS) and coarse stems (CS)) harvested in January 2015 (dried samples, 1) and in September 2015 (fresh samples, 2).

	Guayulins (mg kg^{-1} ± SD)							
Sample (Solvent)	A (A)	A (E)	B (A)	B (E)	C (A)	C (E)	D (A)	D (E)
L1	2080 ± 40 [a]	2140 ± 40 [a]	444 ± 9 [a]	449 ± 4 [a]	1638 ± 8 [a]	1400 ± 45 [b]	360 ± 30 [a]	330 ± 30 [a]
L2	2030 ± 65 [a]	2190 ± 20 [a]	428 ± 6 [a]	460 ± 9 [a]	1650 ± 7 [a]	1560 ± 45 [b]	370 ± 20 [a]	330 ± 20 [a]
FS1	3810 ± 30 [a]	3230 ± 40 [a]	615 ± 9 [a]	500 ± 20 [a]	372 ± 6 [a]	204 ± 4 [a]	8.4 ± 0.7 [c]	6.52 ± 0.07 [a]
FS2	4190 ± 40 [a]	3640 ± 10 [a]	1190 ± 20 [a]	1150 ± 9 [a]	61.5 ± 0.1 [a]	26.4 ± 0.7 [a]	<0.32 *	<0.32 *
CS1	8160 ± 90 [a]	9830 ± 20 [a]	700 ± 10 [d]	884.8 ± 0.7 [a]	1010 ± 4 [a]	1044 ± 7 [a]	25 ± 2 [a]	55 ± 2 [a]
CS2	8800 ± 100 [a]	7800 ± 100 [a]	3090 ± 20 [a]	2810 ± 40 [a]	249 ± 4 [a]	199 ± 4 [a]	23 ± 2 [a]	16 ± 2 [a]

SD = standard deviation; n = 3; * LoQ of guayulin D; paired values (i.e., same sample, same guayulins, different extraction solvent) followed by the same letter (from a to d) are not significantly different according to ANOVA test at $p = 0.05$.

While data relative to guayulins A and B are roughly comparable with those previously reported in literature [4,6,10,11], those relative to the amounts of guayulins C and D in acetone (or ethanol) extracts by different parts of the Guayule shrub are, to the best of our knowledge, unprecedented at all. Firstly, it is very important to highlight that these data are relative to a spontaneous Guayule shrub, whereas almost all literature data are relative to specific and well-known Guayule genotypes coming from cultivation.

3.5.1. Guayulins in Guayule Leaves

As a first remark, it is interesting to note that the relative composition in guayulins on leaves from the first and the second harvest is roughly constant, and this is probably due to the fact that the leaves of the shrubs harvested in September 2015 also arrived not entirely fresh in the laboratory, like those harvested in January 2015. The relative composition of both guayulins is almost the same as the nature of the solvent varies, with only the exception of a slight increase of the concentration of guayulins C and D and a faint decrease of the guayulin A in the acetone extract in comparison to those obtained by ethanol. The ratio between guayulin A and B in leaves is roughly close to 5:1. This data is significantly higher than that (i.e., 1.3) measured by Sidhu et al. for a Gila cultivar [10], but is within the range (between 2.5 and 7) measured by Teetor et al. [11] in brown leaves of three Guayule cultivars. On the other hand, leaves are very rich in guayulins C and D. In particular, the amount of guayulin D in leaves is more than ten times higher than those found in the richest remaining part of Guayule shrub (i.e., the coarse stems from the January harvest, CS1), whereas the concentration of guayulin C in leaves is only ca. 50% higher than that measured in CS1. The significant increase of concentration of guayulins C and D is likely a consequence of the drying process of the guayule biomass. This fact confirms the hypotheses that these compounds could derive from oxidation/degradation pathways from guayulins A and B, respectively [6,8]. The increase of concentration is more evident for guayulin C rather than for guayulin D, but it is possible that this could be caused by the difference between the relative abundance of guayulin A and guayulin B, respectively.

3.5.2. Guayulins in Guayule Stems

The highest amounts of guayulins A and B were found in the coarse stems (CS samples). Depending on the nature of the solvent used for the ASE extraction, amounts ranging between 7800 ± 100 and 9830 ± 20 mg kg^{-1} were measured for guayulin A, whereas concentrations ranging between 700 ± 10 and 3090 ± 20 mg kg^{-1} were measured for guayulin B. These concentrations are in partial agreement with those observed by Teetor et al. for guayulin A [11] and by Sidhu et al. for guayulin B [10]. Indeed, the concentrations of guayulin B measured by Teetor et al. [11] and the amounts of guayulin A found by Sidhu et al. [10] were not as high as those measured in this study. It is interesting to note that the concentration ratio between guayulin A and guayulin B, which is roughly constant at varying of the nature of the solvent, increases by more than 400% passing from fresh to aged stems, and this is due to the loss of more than 70% in the average amount of guayulin B during the storage. On the other hand, data show that changes of the concentration of the guayulin A, albeit much lower than those shown by guayulin B, depend by both the nature of the extraction solvent and the aging of the sample in non-trivial way. Furthermore, the amount of guayulin C in these samples tends to sharply increase (up to five times higher) with ageing, whereas the concentration of this analyte is invariant towards the nature of the extraction solvent. Finally, only a few tens of mg kg^{-1} of guayulin D were found in the coarse stem samples. A moderate increase in concentration of this analyte, likely due to aging effects, was observed only in ethanol extracts.

The amounts of guayulins in fine stems are intermediate between those measured in coarse stems and leaves. The amounts of guayulins A and B in these samples are normally only 30–50% of the relevant values measured for coarse stems, whereas the loss of guayulin B during the ageing is only between 50% and 55% the amount measured in fresh samples. On the other hand, very high increases of concentration of guayulin C were found (ranging between 600% and 770% in samples extracted with acetone and ethanol, respectively) passing from fresh to aged samples, whereas just a few mg kg^{-1} of guayulin D were measured only in FS1 samples, being the remaining samples below the relevant LoQ for this analyte.

4. Conclusions

To the best of our knowledge, for the first time a RP-HPLC method specifically aimed to simultaneously measure the concentration of the known guayulins in resin extracts from parts of Guayule shrub has been developed, validated and applied to real samples. The development of the method has first involved the optimization of the resin extraction procedure from the Guayule biomass among three different approaches (warm solvent extraction, ultrasonic assisted extraction, accelerated solvent extraction). The latter technique provided the best performances and was chosen to be used in the analytical method. The HPLC separation was accomplished by means of an isocratic elution on an Ascentis® C18 column with an 80:20 (*v:v*) mixture of acetonitrile and water. In this way, the separation at the baseline level of all analytes in less than twenty minutes was obtained. Low LoD and LoQ values, a very good linearity over more than three orders of magnitude and a very good precision characterized the proposed method. On the other hand, a slight underestimation bias was observed in all analytes. The method has been successfully tested to fresh and aged samples of spontaneous Guayule shrub, harvested in the Chihuhaua desert (Laredo, Mexico). The guayuline amount was measured as a function of the ageing level of the sample (fresh or air-exposed for eight months), the parts of the plant (leaves, fine stems, coarse stems) and the nature of the extraction solvent (acetone or ethanol). The data obtained are in substantial agreement with those described in the literature for guayulins A and B, while there are no reference data for guayulins C and D. The analysis of the stems revealed evident differences in the concentration of guayulins as a function of their diameter, of the ageing level and, to a lesser extent, of the nature of the extraction solvent. On the other hand, the analysis of the leaves showed no meaningful differences among data obtained from samples obtained in the two crops or by variation of the solvent nature. The leaves were the richest fractions in guayulines C and D, while the coarse stems were the richest in guayulines A and B. The obtained data showed that leaves constitute the fraction of Guayule that is probably most is prone to the effect of oxidative degradation of guayulins A (and B) and the consequent transformation in guayulins C (and D). This behavior is much less evident, where it is present, on the stems. This analytical method is expected to be useful in any further qualitative and quantitative evaluation of guayulins in different parts of Guayule shrub as a function of the nature of the genotype, or the adoption of different cultivation practices.

Acknowledgments: This work was financed by the Italian "Ministero dell'Istruzione, dell'Università e della Ricerca" (CTN-001-00063-46446 ALBE BIOMASSE Lignocellulosiche alternative per lo sviluppo di elastomeri).

Author Contributions: N.S., G.S. and A.M. conceived and designed the experiments; P.M. and I.I. performed the experiments; M.I.P., V.M.N. and J.I.L. analyzed the data; E.R. and A.O.-E. provided Guayule samples; G.S. and N.S. wrote the paper.

Conflicts of Interest: The authors declare no conflict of interest.

References

1. McIntyre, D.; Stephens, H.L.; Schloman, W.W., Jr.; Bhowmick, A.K. Guayule rubber. In *Handbook of Elastomers*, 2nd ed.; Bhowmick, A.K., Stephens, H., Eds.; CRC Press: Boca Raton, FL, USA, 2000; ISBN 9780824703837.
2. Van Beilen, J.B.; Poirier, Y. Establishment of new crops for the production of natural rubber. *Trends Biotechnol.* **2007**, *25*, 522–529. [CrossRef] [PubMed]
3. Van Beilen, J.B.; Poirier, Y. Guayule and Russian Dandelion as Alternative Sources of Natural Rubber. *Crit. Rev. Biotechnol.* **2007**, *27*, 217–231. [CrossRef] [PubMed]
4. Schloman, W.W., Jr.; Garrot, D.J., Jr.; Ray, D.T.; Bennett, D.J. Seasonal Effects on Guayule Resin Composition. *J. Agric. Food Chem.* **1986**, *34*, 177–179. [CrossRef]
5. Eagle, F.A. Guayule. *Rubber Chem. Technol.* **1981**, *54*, 662–684. [CrossRef]
6. Schloman, W.W., Jr.; Hively, R.A.; Krishen, A.; Andrews, A.M. Guayule Byproduct Evaluation: Extract Characterization. *J. Agric. Food Chem.* **1983**, *31*, 873–876. [CrossRef]
7. Romo, J.; Romo De Vivar, A.; Ortega, A.; Diaz, E. Guayulins A and B, new sesquiterpenes isolated from guayule. *Rev. Latinam. Quim.* **1970**, *1*, 132–135.
8. Martinez, M.; Flores, G.; Romo de Vivar, A.; Reynolds, G.; Rodriguez, E. Guayulins C and D from Guayule (Parthenium Argentatum). *J. Nat. Prod.* **1986**, *49*, 1102–1103. [CrossRef] [PubMed]
9. Watkins, S.F.; Fronczek, F.R.; Chiari, G.; Reynolds, G.W.; Rodriguez, E. Molecular structure of guayulin A. *J. Nat. Prod.* **1985**, *48*, 631–633. [CrossRef] [PubMed]
10. Sidhu, O.P.; Ratti, N.; Behl, H.M. Quantitative and Qualitative Variations in Resin Content and Guayulins (A and B) among Different Guayule Cultivars. *J. Agric. Food Chem.* **1995**, *43*, 2012–2015. [CrossRef]
11. Teetor, V.H.; Ray, D.T.; Schloman, W.W., Jr. Evaluating chemical indices of guayule rubber content: Guayulins A and B. *Ind. Crop. Prod.* **2009**, *29*, 590–598. [CrossRef]
12. Mau, C.; Cornish, K. Guayule Plants, Products, and Derivatives. US Patent 8013213, 6 September 2011.
13. Mohan Behl, H.; Marchand, B.; Rodriguez, E. Inheritance of Sesquiterpenoid Phenolic Acid Esters (Guayulins) in F_1 Hybrids of Parthenium (Asteraceae). *Z. Naturforsch.* **1983**, *38*, 494–496. [CrossRef]
14. Rodriguez, E.; Towers, G.H.N.; Mitchell, J.C. Biological activities of sesquiterpene lactones. *Phytochemistry* **1976**, *15*, 1573–1580. [CrossRef]
15. Rodriguez, E.; Reynolds, G.W.; Thompson, J.A. Potent Contact Allergen in the Rubber Plant Guayule (Parthenium argentatum). *Science* **1981**, *211*, 1444–1445. [CrossRef] [PubMed]

16. Cornish, K.; Williams, J.L.; Kirk, M.; Teetor, V.H. Evaluation & control of potential sensitizing & irritating chemical components in natural rubber latex extracted from the industrial crop guayule. *Ind. Biotechnol.* **2009**, *5*, 245–252. [CrossRef]
17. Coffelt, T.A.; Nakayama, F.S.; Ray, D.T.; Cornish, K.; McMahan, C.M. Post-harvest storage effects on guayule latex, rubber, and resin contents and yields. *Ind. Crop. Prod.* **2009**, *29*, 326–335. [CrossRef]
18. Proksch, P.; Mohan Behl, H.; Rodriguez, E. Detection and quantification of guayulins A and B in Parthenium argentaturn (guayule) and F_1 hybrids by high-performance liquid chromatography. *J. Chromatogr. A* **1981**, *213*, 345–348. [CrossRef]
19. Ray, D.T.; Veatch-Blohm, M.E.; Teetor, V.H.; Walsh, B. Upper and Lower Heritability Estimates in Guayule Based on Mode of Reproduction. *J. Am. Soc. Hort. Sci.* **2007**, *132*, 213–218.
20. Lloyd, F.E. *Guayule (Parthenium Argentatum Gray), a Rubber Plant of the Chihuahuan Desert*; Carnegie Institution of Washington: Washington, DC, USA, 1911; pp. 57–58.
21. Norma Oficial Mexicana NOM-007-RECNAT-1997. *Procedimientos, Criterios y Especificaciones Para Realizar el Aprovechamiento, Transporte y Almacenamiento de Ramas, Hojas o Pencas, Flores, Frutos y Semillas*; Diario Oficial de la Federaciòn de los Estados Unidos Mexicanos: Ciudad de Mexico, Mexico, 1997.
22. Mocak, J.; Bond, A.M.; Mitchell, S.; Schollary, G. A statistical overview of standard (IUPAC and ACS) and new procedures for determining the limits of detection and quantification: Application to voltammetric and stripping techniques. *Pure Appl. Chem.* **1997**, *69*, 297–328. [CrossRef]
23. Horwitz, W. Evaluation of analytical methods used for regulation of foods and drugs. *Anal. Chem.* **1982**, *54*, 67A–76A. [CrossRef]
24. AOAC International. *AOAC Peer Verified Methods Programs, Manual on Policies and Procedures*; AOAC International: Arlington, VA, USA, 1998.
25. Salvucci, M.E.; Coffelt, T.A.; Cornish, K. Improved methods for extraction and quantification of resin and rubber from guayule. *Ind. Crop. Prod.* **2009**, *30*, 9–16. [CrossRef]
26. Yeom, H.; Suh, J.H.; Youm, J.R.; Han, S.B. Simultaneous Determination of Triterpenoid Saponins from *Pulsatilla koreana* using High Performance Liquid Chromatography Coupled with a Charged Aerosol Detector (HPLC-CAD). *Bull. Korean. Chem. Soc.* **2010**, *31*, 1159–1164. [CrossRef]

13

Food Sample Preparation for the Determination of Sulfonamides by High-Performance Liquid Chromatography: State-of-the-Art

Dimitrios Bitas [1], Abuzar Kabir [2], Marcello Locatelli [3] and Victoria Samanidou [1,*]

1 Laboratory of Analytical Chemistry, Department of Chemistry, Aristotle University of Thessaloniki, 54124 Thessaloniki, Greece; dimitriosbitas@gmail.com
2 International Forensic Research Institute, Department of Chemistry and Biochemistry, Florida International University, 11200 SW 8th St, Miami, FL 33199, USA; akabir@fiu.edu
3 Department of Pharmacy, University "G. d'Annunzio" of Chieti-Pescara, 66100 Chieti, Italy; m.locatelli@unich.it
* Correspondence: samanidu@chem.auth.gr

Abstract: Antibiotics are a common practice in veterinary medicine, mainly for therapeutic purposes. Sectors of application include livestock farming, aquacultures, and bee-keeping, where bacterial infections are frequent and can be economically damaging. However, antibiotics are usually administered in sub-therapeutic doses as prophylactic and growth promoting agents. Due to their excessive use, antibiotic residues can be present in foods of animal origin, which include meat, fish, milk, eggs, and honey, posing health risks to consumers. For this reason, authorities have set maximum residue limits (MRLs) of certain antibiotics in food matrices, while analytical methods for their determination have been developed. This work focuses on antibiotic extraction and determination, part of which was presented at the "1st Conference in Chemistry for Graduate, Postgraduate Students and PhD Candidates at the Aristotle University of Thessaloniki". Taking a step further, this paper is a review of the most recent sample preparation protocols applied for the extraction of sulfonamide antibiotics from food samples and their determination with high-performance liquid chromatography (HPLC), covering a five-year period.

Keywords: food; sample preparation techniques; sulfonamides; high-performance liquid chromatography; HPLC; ultra-high-performance liquid chromatography; UHPLC

---

1. Introduction

The scope of this state-of-the-art review is to cover the literature regarding the sample preparation protocols developed for the extraction of sulfonamides (SAs) from food samples followed by high-performance liquid chromatography (HPLC) determination, covering the last five years. The review is divided in two main sections. In this first main section, a theoretical background on veterinary drugs and antibiotic use, sulfonamides and their applications, the reported sample preparation techniques, as well as the chemical composition of the reported food matrices and official methods for the determination of antibiotics/sulfonamides in foodstuff samples are provided for the reader to have a prompt introduction to basic terminology. More details can be found in the cited review articles and book sections. In the second main section, the reported sample preparation protocols are provided in full detail for each food matrix.

1.1. Veterinary Drugs—Antibiotics

The European Union Council Directive 96/23/EC "on measures to monitor certain substances and residues thereof in live animals and animal products" divides all pharmacological substances used for veterinary purposes and their corresponding residues into two main groups, A and B. Group A includes substances with anabolic effect, such as antithyroid agents, steroids, and β-agonists, as well as unauthorized substances, such as chloramphenicol, chlorpromazine, metronidazole, and nitrofurans. Group B includes veterinary drugs, such as antibiotics, anthelmintics, anticoccidials and non-steroidal anti-inflammatory drugs, and environmental contaminants, such as organochlorine and organophosphorus compounds [1].

Antibiotics are used in veterinary medicine in order to improve the health and increase the productivity of the food-producing animals and at the same time reduce the morbidity and mortality rates among the livestock. Animals are administered with antibiotics not only for therapeutic purposes but also as prophylactic and metaphylactic measures [2]. Prophylaxis is a preventative measure, where animals are administered with sub-therapeutic doses and in some cases full doses of antibiotics through feed or water and is a common practice in massive livestock production. Metaphylaxis is a measure taken when a number of animals exhibits some of the disease symptoms, and all the animals are administered in order to prevent the disease from spreading. However, both practices are not always effective due to the fact that some antibiotic groups are active during bacterial cell proliferation [3]. The most common routes of administration are through water and feed medication, followed by injection [2]. Antibiotics can also be used as feed additives. Their intake favors the natural intestinal flora of the animals by inhibiting the harmful microorganisms, and as a result, nutrient absorption and assimilation are increased, thus providing a growth promoting effect [3]. Aminoglycosides, β-lactams (penicillins and cephalosporins), macrolides, phenicols, quinolones, SAs, and tetracyclines are among the most common antibiotic substances in veterinary medicine [4].

The presence of veterinary drug residues in animal-originated products relies on the physicochemical properties of each compound that affect the absorption, distribution, metabolism, and excretion of the drug from the animal body. For this reason, a withdrawal period, between the last drug administration and the slaughter or milk, egg, and honey collection, is established in order to ensure the existence of animal-originated products with low drug residues [4]. However, the extensive use of antibiotics besides therapeutic purposes has led to the presence of antibiotic residues in animal originated food products and the development of bacterial drug resistance [3]. The main reasons for antibiotic residues is the illegal use and uncontrolled administration of veterinary drugs to healthy animals, as well as not taking into consideration the withdrawal period, the lack of professional advice, extra label use, frequent drug administration, and increased dosing, as well as contaminated housing, water, and animal feed [4–6]. Antibiotic residues can cause adverse effects to consumers, such as acute allergic or toxic reactions, chronic toxic effects from prolonged exposure to antibiotic residues, and natural intestinal flora disruption [5].

Liver, kidney, and fat are animal tissues with high antibiotic residue concentration, while muscle tissue has relatively lower residues [5]. The European Union Commission Regulation No 37/2010 sets the legislative basis for the "pharmacologically active substances and their classification regarding maximum residue limits in foodstuffs of animal origin", by classifying the substances into the categories of allowed and prohibited. For the allowed substances, a maximum residue limit (MRL) is provided for each target tissue expressed in micrograms per kilogram (μg/kg) of fresh target tissue. Target tissues include muscle, fat, liver, and kidney, as well as milk, eggs, and, in a few cases, honey. The MRLs usually refer to a specific compound, a metabolite, or a compound mixture [7].

1.2. Sulfonamides

SAs constitute a wide-spectrum synthetic antibiotic category effective against a wide range of bacterial species, such as *Bacillus* spp., *Brucella* spp., *Streptococcus* spp., staphylococci gram-positive aerobes, and enterobacteriaceae gram-negative aerobes, as well as protozoa, parasites, and fungi.

SAs are derivatives of sulfanilamide and the various SA analogues, that result from the various **R** radicals of the $-SO_2NH$**R** group (Figure 1). Each SA analogue has different physicochemical and pharmacokinetic properties and as a result antibiotic effect. SAs are relatively insoluble with solubility increasing in alkaline pH, which is an important factor for the type of administration and disease treatment they are intended for [8].

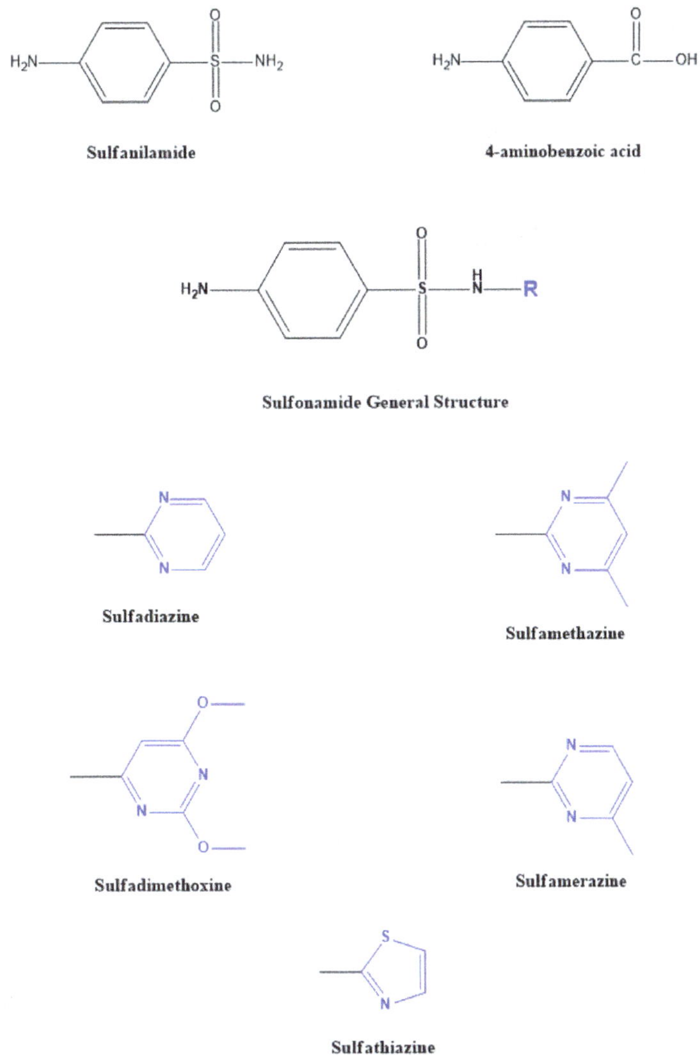

Figure 1. Diagrams of the chemical structure of sulfonamide, sulfanilamide, 4-aminobenzoic acid, and the chemical structure of the most common SA analogues in veterinary medicine.

The SA antibacterial effect relies on their ability to inhibit the conversion of folic acid to tetrahydrofolic acid by competitively antagonizing 4-aminobenzoic acid (Figure 1) for the dihydropteroate synthetase enzyme. Tetrahydrofolic acid is essential for the nucleic acid synthesis, thus SAs inhibit the bacterial DNA and RNA synthesis and subsequently the protein synthesis. Additionally, SAs reduce the bacterial cell permeability for glutamic acid that is essential for the folic acid synthesis. SAs are effective against bacterial species that synthesize the required 4-aminobenzoic acid, but ineffective in the presence of increased amounts of 4-aminobenzoic acid and against species that receive the required 4-aminobenzoic acid from other sources or have antibiotic resistance [6].

SA formulations usually consist of a SA analogue and a diaminopyrimidine, such as aditoprim, baquiloprim, ormetoprim, or trimethoprim, that have a synergistic interaction. The liver and kidney are animal tissues with the highest concentrations of SAs and their metabolites [4].

Sulfamethoxazole, sulfacetamide, and sulfasalazine are SA analogues commonly used in human medicine, while sulfadiazine (SDZ), sulfamethazine (SMZ), sulfadimethoxine (SDMX), sulfamerazine, and sulfathiazole are commonly used in veterinary medicine [6]. The European Union Commission Regulation No 37/2010 sets the MRL for all SA analogues to 100 µg/kg for muscle, fat, liver, and kidney from all food-producing species and bovine, ovine, and caprine milk, while their use is prohibited for animals that produce eggs for human consumption. In the case of more than one SA analogue, the sum of the SA residues should not exceed the provided MRL value (Table 1) [7]. Honeybees are also considered food-producing species and antibiotics are used in beekeeping in order to treat two of the most severe diseases for bees, the American/European foulbrood, and nosemosis. More specifically, the SA analogue sulfothiazole is used against the American foulbrood caused by *Paenibacillus larvae* in order to prevent the spread of the disease, suppress the symptoms, and inhibit the spore germination. SAs can also be used prophylactically against nosemosis. However, a beehive lacks in metabolic pathways for the elimination of antibiotic residues and a withdrawal period does not apply in the case of bees. For this reason, the European Union has not established MRLs for antibiotic residues in honey and only veterinary drugs with zero residues are authorized for beekeeping [9].

Table 1. MRLs for sulfonamide antibiotics, provided by the EU, Codex Alimentarius and FDA.

Sulfonamide	Animal Species	MRL	Target Tissue
Commission Regulation (EU) No 37/2010 [7]			
Sulfonamides	All food producing species	100 µg/kg	Muscle Fat Liver Kidney
	Bovine Ovine Caprine	100 µg/kg	Milk
Codex Alimentarius [10]			
Sulfadimidine	Cattle	25 µg/L	Milk
	Not specified	100 µg/kg	Muscle Liver Kidney Fat
CFR—Code of Federal Regulations—U.S. Food & Drug Administration (FDA) [11]			
Sulfabromomethazine sodium	Cattle Not specified	100 µg/kg 10 µg/L	Uncooked edible tissue Milk
Sodium sulfachloropyrazine monohydrate	Chicken	0	Uncooked edible tissue
Sulfachlorpyridazine	Calves Swine	100 µg/kg	Uncooked edible tissue
Sulfadimethoxine	Chickens Turkeys Cattle Ducks Salmonids Catfish Chukar partridges	100 µg/kg	Uncooked edible tissue
	Not specified	10 µg/L	Milk
Sulfaethoxypyridazine	Cattle Swine Not specified	100 µg/kg 0 0	Uncooked edible tissue Uncooked edible tissue Milk
Sulfamerazine	Trout	0	Uncooked edible tissue
Sulfamethazine	Chickens Turkeys Cattle Swine	100 µg/kg	Uncooked edible tissue
Sulfaquinoxaline	Chickens Turkeys Calves Cattle	100 µg/kg	Uncooked edible tissue

The most crucial problem arising from the uncontrolled use of SAs is the development and spreading of drug resistance, rather than the presence of SA residues in animal originated products itself. SA drug resistance derives from mutations in the dihydropteroate synthase gene that results in enzymes with structural alterations with decreased affinity towards the SAs. Drug resistance genes can be transferred between bacterial strains or genera during a horizontal gene transfer in plasmids, transposons, or integrons. For this reason, SA resistant strains are higher than tetracyclines and other antibiotics, while bacterial species with more than one gene for SA resistance have been observed. Furthermore, drug resistance in an antibiotic group can favor the development of cross-resistance [6].

1.3. Sample Preparation

Food samples are complex heterogenous matrices, where all analytes are distributed in a random manner. Food analysis involves sampling, homogenization, and sample preparation that increase the analytical accuracy and precision. Focusing on sample preparation it usually involves storage, particle size reduction, homogenization, weighting, dilution, filtration, extraction, clean-up, and derivatization. Proper sample preparation protocols result in matrix interference elimination and analyte preconcentration, thus affecting the selectivity, sensitivity, detection capability, and the overall performance of an analytical technique. The most time-consuming step in analytical method development is the optimization of the sample preparation protocol that includes analyte extraction and clean-up. Some of the most common sample preparation techniques used in food analysis are liquid-liquid extraction (LLE), solid-liquid extraction (SLE), solid-phase extraction (SPE), solid-phase microextraction (SPME), stir bar sorptive extraction (SBSE), supercritical fluid extraction (SFE), accelerated solvent extraction (ASE), ultrasonic, and Soxhlet extraction [12].

1.3.1. LLE

LLE is one of the most used sample preparation techniques, along with SPE, and probably the oldest. In LLE the analytes are extracted from an aqueous sample into a water immiscible solvent, according to relative solubility. Despite the wide use of LLE, there are many disadvantages, such as increased time and solvent requirements, analyte lose, sample contamination, and low sensitivity, that reduce LLE applications in modern analytical chemistry. Liquid-phase microextraction (LPME) is an LLE-based microextraction technique that has been developed in order to overcome LLE disadvantages. Extraction takes place between the aqueous sample phase (donor phase) and a water immiscible solvent extraction phase (acceptor phase). LPME can be divided into single-drop microextraction (SMDE), hollow fiber liquid-phase microextraction (HF-LPME), and dispersive liquid-liquid microextraction (DLLME). DLLME is a ternary extraction system that involves the aqueous sample, an extraction solvent, and a disperser solvent. Microliters of the organic extraction solvent and the dispersive solvent are injected into the sample solution and extraction solvent droplets are formed with the help of the dispersive solvent, resulting in a cloudy solution. After the extraction equilibrium between the sample and the extraction solvent is achieved, the cloudy solution is centrifuged and the lower organic phase is collected for analysis. The extraction solvent should be water immiscible, such as chloroform, carbon tetrachloride, and dichloromethane, while the dispersive solvent should be miscible in both aqueous solution and organic solvent, such as ethanol, methanol, acetonitrile, and acetone. Following the principles of green analytical chemistry, conventional extraction solvent can be replaced by ionic liquids that are liquid organic salts, combinations of organic cations with inorganic anions with unique physicochemical properties. Ionic liquids are characterized by low volatility and high density, thus forming more stable droplets and better phase separation than the conventional organic solvents [13].

1.3.2. SLE

The principals of SLE are similar to LLE, except that the analytes are extracted from solid samples. The solid sample is mixed with extraction solvent, the two phases interact, and the soluble sample components diffuse into the extraction phase. Various organic solvents can be used in SLE, however, SLE is usually laborious and has analogous disadvantages as LLE, such as increased solvent requirements, partial extraction, solvent impurities, and emulsion formation. In order to increase the extraction efficiency of the organic solvent, heating and pressure or ultrasounds can be applied during the extraction. In pressurized liquid extraction (PLE), also known as ASE, the solid sample and the extraction solvent are transferred into an extraction cell and the extraction takes place under high temperature (40–200 °C) and pressure (500–3000 psi) for 5–15 min. After the extraction is complete, the sample extract is collected and purged. The increased temperature and pressure applied in PLE result in reduced extraction time, enhanced analyte's solubility and mass transfer, better solvent penetration, and overall improved extraction yields. However, PLE requires expensive equipment and high temperature solvent reduces the extraction selectivity and applicability of PLE for the extraction of less thermal stable compounds. Ultrasound-assisted extraction (UAE) is a less extreme extraction approach in which extraction is assisted by the application of ultrasounds. Compared with other sample preparation techniques, UAE is relatively faster with reduced solvent requirements, and at the same time enables the extraction of analytes in room temperature. However, this technique lacks selectivity and enrichment capability when extraction of trace amounts is required and is usually combined with other clean-up techniques for improved extraction efficiency [14].

1.3.3. Salting-Out Extraction

The salting-out effect has been exploited in sample preparation techniques, such as Quick Easy Cheap Effective Rugged Safe (QuEChERS) and salting-out liquid-liquid extraction (SALLE), in order to improve analyte's extraction from aqueous samples. The salting-out effect is based on the decrease of solubility of water soluble organic analytes when salt concentration in the aqueous sample solution is increased. This effect favors the partition of the analytes into a water-miscible organic solvent and separation between the two phases. Various water-miscible solvents can be used, but acetonitrile is the most convenient water-miscible organic solvent for the application of the salting-out effect due to being chemically inert with organic analytes and the most common mobile phase component in liquid chromatography, and at the same time, acetonitrile has the ability to precipitate matrix proteins. Salting-out agents are usually inorganic and organic salts that provide cations (Mg^{2+}, Sr^{2+}, Ca^{2+}, Ba^{2+}, K^+, Na^+, NH_4^+, Li^+) and anions (SO_4^{2-}, CH_3COO^-, Cl^-, NO_3^-, Br^-, I^-, CNS^-), such as $MgSO_4$, NaCl, CH_3COONH_4, CH_3COONa, and $(NH_4)_2SO_4$, which promote the transfer of the hydrophilic compounds to the organic phase. These salts should be soluble in the aqueous sample but have negligible solubility in the organic phase [15]. In either QuEChERS and SALLE procedures, a water-miscible solvent and a salting-out agent are sequentially added to the sample and the mixture is shaken and centrifuged. The formed organic phase is then collected and can be either injected directly for analysis or further treated for water removal and analyte isolation [14]. Aqueous two-phase system (ATPS) extraction is another alternative based on the partitioning of the analytes between two phases and is used for the extraction of analytes from aqueous samples. The most common biphasic system in food analysis is polymer/salt, while ionic liquid/salt systems were also reported. Polyethylene glycol is usually the polymer of choice combined with phosphate, sulfate, or citrate salts. Parameters that affect the extraction include the molecular weight and concentration of the polymeric phase, the solution pH, and temperature. ATPS does not require the use of organic solvents in contrast with other sample preparation techniques, while sample deproteinization or defatting is not always required [16].

1.3.4. SPE

SPE is a widely used sample preparation technique that provides high enrichment factors and recoveries with reduced sample volume requirements and automation capability. SPE can be used for the daily laboratory routine and in many cases SPE has replaced LLE. In SPE the components of the sample solution interfere with a solid phase (sorbent material) and separation between the target analytes and the matrix interferences can be achieved. The sorbent is usually packed into SPE cartridges between two frits. SPE cartridges are commercially available for specific applications or can be manually prepared with the selected sorbents. In a typical SPE procedure, the SPE cartridges are conditioned/activated with a solvent or a solvent mixture, the sample solution is loaded into the cartridge, the loaded cartridge is washed to remove the retained interreferences, and the retained analytes are eluted with an appropriate solvent (Figure 2a). However, SPE applications are restricted by the type of the sorbent and the characteristics of the sample components, while the tightly packed SPE cartridges increase the extraction time and cause backpressure [13]. These problems can be eliminated with dispersive solid-phase extraction (DSPE) where the sorbent is dispersed in the sample solution and not packed into a cartridge. After the dispersion, the solution is shaken, and when the extraction is complete, the sorbent is collected by centrifugation or filtration and the analytes are eluted by repeating the previous step with the use of an appropriate solvent (Figure 2b) [17]. C18 and OASIS® HLB are two of the most widely applied commercially available materials for SPE. C18 is a nonpolar sorbent that consists of octadecylsilane bonded to silica particles and is suitable for the reversed-phase binding of hydrophobic analytes. OASIS® HLB is a water-wettable hydrophilic-lipophilic-balanced polymeric reversed-phase sorbent that consists of N-vinylpyrrolidone and divinylbenzene and can be used for the binding of acidic, basic, or neutral analytes [14]. Novel SPE sorbents include carbon nanotubes (CNTs) [13], molecularly imprinted polymers (MIPs) [18], and magnetic materials. Magnetic solid-phase extraction (MSPE) is a SPE-based technique that employs magnetic sorbent materials. The magnetic materials comprise of a magnetic metal oxide nanoparticle core, usually Fe_3O_4, coated with inorganic or organic materials, such as silica, alumina, chitosan, or polypyrrole, while the coating can be modified with functional groups for improved sorption capability. A MSPE application is similar to DSPE, with the difference that the sorbent can be collected by means of a magnet (Figure 2b) [19]. While solid-phase techniques can be used for aqueous samples or sample solutions, matrix solid-phase dispersion (MSPD) can be applied directly for the extraction of analytes from solid, semi-solid, and viscous samples. Typically, the sample is mechanically blended with solid support in order to achieve matrix disruption and the development of interactions between the analytes and the sorbent material. The mixture is then transferred and packed into a SPE cartridge and the analytes are eluted with an appropriate sorbent. Apart from the applicability on solid samples, MSDP provides a simple and selective approach for sample extraction and clean-up in a single step [14].

Other solid-phase extraction variations include SPME and SBSE. In both techniques, the sorbent material is coated on a substrate, and extraction/analyte desorption follow the same principles. SPME employs silica or stainless-steel fibers coated with the sorbent material that can be used for the extraction of analytes from gaseous, liquid, and solid samples. A typical SPME procedure involves the partitioning of the analytes between the sorbent and the sample matrix and analyte desorption directly into the analytical instrument. SPME can be coupled with HPLC by means of a six-port injector combined with a desorption chamber, and desorption can be performed with an organic solvent or the mobile phase in static or dynamic mode. SMPE fibers can be coated with various sorbent materials, thus SMPE applicability can be expanded for the extraction of a wide range of analytes and sample matrices [14]. SBSE employs magnetic stir bars coated with polydimethylsiloxane that is a polar polymeric material and develops hydrophilic interactions with the analytes, such as hydrogen bonds and van der Waals forces. In a typical SBSE procedure, a coated stir bar is introduced into the sample solution and the analytes are absorbed onto the coating by continuous stirring. After the extraction, the bar is collected, washed with deionized water, and dried, while analytes can be desorpted, either thermally by thermal or liquid desorption [13]. Both SPME and SBSE are less time-consuming and have reduced sample and solvent requirements in comparison with SPE.

1.3.5. FPSE

Fabric phase sorptive extraction (FPSE) is a recently introduced novel microextraction technique that employs reusable cellulose or polyester fabric substrates homogenously coated with sol-gel hybrid sorbents. In a typical FPSE procedure, the coated fabric, along with a magnetic stir bar, are introduced into a vial that contains the sample or sample solution, and analyte extraction is conducted under stirring. The coated fabric is collected and placed for 4–10 min into a second vial that contains the eluting solvent (Figure 2c). The eluate can be centrifuged prior to analysis. The coated fabric is washed with an appropriate organic solvent and rinsed with deionized water between extractions. Reported FPSE coatings include polydimethylsiloxane, poly(ethyleneglycol), C18, and graphene. The FPSE protocol is simple and fast, with reduced solvent requirements, while the coated fabric can be introduced directly to the liquid samples and is compatible with a wide range of organic solvents. FPSE substrates are characterized by increased sorbent loading and improved adsorption capacity in comparison with SPME, providing high analyte preconcentration [20].

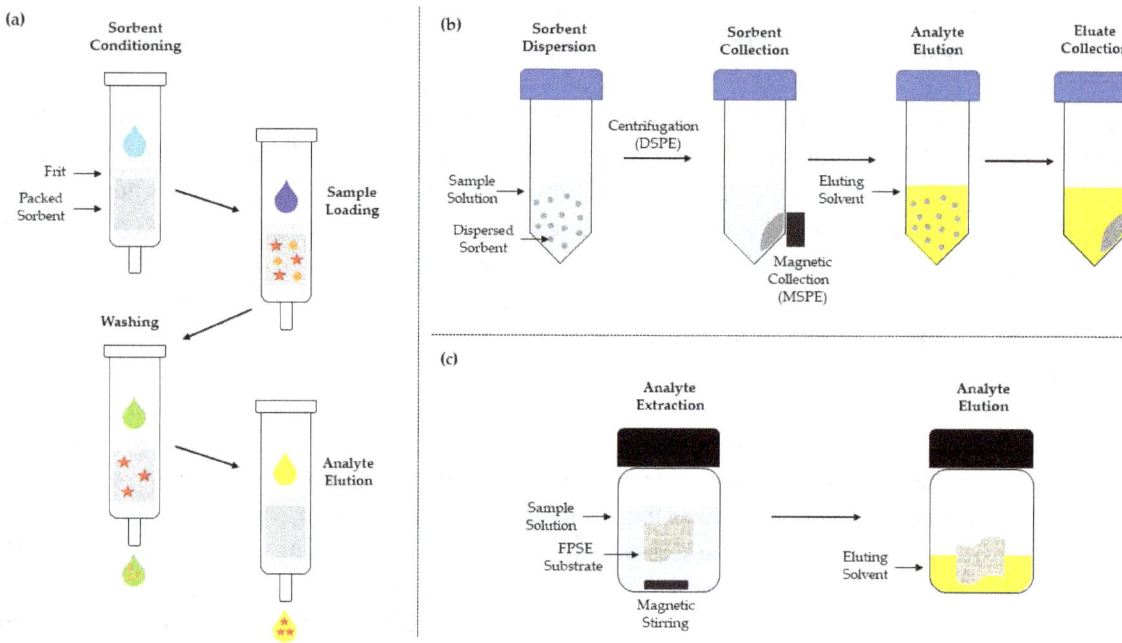

Figure 2. Schematic representation of the basic steps in (**a**) SPE, (**b**) DSPE and MSPE, and (**c**) FPSE.

1.4. Food Composition

Meat contains 72–75% water, 19% proteins, 2.5–5% lipids, 1% vitamins and carbohydrates, and 1% ash. Lipids can vary between 1% and 15%. The main edible animal tissue is the muscle, while other edible parts include organs, such as liver and kidneys, fat, and blood. Edible animal livestock species include bovine and porcine species and poultry. Animal lipids are mostly deposited under the skin (subcutaneous fat), between the muscles (intermuscular fat), and around organs, such as kidneys and heart, but varies between animal species [21]. The chemical composition of fish varies among species, with 50–60% of fish weight being muscle. Fish muscle contains 52–82% water, 16–21% proteins (or 10–25% for farmed species), 0.5–2.3% lipids, 1.2–1.5% ash, and 0.5% carbohydrate content. In comparison with red meat, fish lipid content is lower and ranges between 0.2% and 30%, while lipids are deposited in the liver, muscle, perivisceral, and subcutaneous tissues. Fatty fish species deposit fat all over the muscle tissue that is colored grey, yellow, or pink, such as salmon, and more than 50% of the skin consists of lipids (in other species it ranged between 0.2% and 3.9%) [22]. Milk is a heterogenous mixture that consists of water, emulsified fat, caseins and whey proteins, lactose, minerals, and vitamins. The gross cow milk composition is 86.3% water, 4.9% fat, 3.4% proteins, 4.1% lactose, and 0.7% ash. Buffalo, sheep, and goat milk can also be consumed by humans. Fat is the most important milk component, both organoleptically and commercially, and ranges from below 3% to more than 6% [23]. Eggs consist of white and yolk for 60% and around 30–33% of the total egg weight, respectively. The egg white or albumen is a protein solution that contains over 40 different proteins, with ovalbumin constituting the 54% of the total white proteins. The egg yolk contains lipoproteins, with low-density lipoproteins constituting the 65% of the total yolk proteins. Egg fat is mainly present in egg yolk as triacylglycerol and phospholipids and comprises the 9–10% of the total egg weight. Other egg components include minerals and vitamins that are also located in the egg yolk [24].

1.5. *Official Methods of Analysis*

Official methods for the determination of antibiotics/sulfonamides in food samples are available by regulating agencies. The website of the FDA provides a "Laboratory Methods—Drug & Chemical Residues Methods" section [25] that includes analytical methods for the determination of multiple phenicol residues in honey samples with electrospray liquid chromatography—mass spectrometry (LC-MS) and chloramphenicol residues in crustacean species (shrimp, crab, crawfish) samples with liquid chromatography—tandem mass spectrometry (LC-MS/MS) [26], as well as the determination of fluoroquinolone residues in milk samples with LC-MS/MS [27]. Two multi-class and multi-residue LC-MS/MS methods for the determination of drug residues in milk and aquaculture samples (fish, shrimp) are provided in the "Field Science and Laboratories—Laboratory Information Bulletins" section of the FDA website [28]. A high-performance liquid chromatography—ultraviolet detection (HPLC-UV) method for the determination of sulfamethazine in milk samples is also provided by the FDA [29]. The Association of Official Analytical Chemists (AOAC) provides the "AOAC Official Method 993.32" for the determination of eight sulfonamide residues in raw bovine milk samples with liquid chromatography—ultraviolet detection (LC-UV) [30].

2. Extraction of Sulfonamides from Food Samples

Meat, milk, eggs, and honey are animal-originated products with increased demand. In this section the detailed sample preparation protocols are provided for each reported paper.

2.1. *Animal Tissue Samples*

2.1.1. SLE

The same team developed two SLE protocols for multi-class antibiotic extraction, including 15 SAs, from bovine tissue [31] and fish tissue samples [32] followed by ultra-high-performance liquid chromatography—mass spectrometry (UHPLC-MS/MS). In the first protocol, spiked bovine tissue (2 g) was mixed with acetonitrile (ACN) (10 mL) and ethylenediaminetetraacetic acid (EDTA) (0.1 M, 1 mL) and the mixture was shaken for 20 min and centrifuged at 3100 *g* for 15 min. The supernatant was collected, mixed with *n*-hexane (3 mL), vortexed for 30 s, centrifuged at 3100× *g* for 15 min, and the *n*-hexane layer was discarded. The extract was evaporated under nitrogen stream at 40 °C to 0.5 mL final volume, dissolved in mobile phase (400 μL), and passed through a 0.45 μm filter, prior to analysis. The authors emphasized the simplicity, as well as the reduced cost and time requirements of the developed sample preparation protocol. The addition of EDTA, the sample defatting with *n*-hexane, and the evaporation to 0.5 mL were employed in order to increase the recoveries of all analytes [31]. The same protocol was applied for the fish tissue samples, except for the *n*-hexane treatment step.

Spiked fish tissue (2 g) was mixed with ACN (10 mL) and EDTA (0.1 M, 1 mL), and the mixture was shaken for 20 min and centrifuged at 3100× *g* for 15 min. The supernatant was collected and evaporated under nitrogen stream at 40 °C and the residue was dissolved in mobile phase (400 μL) [32]. In both cases, a SPE sample clean-up step was omitted due to increased selectivity over specific antibiotics that prohibits multi-class extraction, while the sample preparation protocol cost and time requirement were decreased and the number of samples analyzed in a daily routine was increased. Another SLE protocol was reported for the extraction of SDZ, SMZ, SIX, SDMX, and sulfaquinoxaline (SQX) from shrimp tissue samples. Spiked tissue (0.5 g) was mixed with methanol (MeOH)-ACN (50:50, *v*/*v*; 1 mL) and the mixture was vortexed, sonicated for 15 min, and centrifuged at 3500 rpm for 10 min. The extraction step was repeated twice with MeOH-ACN (50:50, *v*/*v*; 1 mL) and twice with 0.1% CH_3COOH-MeOH (60:40, *v*/*v*; 0.5 mL) and the supernatants were collected between extractions. All collected supernatants were combined, evaporated to dryness under nitrogen stream, and the residue was dissolved in MeOH (500 μL) and passed through a 0.20 μm syringe filter. Analysis was carried out by high-performance liquid chromatography—diode array detection (HPLC-DAD). SPE and MSPD were also tested but higher recoveries were achieved with the developed sample preparation protocol [33].

Two ASE protocols were reported for the extraction of 15 SAs and metabolites from baby food samples [34], and multi-class antibiotic extraction, including 9 SAs, from fish tissue samples [35]. In the first protocol, spiked baby food sample (5 g) was transferred into an ASE extraction cell and mixed with 1% CH_3COOH in MeOH-ACN (4:1, *v*/*v*). Extraction was carried out at 70 °C and 1500 psi with 5 min preheating time and 5 min static time, while flush volume was 60% and purge time 60 s. Extraction was repeated for 3 cycles. The extracts (3 × 18 mL) were diluted with extraction solvent (3 × 20 mL), placed in the freezer at −18 °C overnight and centrifuged at 4 °C and 4000 rpm for 5 min. The supernatant was collected and passed through a 0.45 μm filter and analyzed with ultra-high-performance liquid chromatography—orbitrap high-resolution mass spectrometry (UHPLC-Orbitrap-MS). The developed sample preparation protocol provided higher recoveries in comparison with an official AOAC QuEChERS extraction protocol [34]. In the second report, the fish tissue samples were purified with C18 resin inside the ASE extraction cell. The sample extracts were evaporated to dryness, dissolved in mobile phase, and centrifuged at −4 °C and 10,000 rpm prior to high-performance liquid chromatography—mass spectrometry (HPLC-MS/MS) analysis. However, information could be collected only from the abstract and the further details are not available because the rest of the paper is written in Chinese [35].

Two extraction protocols, PLE and USE, were reported for the extraction of 16 SAs and metabolites from chicken, sheep, fish, and horse tissue samples. In both cases, analysis was carried out by high-performance liquid chromatography—quadrupole linear ion trap—mass spectrometry (HPLC-QqLIT-MS/MS). For the PLE protocol, spiked tissue (5 g) was defatted with hexane, mixed with diatomaceous earth, and transferred inside a PLE cell. The extraction solvent was 0.2% CH_3COOH in ACN, the preheating period was 8 min, and extraction was achieved at 90 °C and 1500 psi for 7 min. The total flush volume was 80%, 60 s of purging under nitrogen stream was applied and the extraction step was repeated three times. The collected extracts were placed in the freezer at −18 °C for 1 h and centrifuged at 3500 rpm for 10 min in order to remove the tissue proteins. The supernatant was evaporated to dryness under nitrogen stream at 40 °C and the residue was dissolved in H_2O-ACN (85:15, *v*/*v*; 1 mL). For the UAE protocol, spiked tissue (5 g) was mixed with ACN (10 mL) and the mixture was vortexed for 10 s and placed inside an ultrasonic bath for 60 min. Then, the mixture was placed in the freezer at −18 °C for 1 h and centrifuged at 3500 rpm for 10 min, in order to remove the tissue proteins. The supernatant was collected and evaporated to dryness under nitrogen stream at 40 °C. The residue was dissolved in H_2O-ACN (85:15, *v*/*v*; 2 mL), hexane (2 mL) was added, and the mixture was vortexed for 5 s and centrifuged at 3500 rpm for 10 min. The lower phase was used for analysis. Although both developed protocols could efficiently extract the SAs from the tissue samples, UAE is simpler and less solvent and time-consuming [36].

2.1.2. Salting-Out Extraction

Three QuEChERS extraction protocols were reported in the literature. In the first report, a QuEChERS extraction protocol was developed for the extraction of 22 SAs and metabolites from bovine, chicken, pork, and sheep tissue samples. Spiked tissue (5 g) was mixed with H_2O (5 mL), vortexed for 1 min, and 1% CH_3COOH in ACN (10 mL) was added to the mixture. Sodium hydrogen citrate sesquihydrate (0.5 g), sodium citrate (1 g), $MgSO_4$ (4 g), and NaCl (1 g) were added in sequence and the mixture was shaken and vortexed for 1 min between each addition. The mixture was centrifuged at 3500 rpm for 5 min, the supernatant (6 mL) was collected, mixed with primary secondary amine (PSA) (150 mg) and anhydrous $MgSO_4$ (900 mg), and the mixture was shaken, vortexed, and centrifuged as described. The supernatant (4 mL) was collected and evaporated to dryness under nitrogen stream at 35 °C. The residue was dissolved in 0.01% formic acid in 5% MeOH aqueous solution (500 μL), passed through a 0.22 μm nylon filter, and centrifuged at 13,500 rpm for 10 min prior to high-performance liquid chromatography—high-resolution mass spectrometry (HPLC-HRMS) analysis [37]. In the second report, a modified QuEChERS extraction protocol was developed for the extraction of eight SAs from chicken muscle and egg sample. The extraction protocol employed a commercially available sorbent material Z-Sep$^+$, consisting of C18 and zirconia both bonded to silica particles, for the chicken samples and PSA for the egg samples. Spiked tissue (5 g) was mixed with H_2O (5 mL) and 1% CH_3COOH in ACN (10 mL) and the mixture was shaken for 10 min. $MgSO_4$ (4 g) and CH_3COONa (1 g) were added and the mixture was shaken for 1 min, vortexed for 2 min, and centrifuged at 5000 rpm for 5 min. The supernatant (3 mL) was collected and mixed with C18-zirconia sorbent (300 mg) and the mixture was shaken for 30 s and vortexed for 1 min. The supernatant (2 mL) was collected, evaporated under nitrogen stream, and the residue was dissolved in MeOH-H_2O (50:50, *v*/*v*; 1 mL), vortexed for 1 min and filtered prior to high-performance liquid chromatography—fluorescence detection (HPLC-FLD) analysis. A similar protocol was used for the egg sample (5 g), where it was mixed only with 1% CH_3COOH in ACN (10 mL) and PSA was used instead of the C18-zirconia sorbent [38]. Lastly, a fully automated on-line QuEChERS extraction protocol was reported for the extraction of 27 SAs and metabolites from salmon tissue samples. Spiked tissue (1 g) was mixed with 1% CH_3COOH in ACN-H_2O (84:16, *v*/*v*; 5 mL) and the mixture was vortexed at 3200 rpm for 1 min. $MgSO_4$ (1 g) and CH_3COONa (0.1 g) were added and sample extract (2 mL) was aspirated three times, followed by 30 s equilibrium time and centrifuged at 2264× *g* for 5 min. The supernatant (1 mL) was collected and mixed with Z-Sep$^+$ (45 mg), PSA (32 mg), and Na_2SO_4 (0.25 g) and the mixture was vortexed for 1 min and centrifuged at 2264× *g* for 5 min. The supernatant (200 μL) was collected and mixed with MeOH (300 μL) and ammonium formate solution (0.008 M, 500 μL) prior to ultra-high-performance liquid chromatography—electrospray ionization—quadrupole Orbitrap high-resolution mass spectrometry (UHPLC-ESI-Q-Orbitrap-MS) analysis [39].

Two SALLE protocols were also reported. In the first protocol SALLE was reported in combination with magnetic separation for the extraction of eight SAs from fish tissue samples. Spiked tissue (2 g) was mixed with 0.1% formic acid in ACN (5 mL) and the mixture was vortexed for 1 min. NaCl (0.5 g), $MgSO_4$ (2 g) and Fe_3O_4 (100 mg) were added and the mixture was vortexed for 1 min. The magnetic particles were collected by means of a magnet and the supernatant (1 mL) was collected and evaporated to dryness under nitrogen stream at 45 °C. The residue was dissolved in 0.1% formic acid aqueous solution (1 mL), mixed with *n*-hexane (1 mL), and the mixture was centrifuged at 8000 rpm for 3 min. The aqueous phase was collected and passed through a 0.22 μm syringe filter and analyzed with HPLC-MS/MS. The authors emphasized the good recoveries and the reduced time requirements of the developed sample preparation protocol [40]. In the second report, the samples were treated with ethyl acetate and concentrated under vacuum. The analytes were extracted with HCl solution (2 M), the extract was defatted with *n*-hexane, filtered, and mixed with MeOH-ACN-CH_3COONa (5:5:20, *v*/*v*/*v*). Analysis was carried out by HPLC-FLC. However, information could be collected only from the abstract and the further details are not available because the rest of the paper is written in Chinese [41].

2.1.3. SPE

Two SPE protocols were reported for multi-class antibiotic extraction, including 16 SAs, from bovine liver samples [42] and the extraction of SAs from chicken and pork tissue samples [43]. The first extraction protocol employed Oasis HLB cartridges (3 mL, 200 mg) (Waters, Milford, MA, USA). Spiked liver tissue (2 g) was mixed with ACN (10 mL) and EDTA (0.1 M) and the mixture was shaken for 10 min, sonicated for 20 min, and centrifuged at 4000× *g* for 10 min. The supernatant was collected and evaporated under nitrogen stream to 1 mL final volume. H_2O (5 mL) was added and the mixture was vortexed for 15 s and loaded into a SPE cartridge preconditioned with ACN (10 mL) and H_2O (10 mL). The loaded cartridge was washed with H_2O (5 mL) and dried under reduced pressure for 5 min. The analytes were eluted with ACN (10 mL) and the eluate was evaporated under nitrogen stream to 0.5 mL final volume. The reduced eluate was dissolved in mobile phase (400 μL), *n*-hexane (2 mL) was added, and the mixture was vortexed for 30 s and centrifuged at 4000× *g* for 10 min. The mixture was passed through a 0.45 μm filter and analyzed with UHPLC-MS/MS. The developed sample preparation protocol included a SPE clean-up step in order to reduce the liver tissue interferences and utilized SPE cartridges with wide range selectivity in order to achieve multi-class antibiotic extraction [42]. The second extraction protocol employed multi-walled carbon nanotubes (MWCNTs) as the sorbent material. The samples were treated with ACN and the sample extract was dissolved in Na_2HPO_4 buffer (pH 5.5–6.0) and loaded into the SPE cartridges. The cartridges were washed with acetone-hexane (5:95, *v*/*v*) and the analytes were eluted with acetone-dichloromethane (1:1, *v*/*v*). Analysis was carried out by HPLC-UV. However, information could be collected only from the abstract and the further details are not available because the rest of the paper is written in Chinese [43].

Two variations of SPE, a MSPE and a DSPE protocol, were reported for the extraction of SDZ, sulfathiazole, sulfamerazine, SMZ, and SMP from chicken, pork, and shrimp tissue samples [44] and for multi-class antibiotic extraction, including 21 SAs, from animal tissue samples [45], respectively. The MSPE protocol employed a magnetic Fe_3O_4@JUC-48 nanocomposite as the sorbent material. Spiked tissue (2 g) was mixed with ACN (20 mL) and the mixture was vortexed for 5 min, sonicated for 30 min, and kept overnight. The mixture was centrifuged at 10,000 rpm and the supernatant was collected and stored at 4 °C. Sample extract (8 mL) was mixed with Fe_3O_4@JUC-48 (25 mg) and the mixture was vortexed for 8 min. The magnetic sorbent was collected by means of an external magnet and the supernatant was discarded. The analytes were eluted with MeOH-CH_3COOH (95:5, *v*/*v*; 0.8 mL) and sonication for 10 min. The sorbent was separated by means of an external magnet and the eluate was collected, passed through a 0.22 μm nylon filter, and analyzed with HPLC-DAD. Sorbent reusability was studied by applying the sorbent material in several extraction cycles. Between the extractions, the sorbent was washed with MeOH-CH_3COOH (95:5, *v*/*v*; 3 × 1 mL) and MeOH (3 × 1 mL) and dried at 60 °C. The sorbent could be reused for seven extraction cycles without significant adsorption capacity reduction. The authors emphasized the increased sensitivity and the higher recovery values achieved, as well as the reduced extraction time and sorbent requirements of the developed sample preparation protocol [44]. In the DSPE protocol, the samples were treated with Na_2EDTA (0.1 M) and 1% CH_3COOH in ACN, followed by a DSPE clean-up step. Analysis was carried out by HPLC-MS/MS. However, information could be collected only from the abstract and the further details are not available because the rest of the paper is written in Chinese [45].

Other reported approaches included a combined MSPD-homogeneous ionic liquid microextraction (HILME) protocol for the extraction of seven SAs from bovine, chicken, and pork tissue samples [46], an on-line SPME protocol for the extraction of five antimicrobials, including sulfametoxydiazine, sulfamethoxazole, and SQX from chicken and pork tissue and egg samples [47] and a SBSE protocol for the extraction of ten SAs from chicken and pork tissue samples [48]. For the MSPD-HILME protocol, the ionic liquid employed acted both as elution solvent in MSPD and extraction solvent in HILME. Spiked tissue (0.2 g), silica gel (1 g), and 1-butyl-3-methylimidazolium tetrafluoroborate ($[C_4MIM][BF_4]$) (200 μL) were mixed with a pestle and mortar and the mixture was transferred and placed inside a glass column between absorbent cotton layers. Pure H_2O was passed through the packed mixture and the eluate (3 mL) was mixed with NaCl (0.45 g) and ammonium hexafluorophosphate (2.4 M, 1 mL). The cloudy mixture was centrifuged at 5 °C and 10,000 rpm for 5 min, the supernatant was discarded and the remaining ionic liquid phase was diluted with ACN (300 μL), passed through a 0.22 μm polytetrafluoroethylene membrane filter and analyzed with HPLC-DAD. The authors emphasized the simplicity and the higher extraction recoveries achieved, as well as the reduced reagent requirements of the developed sample preparation protocol [46]. In the on-line SPME protocol, spiked tissue or egg (5 g) was mixed with Na_2SO_4 (5 g) and ACN (10 mL) and the mixture was sonicated for 10 min and centrifuged at 8000 rpm for 5 min. This step was repeated twice and the three sample extracts were combined and evaporated to dryness. The residue was dissolved in ACN-toluene-*n*-hexane (1:4:45, *v*/*v*/*v*; 25 mL) and used for the on-line SPME. The extraction protocol employed molecularly imprinted monolithic capillary columns with SQX as the template molecule. Activated capillaries were treated with propyltrimethoxysilane and filled with the polymerization mixture that consisted of methacrylic acid (functional monomer), ethylene glycol dimethacrylate (cross-linker), *N*,*N*-dimethylformamide, isooctane, and paraxylene (polymerization and porogenic solvents), while polymerization occurred at 60 °C for 70 h. A prepared monolithic capillary column preplaced the sample loop in an on-line HPLC-UV system and extraction consisted of three steps. In the first step, the sample extract passed through the capillary at a flow rate of 0.15 mL/min so that the analytes came in contact with the imprinted polymeric phase and extracted. In the second step, nitrogen was passed through the capillary so that the residual sample solution was completely removed. Finally, the analytes were eluted from the capillary with 400 μL of mobile phase at a flow rate of 0.15 mL/min. The authors emphasized the increased selectivity and sensitivity, as well as the simplicity and the environmental friendliness of the developed sample preparation protocol [47]. Finally, the SBSE protocol employed poly(vinylphthalimide-*co*-*N*,*N*-methylenebisacrylamide) monolith coated stir bars. Spiked tissue (0.5 g) was mixed with ACN (5 mL), the mixture was sonicated for 15 min and centrifuged at 3000 rpm for 5 min and the supernatant was collected and passed through a 0.45 μm filter. This step was repeated and both supernatants were diluted with Milli-Q H_2O (100 mL). For SBSE, sample solution pH was adjusted (pH 4.0) and NaCl (5%, *w*/*v*) was added, while extraction time was 120 min and liquid desorption time was 60 min. Analysis was carried out by HPLC-MS/MS [48].

All reported literature for the extraction of SAs from animal tissue samples is summarized in Table 2, including recoveries, limit of detection (LOD), limit of quantification (LOQ), or/and decision limit (CC_α), decision capability (CC_β) values.

Table 2. Extraction of SAs from animal tissue samples.

Food Sample	Analytes	Sample Preparation	Analytical Technique/Run-Time	LOD-LOQ	Recovery (%)	Ref.
bovine tissue	41 antibiotics (15 SAs)	SLE	UHPLC-MS/MS 12 min	CC_α (µg/kg): 104–132 (SAs) CC_β (µg/kg): 108–164 (SAs)	91–109 (SAs)	[31]
fish tissue	41 antibiotics (15 SAs)	SLE	UHPLC-MS/MS 12 min	CC_α (µg/kg): 110.6–126.9 (SAs) CC_β (µg/kg): 121.2–153.7 (SAs)	92–111 (SAs)	[32]
shrimp tissue	SDZ, SMZ, SIX, SDMX and SQX	SLE	HPLC-DAD 40 min	LOD (µg/kg): 15 LOQ (µg/kg): 50	88.6–108.4	[33]
baby foods (combinations of powdered milk, cereal, vegetable, honey and meat)	15 SAs and metabolites	ASE	UHPLC-Orbitrap-MS 10 min	LOD (µg/kg): 0.03–0.17 LOQ (µg/kg): 0.10–0.55	75.5–96.6	[34]
fish tissue	19 antibiotics (9 SAs)	ASE	HPLC-MS/MS N/A	LOD (ng/g): 0.003–0.6	55.2–113 (all analytes)	[35]
chicken, sheep, fish and horse tissue	16 SAs and metabolites	PLE, USE	HPLC-QqLIT-MS/MS 11 min	CC_α (µg/kg): 111.2–161.4 (PLE), 119.3-142.7 (USE) CC_β (µg/kg): 122.4–222.8 (PLE), 138.6-185.5 (USE)	N/A	[36]
bovine, chicken, pork and sheep tissue	22 SAs and metabolites	QuEChERS extraction	HPLC-HRMS 17 min	LOD (µg/kg): 3–26 LOQ (µg/kg): 11–88 CC_α (µg/kg): 101–111 CC_β (µg/kg): 102–122	88–107 (beef muscle)	[37]
chicken tissue and egg	8 SAs	QuEChERS extraction	HPLC-FLD 23 min	LOD (µg/kg): 5.8–19.9 (chicken muscle), 4.1–25.6 (egg) LOQ (µg/kg): 19.2–66.2 (chicken muscle), 13.6–85.4 (egg)	66.9–86.8 (chicken muscle), 65.9–88.1 (egg)	[38]
salmon tissue	27 SAs and metabolites	on-line QuEChERS extraction	UHPLC-ESI-Q-Orbitrap-MS 7 min	CC_α (µg/kg): 0.04–1.34 CC_β (µg/kg): 0.07–2.33	83–109	[39]
fish tissue	8 SAs	SALLE and magnetic separation	HPLC-MS/MS 8 min	LOD (µg/kg): 2.5–10 LOQ (µg/kg): 5–25	74.87–104.74	[40]
shrimp tissue	14 SAs	SALLE	HPLC-FLD N/A	LOD (µg/kg): 1.0–5.0	77.8–103.6	[41]
bovine tissue	39 antibiotics (16 SAs)	SPE	UHPLC-MS/MS 12 min	CC_α (µg/kg): 65–125 (SAs) CC_β (µg/kg): 81–150 (SAs)	85–110 (SAs)	[42]
chicken and pork tissue	SAs	SPE	HPLC-UV N/A	LOD (mg/L): 0.003 LOQ (mg/L): 0.01	> 70	[43]
chicken, pork and shrimp tissue	SDZ, sulfathiazole, sulfamerazine, SMZ and SMP	MSPE	HPLC-DAD 16 min	LOD (ng/g): 1.73–5.23 LOQ (ng/g): 3.97–15.89	81.4–101.3 (chicken), 76.1–102.6 (pork), 79.2–102.5 (shrimp)	[44]
animal tissue	63 veterinary drugs (21 SAs)	DSPE	HPLC-MS/MS N/A	LOD (µg/kg): 0.1–3.0 (all analytes) LOQ (µg/kg): 0.5–10.0 (all analytes)	62.2–112.0 (all analytes)	[45]
bovine, chicken and pork tissue	7 SAs	MSPD-HILME	HPLC-DAD 30 min	LOD (µg/kg): 4.3–13.4 LOQ (µg/kg): 14.2–44.8	85.4–95.1 (kidney), 104.5–118.0 (liver), 85.5–112.3 (muscle)	[46]
chicken and pork tissue and egg	5 antimicrobials (sulfametoxydiazine, sulfamethoxazole and SQX)	on-line SPME	HPLC-UV 20 min	LOD (µg/L): 0.10–0.14 (SAs) LOQ (µg/L): 0.39–0.47 (SAs)	84.1–99.6 (SAs in chicken), 87.0–108.2 (SAs in pork), 80.1–101.4 (SAs in egg)	[47]
chicken and pork tissue	10 SAs	SBSE	HPLC-MS/MS N/A	LOD (µg/kg): 0.0012–0.0061 (pork), 0.0021–0.0146 (chicken) LOQ (µg/kg): 0.0040–0.0203 (pork), 0.0066–0.0487 (chicken)	62.4–109.9 (pork), 55.2–109.1 (chicken)	[48]

2.2. *Milk Samples*

2.2.1. LLE

An LLE protocol was reported for the extraction of nine SAs from milk samples. Spiked milk (100 μL) was mixed with acidified dichloromethane (800 μL) and the mixture was sonicated for 10 min and centrifuged at 93,000× *g* for 10 min. The organic phase was collected and the step was repeated. The combined organic phases were evaporated to dryness under nitrogen stream at 40 °C and the residue was dissolved in acidified MeOH (100 μL) and filtered prior to HPLC-MS/MS analysis. The authors emphasized the reduced sample and reagent requirements of the developed sample preparation protocol, while a SPE clean-up step was omitted in order to simplify and reduce the cost of the protocol [49]. DLLME and modified QuEChERS extraction, were reported for the extraction of nine SAs from milk samples. In both cases off-line derivatization was conducted with fluorescamine (50 μL) and sonication for 15 min prior to HPLC-FLD analysis. For the DLLME protocol, spiked milk (30 mL) was mixed with 20% trichloroacetic acid aqueous solution (15 mL) and the mixture was vortexed for 10 s and centrifuged at 6000 rpm for 5 min. The supernatant was passed through a 0.2 μm filter and pH was adjusted to pH 4.0–4.5. Treated supernatant (5 mL) was mixed with chloroform (1000 μL, extraction solvent) and ACN (1900 μL, dispersive solvent) and the mixture was shaken until a cloudy solution was formed and centrifuged at 6000 rpm for 5 min. The chloroform phase was collected with a syringe, evaporated under nitrogen stream and the residue was dissolved in Tris buffer (pH 7.0, 1.5 mL) and passed through a 0.2 μm nylon filter. For the QuEChERS protocol, spiked milk (2 mL) was mixed with H_2O (8 mL), vortexed for 10 s, 5% CH_3COOH in ACN (10 mL) was added, and the mixture was shaken for 30 s. The QuEChERS (C18, $MgSO_4$ and PSA) were added and the mixture was shaken for 1 min and centrifuged at 4000 rpm for 5 min. The supernatant (1.5 mL) was collected, evaporated under nitrogen stream and the residue was dissolved in Tris buffer (pH 7.0, 1.5 mL) and passed through a 0.2 μm nylon filter. Both developed sample preparation protocols were simple, fast, and environmentally friendly, providing good recoveries. When compared, DLLME gave lower LOD values and higher recoveries, while QuEChERS extraction was more reproducible with higher throughput [50].

2.2.2. Salting-Out Extraction

SALLE, combined with SPE [51] and a miniaturized SALLE protocol [52], were reported for multi-veterinary drug extraction, including 26 SAs, from milk samples and sulfonamide from tea beverage, water, milk, honey, plasma, blood, and urine samples, respectively. The first protocol employed Oasis HLB Plus cartridges (225 mg) (Waters, Milford, MA, USA). For the SALLE step, spiked milk (5 g) was mixed with oxalic acid-EDTA buffer (pH 3.0, 5 mL) and ACN (10 mL) and the mixture was shaken for 30 s and centrifuged at 3000 rpm for 5 min. The supernatant was collected, $(NH_4)_2SO_4$ (1 g) was added and the mixture was shaken for 2 min, left for 2 min and centrifuged at 3000 rpm for 3 min. Three layers resulted after centrifuging, and the upper ACN and the lower aqueous layer were used for the SPE clean-up step, while the middle layer was the milk fat. The SPE cartridges were preconditioned with MeOH (10 mL), H_2O (10 mL), and oxalic acid-EDTA buffer (pH 3.0, 2 mL), loaded with the aqueous layer, washed with buffer (2 mL) and the analytes were eluted with the ACN layer (10 mL) and MeOH (5 mL). The eluate (3 ML) was evaporated to 0.1–0.2 mL final volume under nitrogen stream at 50 °C for 20 min and the residue was dissolved in CH_3COONH_4 solution (0.1 M, 1 mL), vortexed for 30 s, and passed through a 0.45 μm filter device prior to UHPLC-ESI-Q-Orbitrap-MS analysis [51]. The second protocol employed two 1 mL syringes coupled via their tips that contained the sample solution and extraction solution, respectively. For the preparation of milk samples, spiked milk (1 mL) was mixed with ACN-MeOH-H_2O (40:20:20, *v*/*v*/*v*; 1 mL), the mixture was centrifuged at 3000 rpm for 10 min. For the preparation of honey samples, spiked honey was diluted with H_2O at a concentration of 0.1 g/mL and the mixture was homogenized and centrifuged at 4000 rpm for 20 min. Milk or honey supernatant (0.5 mL) was retracted with the sample syringe A, NaCl (250 mg/mL) was added, and the mixture was vortexed for 20 s and adjusted to pH 7.0 with NaOH solution (0.1 M). ACN (250 mL) was retracted with the extraction solution syringe B and both syringes were coupled and held vertically. The extraction was conducted by injecting the extraction solvent from syringe B into syringe A, forming a cloudy solution and the content was pumped back to syringe B. This step was repeated five times and the mixture was left in syringe B vertically for 2 min. After phase formation, the upper phase was collected and injected for HPLC-UV analysis. The authors emphasized the reduced organic solvent and sample requirements, as well as the simplicity and the improved extraction efficiency of the developed sample preparation protocol [52].

Two ATPS extraction protocols were reported in the literature for the extraction of SAs from milk. A modified ATPS protocol was reported for the extraction of SDZ and SMZ from milk, egg, and water samples. The extraction protocol employed polyoxyethylene lauryl ether and $Na_2C_4H_4O_6$ in order to form the polymer-organic salt extraction system. For the preparation of milk and egg samples, spiked milk (50 mL) or homogenized spiked egg (50 mL) was mixed with 10% trichloroacetic acid solution (20 mL) and H_2O to 100 mL final volume and the mixture was shaken and centrifuged at 4000 rpm for 30 min. The supernatant was collected and passed through a 0.45 μm filter in order to remove the proteins. The filtrate was added to a mixture of polyoxyethylene lauryl ether (0.027 g/mL) and $Na_2C_4H_4O_6$ (0.180 g/mL) and filled with H_2O to 10 mL final volume. The mixture was placed in a heated water bath under continuous stirring for 20 min and after the phase formation, the upper phase was collected and analyzed with HPLC-UV [53]. An ionic liquid ATPS extraction protocol was reported for the extraction of six SAs from milk samples. Spiked milk (5 mL) diluted with pure H_2O (2.5 mL) was mixed with 10% $HClO_4$ aqueous solution (500 μL) and the mixture was shaken for 2 min and centrifuged at 10,000 rpm for 10 min. The supernatant was collected and passed through a 0.45 μm filter. Butyl-3-methylimidazolium tetrafluoroborate ($[C_4MIM][BF_4]$) (300 μL) and $C_6H_5Na_3O_7{\cdot}2H_2O$ (3 g) were added and the mixture was shaken and centrifuged at 10,000 rpm for 10 min. The upper IL phase was collected, diluted with ACN at 1:1 ratio, sonicated, and passed through a 0.22 μm filter. Analysis was carried out by HPLC-UV. The authors emphasized the reduced organic solvent consumption of the developed sample preparation protocol in comparison with classic LLE, as well as the combination of sample preconcentration and clean-up in one step [54].

2.2.3. SPE

Two SPE protocols were reported for multi-veterinary drug extraction, including 18 SAs [55], and the extraction of six SAs [56] from milk samples. The first extraction protocol employed Oasis MCX cartridges (3 mL, 60 mg) (Waters, Milford, MA, USA). Spiked milk (2 g) was mixed with 1% CH_3COOH in ACN (5 mL) and the mixture was vortexed for 30 s and centrifuged at 5000 rpm for 12 min. The supernatant was evaporated to dryness under nitrogen stream at 40 °C and the residue was dissolved in HCl solution (0.1 M, 3 mL) and loaded to a SPE cartridge, conditioned with MeOH (3 mL) and HCl solution (0.1 M, 3 mL). The loaded cartridge was washed with HCl solution (0.1 M, 3 mL) and MeOH (3 mL) and the analytes were eluted with 10% ammonia in ACN (4 mL). The eluate was evaporated to dryness under nitrogen stream at 40 °C and the residue was dissolved in 0.1% CH_3COOH in CH_3COONH_4-MeOH (90:10, *v*/*v*; 1 mL) and passed through a 0.22 μm filter prior to ultra-high-performance liquid chromatography—electrospray ionization—mass spectrometry (UHPLC-ESI-MS/MS) analysis [55]. The second extraction protocol employed multi-template MIPs prepared by sol-gel synthesis. Spiked milk (1 g), deproteinized with ACN, was loaded into MISPE cartridges packed with the prepared MIPs (30 mg) and left for 15 min to equilibrate. The loaded cartridge was washed with MeOH (2 mL) and the analytes were eluted with 1% CH_3COOH-MeOH-ACN (50:10:40, *v*/*v*/*v*; 2 mL) at a flow rate of 1 mL/min and the eluate was evaporated to dryness under nitrogen stream. Analysis was carried out by HPLC-DAD. MISPE cartridge conditioning was omitted as an unnecessary step, thus reducing the time of the developed sample preparation protocol [56].

Three MSPE protocols were reported for the extraction of nine SAs [57], SMP, SMZ, sulfamethoxazole and sulfachloropyridazine [58], and five SAs [59] from milk samples. The first protocol employed silica-based magnetic sorbent material. Magnetic sorbent (0.1 g), conditioned with MeOH (5 mL) and sonication for 5 min and washed with deionized H_2O (2 × 10 mL), was added to spiked milk (10 mL). The mixture was sonicated for 15 min, the magnetic sorbent was collected by means of a magnet and the supernatant was discarded. The sorbent was washed with acetate buffer (pH 4.0, 3 × 5 mL) and the analytes were with 10^{-3} M NaOH in MeOH (5 mL) for 5 min. The eluate was collected, evaporated to dryness under nitrogen stream and the residue was dissolved in 1% formic acid aqueous solution (500 μL) and passed through a 0.2 μm nylon filter prior to HPLC-DAD analysis. The authors emphasized the simplicity, higher recovery values, and the lower organic solvent requirements of the developed sample preparation protocol in comparison with classic SPE [57]. The second protocol employed a magnetic hyper cross-linked polystyrene composite as the sorbent material. Spiked milk (25 mL) was agitated for 15 min and the magnetic sorbent was added (20 mg). The sample pH was adjusted to pH 5.0, extraction/stirring time was 10 min and analytes were eluted with ACN (2 × 1 mL) and sonication for 5 min. Analysis was carried out by high-performance liquid chromatography—amperometric detection (HPLC-AD). The proposed magnetic composite combined large surface area, high adsorption, and magnetic separation, while a small amount could be used for the extraction from large volumes of untreated milk. The authors emphasized the good recovery's simplicity of the sample preparation protocol, as well as the reduced time and solvent requirements in comparison with classic sample preparation techniques [58]. The last MSPE protocol employed a magnetic graphene-based composite ($CoFe_2O_4$-graphene) as the sorbent material. Spiked milk (1.5 mL) was mixed with 15% $HClO_4$ aqueous solution (0.2 mL) and the mixture was vortexed for 30 s and centrifuged at 14,000 rpm for 5 min. The supernatant was collected, diluted with deionized H_2O (100 mL), and the pH was adjusted to pH 4.0 and mixed with the magnetic sorbent (15 mg). The magnetic sorbent was previously conditioned with MeOH (5 mL), H_2O (5 mL) and sonication for 5 min. The mixture was shaken for 20 min and vortexed for 2 min. The magnetic particles were collected by means of a magnet and the supernatant was discarded. The analytes were eluted with 5% CH_3COOH in MeOH (0.5 mL) and the eluate was passed thought a 0.22 μm filter prior to HPLC-UV analysis. The authors emphasized the simplicity, low LOD values, and improved recoveries of the developed sample preparation protocol in comparison with other reported methods. The magnetic sorbent displayed increased extraction efficiency for the analytes, while it could be reused after washing with ACN and ultrapure H_2O [59].

2.2.4. Other Extraction Techniques

Interesting approaches reported for the extraction of SAs from milk samples include FPSE, graphene-modified melamine sponge (GMeS) microextraction and miniaturized syringe assisted extraction (mini-SAE). The FPSE protocol was reported for the extraction of SMZ, SIX, and SDMX from milk samples. The extraction protocol employed highly polar sol-gel poly(ethylene glycol) coated cotton cellulose fabric segments as the sorbent material. FPSE media incubated in MeOH-ACN (50:50, *v*/*v*; 2 mL) for 5 min and rinsed with H_2O (2 mL), was introduced into spiked whole milk (1 g) for 30 min. The fabric-milk system was stirred by means of a magnetic stirrer for 30 min and the extraction media was transferred and incubated in MeOH (250 μL) for 8 min and ACN (250 μL) for 5 min. The extract was filtered prior to HPLC-UV analysis. The coated fabric was washed with ACN-MeOH (50:50, *v*/*v*; 2 mL) for 5 min, left to dry for 5–10 min, and kept in an air-tight container between extractions and could be reused for up to 30 times. The developed sample preparation protocol eliminated deproteinization and evaporation/reconstitution, thus reducing the extraction time and the errors resulting from these steps. The proposed fabric sorbent could be applied directly into the untreated milk sample offering a simpler extraction protocol and higher recoveries. Furthermore, the fabric sorbent displayed high chemical and solvent stability that allows the use of the suitable extraction solvent for sample analysis with multiple chromatographic techniques [60]. The GMeS microextraction protocol was reported for the extraction of eight SAs from milk, egg, and water. The extraction protocol employed novel graphene-modified melamine sponges as the sorbent material. For the preparation of milk samples, spiked milk (15 mL) was defatted with centrifuging at 4000 rpm and 4 °C for 10 min and deproteinized with 15% trichloroacetic acid solution (1 mL for every 10 mL defatted sample solution), vortexing for 1 min, and centrifuging at 4000 rpm for 5 min. The supernatant was collected and mixed with NaCl (6% *w*/*v*) and centrifuged. The supernatant was used for the GMeS extraction. For the preparation of egg samples, homogenized spiked egg (1 g) was mixed with double distilled H_2O (8.7 mL), 15% trichloroacetic acid solution (0.3 mL), and NaCl (6% *w*/*v*) and stirred for 1 min prior to extraction. GMeS cubes conditioned with MeOH and distilled H_2O were placed inside the sample solution (10 mL) and stirred at 600 rpm for 30 min. The cube was collected, placed into a syringe cartridge, rinsed with H_2O, and squeezed in order to remove the absorbed sample. The analytes were eluted with 5% ammonia in ACN (2 × 1 mL), the eluate was evaporated to dryness under nitrogen stream and the residue was dissolved in H_2O-ACN (70:30, *v*/*v*; 100 μL) and sonicated for 1 min prior to HPLC-DAD analysis. The authors emphasized the simple and rapid preparation and easier handling, as well as the improved recoveries and environmental friendliness of developed GMeS material in comparison with other sorbents found in the literature [61]. The mini-SAE protocol was reported for the extraction of SDZ and sulfamonomethoxine from milk samples. The extraction protocol employed a poly (hydroxyethyl methacrylate) polymer as the sorbent material. Spiked milk (50 g) was mixed with 16% lead acetate aqueous solution (3 mL) and the mixture was stirred for 5 min and centrifuged at 4000 rpm for 4 min. The supernatant was collected, 16% lead acetate aqueous solution (2 mL) was added, and the mixture was centrifuged at 4000 rpm for 4 min. The supernatant (1 mL) was loaded into the mini-SAE device packed with the polymer sorbent (50 mg) and conditioned with MeOH (2 mL) and H_2O (2 mL). The device was washed with H_2O (1 mL) and the analytes were eluted with 5% CH_3COOH in MeOH (3 mL). The eluate was evaporated to dryness under nitrogen stream and the residue was dissolved in phosphate buffer (pH 4.0, 1 mL) and derivatized with fluorescamine prior to HPLC-FLD analysis [62].

2.3. Milk Product Samples

Three protocols were also reported for the extraction of SAs from milk products (baby formula, cheese, and butter). Firstly, a SLE protocol was reported for multi-veterinary drug extraction, including 24 SAs, from baby formula samples. Spiked formula (1 g) was mixed with EDTA aqueous solution (0.05 M, 10 mL), the mixture was vortexed, 0.1% formic acid in ACN (10 mL) was added, and the mixture was vortexed, shaken for 15 min and centrifuged at 2000 rcf for 10 min. The supernatant (2 mL) was collected, evaporated to dryness under nitrogen stream at 40 °C, and the residue was dissolved in H_2O-ACN (75:25, *v*/*v*; 1 mL). Analysis was carried out by UHPLC-MS/MS. A clean-up step was omitted due to low recoveries for β-lactams, tetracyclines and dyes, and variable recoveries for the other analytes [63]. A QuEChERS extraction protocol was reported for multi-veterinary drug extraction, including sulfachloropyridazine, sulfadimidine, SDMX, and SQX, from cheese samples. Spiked cheese (10 g) was mixed with 1% CH_3COOH in ACN (10 mL) and Na_2EDTA solution (0.1 M, 10 mL) and the mixture was vortexed for 1 min. $MgSO_4$ (4 g) and CH_3COONa (1 g) were added and the mixture was stirred for 1 min and centrifuged at 4500 *g* for 5 min. The supernatant (2 mL) was collected and passed through a 0.2 μm nylon filter and the filtrate (1 mL) was diluted with 0.01% formic acid solution-MeOH (50:50, *v*/*v*; 1 mL) prior to analysis with UHPLC-MS/MS. The developed sample preparation protocol enabled the extraction of multiple veterinary drugs, in comparison with other similar protocols, that were used for the extraction of a single antibiotic or antibiotic group [64]. An ionic liquid—magnetic bar—liquid-phase microextraction (IL-MB-LPME) was reported for the extraction of eight SAs from butter samples. The extraction protocol employed magnetic hollow fibers as the extraction configuration and 1-octyl-3-methylimidazolium hexafluorophosphate ([C_8MIM][PF_6]) immobilized on the hollow fiber micropores as the extraction solvent. Spiked butter (30 g) was added into a vessel containing eight magnetic fibers and Na_2SO_4 aqueous solution (3 M, 6 mL) and the vessel was sealed and placed into a water bath at 45 °C and magnetic stirring at 500 rpm for 25 min. The magnetic fibers were collected by means of a magnet, washed with hexane (1 mL), and the analytes were eluted with MeOH (200 μL) and sonication for 3 min. The eluate was collected, mixed with Na_2SO_4 (100 mg) and the supernatant was passed through a 0.22 μm filter. Analysis was carried out by HPLC-UV. The Na_2SO_4 aqueous solution acted both as the extraction solvent for the extraction of the SAs from the butter and as the sample solution for the IL magnetic hollow fibers, thus the developed sample preparation protocol combined analyte extraction, clean-up, and preconcentration in one step [65].

2.4. *Egg Samples*

Two SPE protocols were reported for the extraction of SAs from egg samples. The first SPE protocol was reported for the extraction of 13 SAs from egg samples. The extraction protocol employed Strata-X SCX cartridges (Phenomenex, Macclesfield, UK). Homogenized spiked egg (10 g) was adjusted to pH 5.0–6.0 with 10% CH_3COOH solution (900 μL) for 15 min. Chloroform-acetone (50:50, *v*/*v*; 30 mL) was added and the mixture was shaken for 10 min and sonicated for 20 min. NaCl (3 g) and Na_2SO_4 (3 g) were added and the mixture was centrifuged at 2209 *g* and 10 °C for 10 min and placed at −70 °C for 30 min. The organic phase was collected (25 mL), mixed with CH_3COOH (2.5 mL), and loaded into the SPE cartridge conditioned with *n*-hexane (2 × 3 mL) and acetone-5% CH_3COOH in chloroform (50:50, *v*/*v*; 2 × 3 mL). The loaded cartridge was washed with H_2O (5 mL) and MeOH (5 mL) and the analytes were eluted with MeOH-ammonia solution (97.5:2.5, *v*/*v*; 13 mL). The eluate was evaporated to dryness under nitrogen stream at 45 °C and the residue was dissolved in mobile phase (0.5 mL), mixed with *n*-hexane (0.5 mL), and centrifuged at 2209× *g* and 20 °C for 10 min. The lower phase was collected, centrifuged for another 10 min, and the supernatant was analyzed with HPLC-DAD [66]. The second SPE protocol was reported for the extraction of SDZ from egg samples. The extraction protocol employed SDZ imprinted microspheres (100 mg) packed into a glass syringe conditioned with MeOH (5 mL) and Milli-Q H_2O (5 mL). Spiked egg yolk (2 g) and white (2 g) were respectively mixed with MeOH (10 mL) and the mixture was sonicated for 10 min. The supernatants were collected and the step was repeated for both egg yolk and white. All collected supernatants were combined, concentrated to 10 mL final volume, and loaded to the MISPE cartridge. The loaded cartridge was washed with MeOH-H_2O (30:70, *v*/*v*; 1 mL) and the analyte was eluted with MeOH (1 mL). The eluate was analyzed directly with HPLC-DAD. The authors emphasized the clean-up efficiency and analyte preconcentration achieved by the developed extraction protocol, while sample defatting was not necessary [67]. Additionally, QuEChERS extraction [38], on-line SPME [47], ATPS [53], and GMeS microextraction [61] were reported for the extraction of SAs from egg samples and sample preparation protocols are given in detail in Sections 2.1 and 2.2.

2.5. *Honey Samples*

An on-line SPE protocol was reported for the extraction of 15 SAs from honey samples. The extraction was achieved in a Zorbax Extended C-18 (12 mm × 4.6 mm; 5 μm) column (Agilent, Santa Clata, CA, USA). Spiked honey (1 g) was hydrolyzed with HCl solution (3 M, 800 μL) for 90 min and neutralized with citrate buffer (pH 3.5, 200 μL) and NaOH solution (10 M, 240 μL). The SAs were derivatized with 0.2% fluorescamine (200 μL) and the sample solution was passed through a 0.22 μm filter and injected to the on-line SPE-HPLC-FLD system. The authors emphasized the reduced organic solvent and sample requirements, as well as the simplicity, environmental friendliness, and increased selectivity and sensitivity of the automated SPE protocol [68]. Additionally, a miniaturized SALLE [52] protocol was reported for the extraction of sulfonamide from honey samples and details are provided in Section 2.2.

All reported literature for the extraction of SAs from milk and milk product, egg, and honey samples is summarized in Table 3.

Table 3. Extraction of SAs from milk and milk product, egg, and honey samples.

Food Sample	Analytes	Sample Preparation	Analytical Technique/Run-Time	LOD-LOQ	Recovery (%)	Ref.
milk	9 SAs	LLE	HPLC-MS/MS 30 min	LOQ (μg/kg): 12.5–45 CC_α (μg/kg): 106–122 CC_β (μg/kg): 112–145	89–105	[49]
milk	9 SAs	DLLME, QuEChERS extraction	HPLC-FLD 15 min	LOD (μg/L): 0.60–1.21 (DLLME), 1.15–2.73 (QuEChERS) LOQ (μg/L): 2.01–4.02 (DLLME), 3.85–9.09 (QuEChERS)	90.8–104.7 (DLLME), 83.6–104.8 (QuEChERS)	[50]
milk	105 veterinary drugs (26 SAs)	SALLE and SPE	UHPLC-ESI-Q-Orbitrap-MS 14 min	LOQ (μg/kg): 1.0 (all analytes)	71–120 (all analytes)	[51]
tea beverage, water, milk, honey, plasma, blood and urine	sulfonamide	miniaturized SALLE	HPLC-UV 7 min	LOD (ng/mL): 0.3 LOQ (ng/mL): 1.0	96.66 (tea beverage), 76.67 (milk), 43.33 (honey)	[52]
milk, egg and water	SDZ and SMZ	ATPS extraction	HPLC-UV N/A	LOD (pg/mL): 2.92–3.64 (milk), 2.90–3.49 (egg) LOQ (pg/mL): 9.73–12.15 (milk), 9.66–11.62 (egg)	97.14–99.52 (milk), 96.90–99.30 (egg)	[53]
milk	6 SAs	ATPS extraction	HPLC-UV 30 min	LOD (ng/mL): 2.04–2.84 LOQ (ng/mL): 6.73–9.37	72.32–108.96	[54]
milk	38 veterinary drugs (18 SAs)	SPE	UHPLC-ESI-MS/MS 13 min	CC_α (μg/kg): 109–114 (SAs) CC_β (μg/kg): 116–123 (SAs)	87–119 (all analytes)	[55]
milk	6 SAs	SPE	HPLC-DAD 15.3 min	LOD (μg/kg): 1.9–13.3 LOQ (μg/kg): 5.6–42.2 CC_α (μg/kg): 101.9–113.5 CC_β (μg/kg): 114.4–135.4	N/A	[56]
milk	9 SAs	MSPE	HPLC-DAD 35 min	LOD (μg/L): 7–14 CC_α (μg/kg): 108.86–117.16 CC_β (μg/kg): 117.73–134.32	81.88–114.98	[57]
milk and water	SMP, SMZ, sulfamethoxazole and sulfachloropyridazine	MSPE	HPLC-AD N/A	LOD (ng/mL): 2.0–2.5 (milk) LOQ (ng/mL): 6.0–7.5 (milk)	92–105 (milk)	[58]
milk	5 SAs	MSPE	HPLC-UV 8 min	LOD (μg/L): 1.16–1.59 LOQ (μg/L): 3.52–4.81	62.0–104.3	[59]
milk	SMZ, SIX and SDMX	FPSE	HPLC-UV 6.5 min	CC_α (μg/kg): 114.4–116.5 CC_β (μg/kg): 104.1–118.5	93–107	[60]
milk, egg and water	8 SAs	GMeS microextraction	HPLC-DAD 30 min	LOQ (μg/kg): 0.31–0.91 (milk), 0.96–1.32 (egg)	90–105 (milk), 90–108 (egg)	[61]

Table 3. *Cont.*

Food Sample	Analytes	Sample Preparation	Analytical Technique/Run-Time	LOD-LOQ	Recovery (%)	Ref.
milk	SDZ and sulfamonomethoxine	mini-SAE	HPLC-FLD N/A	LOD (ng/g): 0.19–0.26 LOQ (ng/g): 0.67–0.87	85.6–100.3	[62]
baby formula	150 veterinary drugs (24 SAs)	SLE	UHPLC-MS/MS 17.5 min	LOQ (ng/g): 1–10 (all analytes)	50–120 (all analytes)	[63]
cheese	17 veterinary drugs (sulfachloropyridazine, sulfadimidine, SDMX and SQX)	QuEChERS extraction	UHPLC-MS/MS 8.5 min	LOD (μg/kg): 0.2–1.7 (SAs) LOQ (μg/kg): 0.7–5.5 (SAs) CC_{α} (μg/kg): 3.4–5.8 (SAs) CC_{β} (μg/kg): 5.7–10.2 (SAs)	72.5–106.3 (SAs)	[64]
butter	8 SAs	IL-MB-LPME	HPLC-UV 30 min	LOD (μg/kg): 1.20–2.17 LOQ (μg/kg): 4.00–7.25	73.25–103.85	[65]
egg	13 SAs	SPE	HPLC-DAD 45 min	LOD (μg/kg): 0.30–1.29 LOQ (μg/kg): 0.92–3.92 CC_{α} (μg/kg): 11.3–18.5 CC_{β} (μg/kg): 13.2–27.3	45.2–87.5	[66]
egg	SDZ	SPE	HPLC-DAD N/A	LOD (μg/L): 0.06 (egg yolk), 0.05 (egg white) LOQ (μg/L): 0.20 (egg yolk), 0.17 (egg white)	78.22–86.10	[67]
honey	15 SAs	on-line SPE	HPLC-FLD 30 min	LOD (ng/g): 0.1–1.0	76–108	[68]

3. Conclusions

The most recent literature regarding the extraction of SAs from food samples was successfully reported. In the case of animal tissue samples, SLE [31–33], ASE [34–36], QuEChERS extraction [37,38], SALLE [40,41], and SPE [42] were the most reported methodologies for the extraction of SAs, along with other antibiotic or veterinary drugs, followed by reports of USE [36], DSPE [45], and SBSE [48]. Reports of on-line techniques include a fully automated on-line QuEChERS extraction protocol [39] and an on-line SPME protocol that utilized molecularly imprinted monolithic capillary columns [47]. Reported novel solid-phase sorbent materials include MWCNTs utilized in a SPE protocol [43], and Fe_3O_4@JUC-48 nanocomposite utilized in a MSPE protocol [44]. An ionic liquid application was also reported in a MSPD-HILME protocol [46]. SPE [55], MSPE [57], and ATPS extraction [53,54] were the most reported approaches for the extraction of SAs from milk samples. Other approaches include LLE [49], DLLME [50], as well as a modified QuEChERS extraction protocol that employed C18, $MgSO_4$ and PSA [50], SALLE combined with SPE [51], and a miniaturized SALLE protocol [52]. Interesting approaches include FPSE [60], GMeS microextraction [61], and mini-SAE [62]. Reported novel materials include a magnetic hyper cross-linked polystyrene composite [58] and a magnetic graphene-based composite [59] utilized in MSPE protocols and multi-template MIPs prepared by sol-gel synthesis utilized in a SPE protocol [56]. Furthermore, SLE [63], QuEChERS extraction [64], and IL-MB-LPME [65] were reported for the extraction of SAs from milk products. In the case of egg samples, SPE [66,67] was the main reported technique, while QuEChERS extraction [38], on-line SPME [47], ATPS [53], and GMeS microextraction [61] protocols included eggs, along with other food matrices. Finally, for honey samples only an on-line SPE [68] and a miniaturized SALLE [52] protocol were reported in the recent literature.

Author Contributions: The authors have equally contributed to the manuscript.

Funding: This research received no external funding.

Conflicts of Interest: The authors declare no conflict of interest.

References

1. The Council of The European Union. Council Directive 96/23/EC. *Off. J.* **1996**, *125*, 10–32.
2. PAGE, S.W.; GAUTIER, P. Use of antimicrobial agents in livestock. *Rev. Sci. Tech. OIE* **2012**, *31*, 145–188. [CrossRef]
3. Moreno, L.; Lanusse, C. Specific Veterinary Drug Residues of Concern in Meat Production. In *New Aspects of Meat Quality*; Elsevier: New York, NY, USA, 2017; pp. 605–627, ISBN 9780081005934.
4. Croubels, S.; Daeseleire, E. *Veterinary Drug Residues in Foods*; Woodhead Publishing Limited: Sawston, UK, 2012; ISBN 9780857090584.
5. Moreno, L.; Lanusse, C. Veterinary Drug Residues in Meat-Related Edible Tissues. In *New Aspects of Meat Quality*; Elsevier: New York, NY, USA, 2017; pp. 581–603, ISBN 9780081005934.
6. Baran, W.; Adamek, E.; Ziemiańska, J.; Sobczak, A. Effects of the presence of sulfonamides in the environment and their influence on human health. *J. Hazard. Mater.* **2011**, *196*, 1–15. [CrossRef] [PubMed]
7. EUR-Lex. Available online: http://eur-lex.europa.eu/legal-content/EN (accessed on 14 January 2017).
8. Prescott, J.F. Sulfonamides, Diaminopyrimidines, and Their Combinations. In *Antimicrobial Therapy in Veterinary Medicine*; John Wiley & Sons, Inc: Hoboken, NJ, USA, 2013; pp. 279–294.
9. Reybroeck, W.; Daeseleire, E.; De Brabander, H.F.; Herman, L. Antimicrobials in beekeeping. *Vet. Microbiol.* **2012**, *158*, 1–11. [CrossRef] [PubMed]
10. Maximum Residue Limits, Codex Alimentarius. Available online: http://www.fao.org/fao-who-codexalimentarius/codex-texts/maximum-residue-limits/pt/ (accessed on 15 May 2018).
11. U.S. Food & Drug Administration CFR, Code of Federal Regulations Title 21. Available online: https://www.accessdata.fda.gov/scripts/cdrh/cfdocs/cfcfr/cfrsearch.cfm?cfrpart=556 (accessed on 15 May 2018).
12. Otles, S.; Ozyurt, V.H. Sampling and Sample Preparation. In *Handbook of Food Chemistry*; Springer: Berlin/Heidelberg, Germnay, 2015; pp. 151–164.

13. Saha, S.; Singh, A.K.; Keshari, A.K.; Raj, V.; Rai, A.; Maity, S. Modern Extraction Techniques for Drugs and Medicinal Agents. In *Ingredients Extraction by Physicochemical Methods in Food*; Elsevier: New York, NY, USA, 2018; pp. 65–106, ISBN 9780128115213.
14. Pérez-Rodríguez, M.; Pellerano, R.G.; Pezza, L.; Pezza, H.R. An overview of the main foodstuff sample preparation technologies for tetracycline residue determination. *Talanta* **2018**, *182*, 1–21. [CrossRef] [PubMed]
15. Tang, Y.Q.; Weng, N. Salting-out assisted liquid-liquid extraction for bioanalysis. *Bioanalysis* **2013**, *5*, 1583–1598. [CrossRef] [PubMed]
16. Iqbal, M.; Tao, Y.; Xie, S.; Zhu, Y.; Chen, D.; Wang, X.; Huang, L.; Peng, D.; Sattar, A.; Shabbir, M.A.B.; et al. Aqueous two-phase system (ATPS): An overview and advances in its applications. *Biol. Proc. Online* **2016**, *18*, 18. [CrossRef] [PubMed]
17. Maciel, E.V.S.; de Toffoli, A.L.; Lanças, F.M. Recent trends in sorption-based sample preparation and liquid chromatography techniques for food analysis. *Electrophoresis* **2018**, 1–48. [CrossRef] [PubMed]
18. Bitas, D.; Samanidou, V. Molecularly imprinted polymers as extracting media for the chromatographic determination of antibiotics in milk. *Molecules* **2018**, *23*, 316. [CrossRef] [PubMed]
19. Vasconcelos, I.; Fernandes, C. Magnetic solid phase extraction for determination of drugs in biological matrices. *TrAC Trends Anal. Chem.* **2017**, *89*, 41–52. [CrossRef]
20. Kazantzi, V.; Anthemidis, A. Fabric Sol–gel Phase Sorptive Extraction Technique: A Review. *Separations* **2017**, *4*, 20. [CrossRef]
21. Cobos, Á.; Díaz, O. Chemical Composition of Meat and Meat Products. In *Handbook of Food Chemistry*; Springer: Berlin/Heidelberg, Germnay, 2015; pp. 471–510.
22. Coppes Petricorena, Z. Chemical Composition of Fish and Fishery Products. In *Handbook of Food Chemistry*; Springer: Berlin/Heidelberg, Germnay, 2015; pp. 403–435.
23. Mehta, B.M. Chemical Composition of Milk and Milk Products. In *Handbook of Food Chemistry*; Springer: Berlin/Heidelberg, Germnay, 2015; pp. 511–553.
24. Sunwoo, H.H.; Gujral, N. Chemical Composition of Eggs and Egg Products. In *Handbook of Food Chemistry*; Springer: Berlin/Heidelberg, Germnay, 2015; pp. 331–363.
25. U.S. Food and Drug Administration. Laboratory Methods, Drug & Chemical Residues Methods. Available online: https://www.fda.gov/Food/FoodScienceResearch/LaboratoryMethods/ucm2006950.htm (accessed on 17 May 2018).
26. U.S. Food and Drug Administration. Laboratory Methods, Analytical Methods for Residues of Chloramphenicol & Related Compounds in Foods. Available online: https://www.fda.gov/Food/FoodScienceResearch/LaboratoryMethods/ucm113126.htm (accessed on 17 May 2018).
27. FDA's Center for Food Safety and Applied Nutrition. Laboratory Methods, Preparation and LC/MS/MS Analysis of Honey for Fluoroquinolone Residues. Available online: https://www.fda.gov/Food/FoodScienceResearch/LaboratoryMethods/ucm071495.htm (accessed on 17 May 2018).
28. U.S. Food and Drug Administration. Field Science and Laboratories, Laboratory Information Bulletins. Available online: https://www.fda.gov/ScienceResearch/FieldScience/ucm231463.htm (accessed on 17 May 2018).
29. FDA's Center for Food Safety and Applied Nutrition. Laboratory Methods, Bacteriological Analytical Manual (BAM). Available online: https://www.fda.gov/Food/FoodScienceResearch/LaboratoryMethods/ucm2006949.htm (accessed on 17 May 2018).
30. Horwitz, W. *Official Methods of Analysis of AOAC International*, 17th ed.; AOAC International: Gaithersburg, MD, USA, 2000.
31. Freitas, A.; Barbosa, J.; Ramos, F. Multi-residue and multi-class method for the determination of antibiotics in bovine muscle by ultra-high-performance liquid chromatography tandem mass spectrometry. *Meat Sci.* **2014**, *98*, 58–64. [CrossRef] [PubMed]
32. Freitas, A.; Leston, S.; Rosa, J.J.J.; Castilho, M.; Barbosa, J.; Rema, P.; Pardal, M.Â.A.; Ramos, F. Multi-residue and multi-class determination of antibiotics in gilthead sea bream (*Sparus aurata*) by ultra high-performance liquid chromatography-tandem mass spectrometry. *Food Addit. Contam. Part A* **2014**, *31*, 817–826. [CrossRef] [PubMed]
33. Charitonos, S.; Samanidou, V.F.; Papadoyannis, I. Development of an HPLC-DAD Method for the Determination of Five Sulfonamides in Shrimps and Validation According to the European Decision 657/2002/EC. *Food Anal. Methods* **2017**, *10*, 2011–2017. [CrossRef]

34. Konak, Ü.; Certel, M.; Şık, B.; Tongur, T. Development of an analysis method for determination of sulfonamides and their five acetylated metabolites in baby foods by ultra-high performance liquid chromatography coupled to high-resolution mass spectrometry (Orbitrap-MS). *J. Chromatogr. B Anal. Technol. Biomed. Life Sci.* **2017**, *1057*, 81–91. [CrossRef] [PubMed]
35. Liu, S.; Du, J.; Chen, J.; Zhao, H. Determination of 19 antibiotic and 2 sulfonamide metabolite residues in wild fish muscle in mariculture areas of Laizhou Bay using accelerated solvent extraction and high performance liquid chromatography-tandem mass spectrometry. *Se Pu Chin. J. Chromatogr.* **2014**, *32*, 1320–1325. [CrossRef]
36. Hoff, R.B.; Pizzolato, T.M.; Peralba, M.D.C.R.; Díaz-Cruz, M.S.; Barceló, D. Determination of sulfonamide antibiotics and metabolites in liver, muscle and kidney samples by pressurized liquid extraction or ultrasound-assisted extraction followed by liquid chromatography-quadrupole linear ion trap-tandem mass spectrometry (HPLC-QqL). *Talanta* **2015**, *134*, 768–778. [CrossRef] [PubMed]
37. Abdallah, H.; Arnaudguilhem, C.; Jaber, F.; Lobinski, R. Multiresidue analysis of 22 sulfonamides and their metabolites in animal tissues using quick, easy, cheap, effective, rugged, and safe extraction and high resolution mass spectrometry (hybrid linear ion trap-Orbitrap). *J. Chromatogr. A* **2014**, *1355*, 61–72. [CrossRef] [PubMed]
38. Huertas-Pérez, J.F.; Arroyo-Manzanares, N.; Havlíková, L.; Gámiz-Gracia, L.; Solich, P.; García-Campaña, A.M. Method optimization and validation for the determination of eight sulfonamides in chicken muscle and eggs by modified QuEChERS and liquid chromatography with fluorescence detection. *J. Pharm. Biomed. Anal.* **2016**, *124*, 261–266. [CrossRef] [PubMed]
39. Jia, W.; Shi, L.; Chu, X. Untargeted screening of sulfonamides and their metabolites in salmon using liquid chromatography coupled to quadrupole Orbitrap mass spectrometry. *Food Chem.* **2018**, *239*, 427–433. [CrossRef] [PubMed]
40. Li, J.; Liu, H.; Zhang, J.; Liu, Y.; Wu, L. A novelty strategy for the fast analysis of sulfonamide antibiotics in fish tissue using magnetic separation with high-performance liquid chromatography–tandem mass spectrometry. *Biomed. Chromatogr.* **2016**, *30*, 1331–1337. [CrossRef] [PubMed]
41. Huang, D.; Huang, X.; Gu, R.; Hui, Y.; Tian, L.; Feng, B.; Zhang, X.; Yu, H. Determination of 14 sulfonamide residues in shrimps by high performance liquid chromatography coupled with post-column derivatization. *Se Pu Chin. J. Chromatogr.* **2014**, *32*, 874–879. [CrossRef]
42. Freitas, A.; Barbosa, J.; Ramos, F. Multidetection of antibiotics in liver tissue by ultra-high-pressure-liquid-chromatography-tandem mass spectrometry. *J. Chromatogr. B Anal. Technol. Biomed. Life Sci.* **2015**, *976*, 49–54. [CrossRef] [PubMed]
43. Zhao, H.; Liu, H.; Yan, Z. Analysis of sulfonamide residues in pork and chicken by high performance liquid chromatography coupled with solid-phase extraction using multiwalled carbon nanotubes as adsorbent. *Se Pu Chin. J. Chromatogr.* **2014**, *32*, 294–298. [CrossRef]
44. Xia, L.; Liu, L.; Lv, X.; Qu, F.; Li, G.; You, J. Towards the determination of sulfonamides in meat samples: A magnetic and mesoporous metal-organic framework as an efficient sorbent for magnetic solid phase extraction combined with high-performance liquid chromatography. *J. Chromatogr. A* **2017**, *1500*, 24–31. [CrossRef] [PubMed]
45. Luo, H.; Xie, M.; Huang, X.; Wu, H.; Zhu, Z.; Huang, F.; Lin, X.; Ouyang, G. Multiresidue analysis of 63 veterinary drugs in meat by dispersive solid-phase extraction and high performance liquid chromatography-tandem mass spectrometry. *Chin. J. Chromatogr.* **2015**, *33*, 354–362. [CrossRef]
46. Wang, Z.; He, M.; Jiang, C.; Zhang, F.; Du, S.; Feng, W.; Zhang, H. Matrix solid-phase dispersion coupled with homogeneous ionic liquid microextraction for the determination of sulfonamides in animal tissues using high-performance liquid chromatography. *J. Sep. Sci.* **2015**, *38*, 4127–4135. [CrossRef] [PubMed]
47. Zhang, Q.; Xiao, X.; Li, G. Porous molecularly imprinted monolithic capillary column for on-line extraction coupled to high-performance liquid chromatography for trace analysis of antimicrobials in food samples. *Talanta* **2014**, *123*, 63–70. [CrossRef] [PubMed]
48. Huang, X.; Chen, L.; Yuan, D. Development of monolith-based stir bar sorptive extraction and liquid chromatography tandem mass spectrometry method for sensitive determination of ten sulfonamides in pork and chicken samples. *Anal. Bioanal. Chem.* **2013**, *405*, 6885–6889. [CrossRef] [PubMed]
49. Nebot, C.; Regal, P.; Miranda, J.M.; Fente, C.; Cepeda, A. Rapid method for quantification of nine sulfonamides in bovine milk using HPLC/MS/MS and without using SPE. *Food Chem.* **2013**, *141*, 2294–2299. [CrossRef] [PubMed]

50. Arroyo-Manzanares, N.; Gámiz-Gracia, L.; García-Campaña, A.M. Alternative sample treatments for the determination of sulfonamides in milk by HPLC with fluorescence detection. *Food Chem.* **2014**, *143*, 459–464. [CrossRef] [PubMed]
51. Wang, J.; Leung, D.; Chow, W.; Chang, J.; Wong, J.W. Development and Validation of a Multiclass Method for Analysis of Veterinary Drug Residues in Milk Using Ultrahigh Performance Liquid Chromatography Electrospray Ionization Quadrupole Orbitrap Mass Spectrometry. *J. Agric. Food Chem.* **2015**, *63*, 9175–9187. [CrossRef] [PubMed]
52. Sereshti, H.; Khosraviani, M.; Sadegh Amini-Fazl, M. Miniaturized salting-out liquid-liquid extraction in a coupled-syringe system combined with HPLC-UV for extraction and determination of sulfanilamide. *Talanta* **2014**, *121*, 199–204. [CrossRef] [PubMed]
53. Lu, Y.; Cong, B.; Tan, Z.; Yan, Y. Synchronized separation, concentration and determination of trace sulfadiazine and sulfamethazine in food and environment by using polyoxyethylene lauryl ether-salt aqueous two-phase system coupled to high-performance liquid chromatography. *Ecotoxicol. Environ. Saf.* **2016**, *133*, 105–113. [CrossRef] [PubMed]
54. Shao, M.; Zhang, X.; Li, N.; Shi, J.; Zhang, H.; Wang, Z.; Zhang, H.; Yu, A.; Yu, Y. Ionic liquid-based aqueous two-phase system extraction of sulfonamides in milk. *J. Chromatogr. B Anal. Technol. Biomed. Life Sci.* **2014**, *961*, 5–12. [CrossRef] [PubMed]
55. Hou, X.L.; Chen, G.; Zhu, L.; Yang, T.; Zhao, J.; Wang, L.; Wu, Y.L. Development and validation of an ultra high performance liquid chromatography tandem mass spectrometry method for simultaneous determination of sulfonamides, quinolones and benzimidazoles in bovine milk. *J. Chromatogr. B Anal. Technol. Biomed. Life Sci.* **2014**, *962*, 20–29. [CrossRef] [PubMed]
56. Kechagia, M.; Samanidou, V.; Kabir, A.; Furton, K.G. One-pot synthesis of a multi-template molecularly imprinted polymer for the extraction of six sulfonamide residues from milk before high-performance liquid chromatography with diode array detection. *J. Sep. Sci.* **2017**, *41*, 723–731. [CrossRef] [PubMed]
57. Ibarra, I.S.; Miranda, J.M.; Rodriguez, J.A.; Nebot, C.; Cepeda, A. Magnetic solid phase extraction followed by high-performance liquid chromatography for the determination of sulphonamides in milk samples. *Food Chem.* **2014**, *157*, 511–517. [CrossRef] [PubMed]
58. Tolmacheva, V.V.; Apyari, V.V.; Furletov, A.A.; Dmitrienko, S.G.; Zolotov, Y.A. Facile synthesis of magnetic hypercrosslinked polystyrene and its application in the magnetic solid-phase extraction of sulfonamides from water and milk samples before their HPLC determination. *Talanta* **2016**, *152*, 203–210. [CrossRef] [PubMed]
59. Li, Y.; Wu, X.; Li, Z.; Zhong, S.; Wang, W.; Wang, A.; Chen, J. Fabrication of $CoFe_2O_4$-graphene nanocomposite and its application in the magnetic solid phase extraction of sulfonamides from milk samples. *Talanta* **2015**, *144*, 1279–1286. [CrossRef] [PubMed]
60. Karageorgou, E.; Manousi, N.; Samanidou, V.; Kabir, A.; Furton, K.G. Fabric phase sorptive extraction for the fast isolation of sulfonamides residues from raw milk followed by high performance liquid chromatography with ultraviolet detection. *Food Chem.* **2016**, *196*, 428–436. [CrossRef] [PubMed]
61. Chatzimitakos, T.; Samanidou, V.; Stalikas, C.D. Graphene-functionalized melamine sponges for microextraction of sulfonamides from food and environmental samples. *J. Chromatogr. A* **2017**, *1522*, 1–8. [CrossRef] [PubMed]
62. Gao, M.; Yan, H.; Sun, N. Water-compatible poly (hydroxyethyl methacrylate) polymer sorbent for miniaturized syringe assisted extraction of sulfonamides in milk. *Anal. Chim. Acta* **2013**, *800*, 43–49. [CrossRef] [PubMed]
63. Zhao, H.; Zulkoski, J.; Mastovska, K. Development and Validation of a Multiclass, Multiresidue Method for Veterinary Drug Analysis in Infant Formula and Related Ingredients Using UHPLC-MS/MS. *J. Agric. Food Chem.* **2017**, *65*, 7268–7287. [CrossRef] [PubMed]
64. Pérez, M.L.G.; Romero-González, R.; Vidal, J.L.M.; Frenich, A.G. Analysis of veterinary drug residues in cheese by ultra-high-performance LC coupled to triple quadrupole MS/MS. *J. Sep. Sci.* **2013**, *36*, 1223–1230. [CrossRef] [PubMed]
65. Wu, L.; Song, Y.; Hu, M.; Xu, X.; Zhang, H.; Yu, A.; Ma, Q.; Wang, Z. Determination of sulfonamides in butter samples by ionic liquid magnetic bar liquid-phase microextraction high-performance liquid chromatography. *Anal. Bioanal. Chem.* **2015**, *407*, 569–580. [CrossRef] [PubMed]

66. Summa, S.; Lo Magro, S.; Armentano, A.; Muscarella, M. Development and validation of an HPLC/DAD method for the determination of 13 sulphonamides in eggs. *Food Chem.* **2015**, *187*, 477–484. [CrossRef] [PubMed]
67. He, X.; Tan, L.; Wu, W.; Wang, J. Determination of sulfadiazine in eggs using molecularly imprinted solid-phase extraction coupled with high-performance liquid chromatography. *J. Sep. Sci.* **2016**, *39*, 2204–2212. [CrossRef] [PubMed]
68. Sajid, M.; Na, N.; Safdar, M.; Lu, X.; Ma, L.; He, L.; Ouyang, J. Rapid trace level determination of sulfonamide residues in honey with online extraction using short C-18 column by high-performance liquid chromatography with fluorescence detection. *J. Chromatogr. A* **2013**, *1314*, 173–179. [CrossRef] [PubMed]

14

Effects of Alignment of Weak Interaction Sites in Molecular Shape Recognition High-Performance Liquid Chromatography

Hiroki Noguchi [1], Tianhang Liu [1], Makoto Takafuji [1,2] and Hirotaka Ihara [2,3,*]

1. Department of Applied Chemistry and Biochemistry, Kumamoto University; 2-39-1 Kurokami, Chuo-ku, Kumamoto 860-8555, Japan; 130d9219@st.kumamoto-u.ac.jp (H.N.); 168d8359@st.kumamoto-u.ac.jp (T.L.); takafuji@kumamoto-u.ac.jp (M.T.)
2. Kumamoto Institute for Photo-electro Organics (PHOENICS), Kumamoto 862-0901, Japan
3. Department of New Frontier Sciences, Kumamoto University, 2-39-1 Kurokami, Chuo-ku, Kumamoto 860-8555, Japan

* Correspondence: ihara@kumamoto-u.ac.jp

Academic Editor: Frank L. Dorman

Abstract: This paper introduces organic phases with aligned carbonyl groups derived from L-aspartic acid, L-glutamic acid, and L-α-aminoadipic acid; their stationary phases are denoted as Sil-Asp-2C$_n$, Sil-Glu-2C$_n$, and Sil-Adi-2C$_n$, respectively. The stationary phases were used in high-performance liquid chromatography to investigate the alignment effect of carbonyl groups, which act as π-interaction sites, on molecular shape selectivity. The selectivities of the synthesized organic phases were evaluated using polyaromatic hydrocarbons (PAHs), including geometric isomers (e.g., *cis*- and *trans*-stilbenes). The PAH selectivities of the prepared stationary phases were higher than that of conventional octadecyl silica. Among the stationary phases prepared in this study, Sil-Glu-2C$_n$ (n = 18) showed the highest selectivity toward terphenyl isomers with different twist configurations and the lowest selectivity toward planar PAHs with different aspect ratios. The results show that the molecular shape selectivity of the phases was affected by the alignment of interaction sites. As a practical application of the octadecylated amino acid derivative-bonded stationary phases, we evaluated their selectivity for tocopherol isomers and achieved good separation. Furthermore, Sil-Asp-2C$_n$ (n = 1) showed hydrophilic interaction chromatography mode retention behavior for the separation of polar molecules like nucleosides.

Keywords: aspartic acid; glutamic acid; aminoadipic acid; carbonyl-π interaction; molecular shape selectivity; polyaromatic hydrocarbon; tocopherol; reversed–phase; hydrophilic interaction chromatography

---

1. Introduction

The development of highly selective high-performance liquid chromatography (HPLC) stationary phases is important for progress in academic and industrial fields. Surface modification of inorganic carrier materials, such as porous silica particles, with new organic phases can provide useful tools for separations. Recently, organic phases including ionic liquids [1,2], alkyl chains with embedded polar groups (EPG) [3,4], and bio-inspired phases such as organogels [5–7] and peptides [8,9] have been developed for use in HPLC. These organic phases show unique selectivities in HPLC because of their introduced specific interaction sites. Molecular ordering of the organic phase is also important for achieving a highly selective stationary phase. Although the selectivity of octadecyl silica (ODS) is mainly based on its hydrophobic distribution, dense octadecyl groups on silica increase the molecular shape selectivity of polyaromatic hydrocarbons (PAHs) and their isomers [10,11]. This selectivity improvement is explained by highly ordered alkyl chain conformation. In addition, in the case of the polar group embedded alkyl phase, the polar group that is arranged close to the surface of silica reduces the silanol activity and improves separation of basic compound [12,13]. Furthermore, various organic phases, such as macrocyclic molecules [14,15] and helical oligoamides and -peptides [16,17], achieve specific selectivity by aligning functional groups based on their steric structure. We also found that highly ordered integration of weak interaction sites such as carbonyl groups can enhance the molecular shape selectivity of a stationary phase [18,19]. This was demonstrated by polymerization of carbonyl group-containing monomers and their crystallization on silica. The ordering of interaction sites in an organic phase with a simple chemical structure is an important approach to enhance molecular shape selectivity. In this work, we focus on the effect of the alignment of interaction sites in organic phases. We synthesize three types of carbonyl-containing molecular units derived from L-aspartic acid (Asp), L-glutamic acid (Glu), and L-α-aminoadipic acid (Adi) bonded to silica (Figure 1). The selected amino acids contain two carbonyl groups and have different numbers of methylene groups (x = 1, 2, and 3) between carbonyl groups. The amino acids are hydrophobized by introduction of two octadecyl alkyl chains (C_n). The number of methylene groups affects the alignment of interaction sites. The purpose of this work is to investigate the effect of interaction site alignment on the molecular shape selectivity in reversed-phase HPLC. We also examine the applicability of an amino acid derivative as an organic phase in hydrophilic interaction chromatography (HILIC) by changing the length of the alkyl chains introduced on Asp.

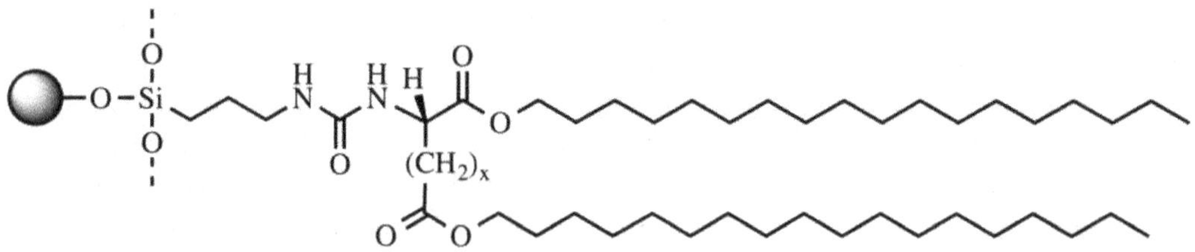

x=1: Sil-Asp-2C_{18}, x=2: Sil-Glu-2C_{18}, x=3: Sil-Adi-2C_{18}

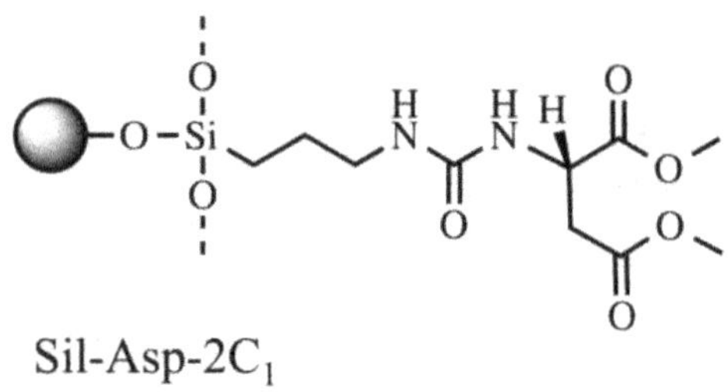

Sil-Asp-2C_1

Figure 1. Chemical structures of amino acid derivative-bonded stationary phases.

2. Materials and Methods

2.1. Preparation of Amino Acid Derivatives-Bonded Stationary Phases

A typical synthetic procedure is as follows. Glu (3.00 g), stearyl alcohol (11.58 g), and *p*-toluenesulfonic acid monohydrate (12.02 g) as a catalyst were dispersed in toluene (100 mL). The mixture was stirred under reflux for 12 h in a flask with an attached Dean-Stark apparatus (FLAT Inc., Chiba, Japan). The solution was concentrated using an evaporator, and the residue was recrystallized from ethanol to give the *p*-toluenesulfonic acid salt of L-glutamic acid dioctadecylate: yield: 12.88 g (78.0%); melting point (mp): 70.6–71.4 °C. Similar procedures gave the corresponding derivatives from Asp (yield: 72.3%; mp: 88.0–90.1 °C) and Adi (yield: 75.6%; mp: 79.5–80.8 °C). Dimethyl esterification of Asp with thionyl chloride in methanol was also performed using a previously reported method [20]. The obtained diester derivatives of the amino acids were dissolved in chloroform, neutralized with sodium hydrogen carbonate, reacted with 3-isocyanatopropyltriethoxysilane (TES) as a silane-coupling agent, and recrystallized from methanol to yield TES-Asp-2C_{18}, TES-Asp-2C_1, TES-Glu-2C_{18}, and TES-Adi-2C_{18}, where 2C_{18} and 2C_1 correspond to dioctadecylated and dimethylated derivatives, respectively. The chemical structures of the obtained compounds were determined using ^{1}H nuclear magnetic resonance (NMR) and Fourier transform infrared (FT-IR) spectroscopies, and elemental analysis (EA) (See Figures S1–S11). The obtained amino acid derivatives were mixed with porous silica particles (YMC GEL SL-12S-05 silica gel, diameter 5 μm, pore size 120 Å, specific surface area 330 $m^2 \cdot g^{-1}$) in toluene solution under reflux for 72 h to bond on the surface of the silica particles, giving Sil-Asp-2C_{18}, Sil-Asp-2C_1, Sil-Glu-2C_{18}, and Sil-Adi-2C_{18}. The amounts of amino acid derivatives that bonded to silica were determined using EA and thermogravimetric analysis (TGA).

2.2. Instrumentations

EA was performed using a Micro Corder JM10 analyzer (J-Science Lab Co., Ltd., Kyoto, Japan). FT-IR spectroscopy was performed using an FT/IR-4100 instrument (JASCO, Tokyo, Japan); a DR PRO410-M accessory (JASCO, Tokyo, Japan) was used for diffuse-reflectance infrared Fourier transform (DRIFT) spectroscopy. ^{1}H-NMR spectra were recorded using a Unity Inova AS400 spectrometer (Varian, Palo Alto, CA, USA) operating at 400 MHz. Chemical shifts are expressed in parts per million (ppm), and tetramethylsilane was used as an internal standard. Transmission electron microscopy (TEM) was performed using a JEM-2000× microscope (JEOL, Tokyo, Japan). TGA was performed using a SII Exstar 6000 TG/DTA 6300 thermobalance (SII, Chiba, Japan) in static air from 100 to 900 °C at a heating rate of 10 $°C \cdot min^{-1}$.

2.3. Liquid Chromatography

The synthesized stationary phases were packed into stainless-steel columns that were 150 mm long with an internal diameter of 4.6 mm. A JASCO 980 pump (JASCO Corporation, Tokyo, Japan) with a Rheodyne 7725 injector (JASCO Corporation, Tokyo, Japan) (10-μL loop) and JASCO MD-2010 plus multiwavelength detector (JASCO Corporation, Tokyo, Japan) were used for liquid chromatography. The column temperature was maintained using a column jacket with a heating and cooling system. ChromNAV (version 1.14) software (JASCO Corporation, Tokyo, Japan) was used for system control and data analysis. Chromatographic-grade solvents purchased from Nacalai Tesque, Inc. (Kyoto, Japan) were used to prepare the mobile phases and samples. A commercial ODS column (GL-Science ODS-3, C: 15%, column length: 150 mm, internal diameter: 4.6 mm) (GL Science Inc., Tokyo, Japan) was used as the reference column. The retention time of methanol was used as the void volume (t_0) marker. The absorption of methanol was measured at 220 nm. The retention factor (k) was calculated from the retention time and t_0. The separation factor (α) was determined as the ratio of k values.

3. Results and Discussions

3.1. Evaluation of the Molecular Orientation of Organic Phases

Figure 2 shows TEM images of TES-Asp-2C_{18}, TES-Glu-2C_{18}, and TES-Adi-2C_{18} dissolved in methanol at a concentration of 0.6 mM and then cast on a carbon-coated copper grid. The amino acid derivatives formed fibrous aggregates with diameters of around 9.5 nm in methanol. This is probably because of the biphasic structure of the 2C_{18} derivatives based on their non-polar properties derived from the long-chain alkyl groups, and the polar urea and ester groups. It is expected that the hydrogen bonding between the carbonyl and urea groups of the amino acid derivatives support the formation of fibrous aggregates. These results suggest that the 2C_{18} derivatives can form ordered structures on the silica particle surfaces in a HPLC mobile phase such as methanol.

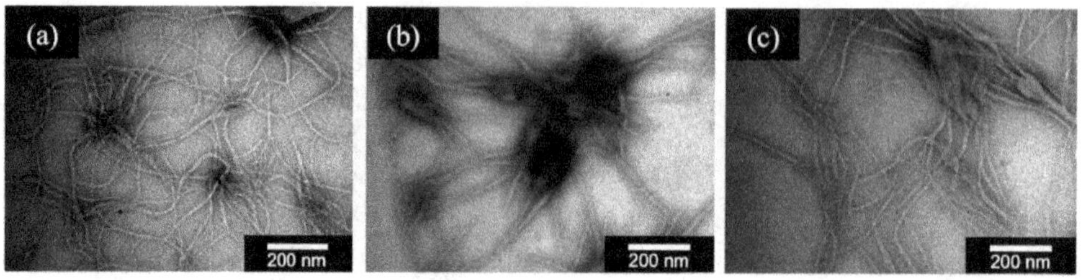

Figure 2. Transmission electron microscopy (TEM) images of nanofibrillar aggregates from (**a**) TES-Asp-2C$_{18}$: dioctadecyl ((3-(triethoxysilyl)propyl)carbamoyl)-L-aspartate; (**b**) TES-Glu-2C$_{18}$: dioctadecyl ((3-(triethoxysilyl)propyl)carbamoyl)-L-glutamate; and (**c**) TES-Adi-2C$_{18}$: dioctadecyl ((3-(triethoxysilyl)propyl)carbamoyl)-L-2-amino adipic acid in MeOH (0.6 mM). These samples were stained by 1.0 wt % uranyl acetate.

3.2. Bonding Density of Amino Acid Derivatives on Silica Surface

Bonding of the amino acid derivatives on the surface of silica was confirmed by DRIFT spectroscopy, EA and TGA. As illustrated in Figure 3, the DRIFT spectra of Sil-Asp-2C$_1$, Sil-Asp-2C$_{18}$, Sil-Glu-2C$_{18}$ and Sil-Adi-2C$_{18}$ contained peaks around 2850 and 2900 cm^{-1}, which are attributed to the C–H stretching of the long alkyl chains, and at 1650 and 1715 cm^{-1} with a shoulder at 1745 cm^{-1}, which are characterized as C=O vibrations of urea and ester bonds, respectively. The EA results indicate that the loading amounts of carbon in the C$_{18}$-amino acid derivative-bonded silica samples were almost same as that of the reference ODS (C: 15%). This suggests that the density of alkyl chains on the silica surface was similar to that on the reference ODS. The C/N values of the samples were close to the theoretical values. No marked weight loss of the samples was observed after successive washing with good solvents such as chloroform. These results indicate that the amino acid derivatives were covalently bonded to the silica surface. The weight percentage (P_w) and surface coverage (N) of the bonded phases summarized in Table 1 were calculated from P_w (wt %) = $P_cM/12n_C$ and N (μmol·m^{-2}) = $10^6\ P_c/[1200n_c - P_cM]S$, where P_c is the percentage of carbon based on the EA results, S is the specific surface area of the porous silica particles used for bonding (330 m^2·g^{-1}), n_c is the number of carbon atoms in a bonded molecule, and M is the molecular weight of the molecules used for bonding [21]. The thermal weight loss of the stationary phases between 100 and 900 °C (Figure S12) agrees with the EA results.

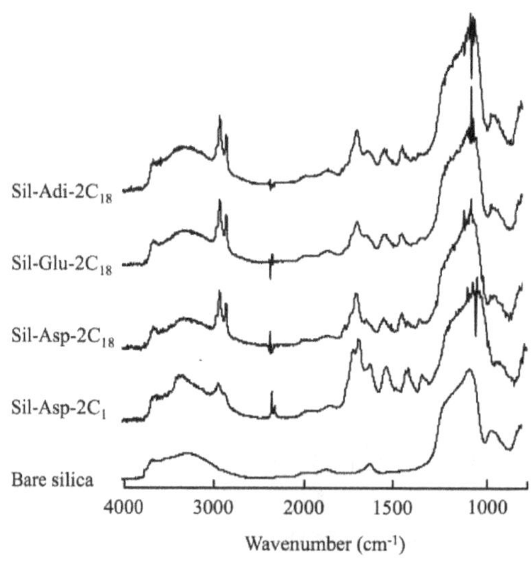

Figure 3. Comparison of diffuse-reflectance infrared Fourier transform (DRIFT) spectra of Sil-Asp-2C$_1$, Sil-Asp-2C$_{18}$, Sil-Glu-2C$_{18}$, Sil-Adi-2C$_{18}$, and bare silica.

Table 1. Bonded amount of amino acid derivatives. calcd.: Calculated value.

	%C	%N	%H	C/N		Bonded Amount		
				Found	calcd.	$\mu mol \cdot m^{-2}$	wt %	TG %
Sil-Asp-2C_{18}	17.05	0.9	2.86	18.94	18.86	0.98	23.32	23.00
Sil-Glu-2C_{18}	13.71	0.7	2.37	19.59	19.29	0.77	18.69	18.20
Sil-Adi-2C_{18}	15.36	0.8	2.66	19.20	19.71	0.84	20.88	22.06
Sil-Asp-2C_1	14.59	3.5	2.28	4.17	4.29	3.68	33.23	26.19

Theoretical C/N values and immobilized amount were calculated from $C_{44}H_{85}N_2O_5$ for Sil-Asp-2C_{18}, $C_{45}H_{87}N_2O_5$ for Sil-Glu-2C_{18}, $C_{46}H_{89}N_2O_5$ for Sil-Adi-2C_{18}, $C_{10}H_{17}N_2O_5$ for Sil-Asp-2C_1.

3.3. Reversed Phase Liquid Chromatography

Chromatographic separations were performed with the Sil-Asp-2C_{18}, Sil-Glu-2C_{18} and Sil-Adi-2C_{18} columns (C_{18}-amino acid derivative columns) using long-chain alkylated benzene derivatives and PAHs as solutes and methanol/water (9:1) as the mobile phase. With this sample set, the elution order of C18-amino acid derivative columns followed log P values of analytes (see Figure S13). In contrast, unique separation behaviors were observed when the analytes were PAHs with characteristic structures. The molecular shape selectivity of ODS is affected by the density of alkyl chains [22]. The elution order of phenanthro[3,4-c]phenanthrene (PhPh), 1,2:3,4:5,6:7,8-tetrabenzonaphthalene (TBN), and benzo[a]pyrene (BaP) with monomeric ODS (C: 15%) is BaP $\leq$ PhPh $<$ TBN, and with highly carbon loaded polymeric ODS (C: 30%) is PhPh $\leq$ TBN $<$ BaP [22]. The elution orders of prepared stationary phases were different from both ODS (PhPh $<$ BaP $<$ TBN, see Figure 4). Because the C18 amino acid derivatives were self-assemble molecules, there were possibilities that C18 amino acid derivatives were bonded on the surface of silica particles locally and densely. It was supposed that the assembled structure affect for the molecular recognition. To investigate this in further detail, we clarified the molecular shape selectivities of the C_{18}-amino acid derivative columns using geometric isomers (*cis-/trans*-stilbenes), positional isomers with different twist angles (*o*-, *m*-, and *p*-terphenyls), and a planar compound (triphenylene), as analytes. The data in Table 2 reveal that the C_{18}-amino acid derivative columns showed higher selectivity toward planar molecules than ODS did. Sil-Glu-2C_{18} showed the highest selectivity for *p*- and *o*-terphenyls (α = 2.46, 2.39 and 2.03 for Sil-Glu-2C_{18}, Sil-Asp-2C_{18} and Sil-Adi-2C_{18}, respectively), but its selectivity for triphenylene and *o*-terphenyls was lower than those of Sil-Asp-2C_{18} and Sil-Adi-2C_{18} (α = 2.72, 3.81 and 2.99 for Sil-Glu-2C_{18}, Sil-Asp-2C_{18} and Sil-Adi-2C_{18}, respectively). The length-to-breadth ratio (L/B) [23,24] is a useful indicator for evaluating molecular shape. In this work, four planar PAHs, triphenylene (L/B = 1.12), benz[a]anthracene (L/B = 1.58), chrysene (L/B = 1.72), and naphthacene (L/B = 1.89), were selected to investigate the molecular shape selectivity of the C_{18}-amino acid derivative columns. Determination of the selectivities of the columns for these molecules showed that the stationary phases with bonded C_{18}-amino acid derivatives selectively recognized molecules with higher L/B values. The naphthacene/triphenylene separation factors were the highest and lowest for Sil-Asp-2C_{18} and Sil-Glu-2C_{18}, respectively. We also evaluated the selectivity of Sil-Asp-2C_1 using PAHs, as summarized in Table 2.

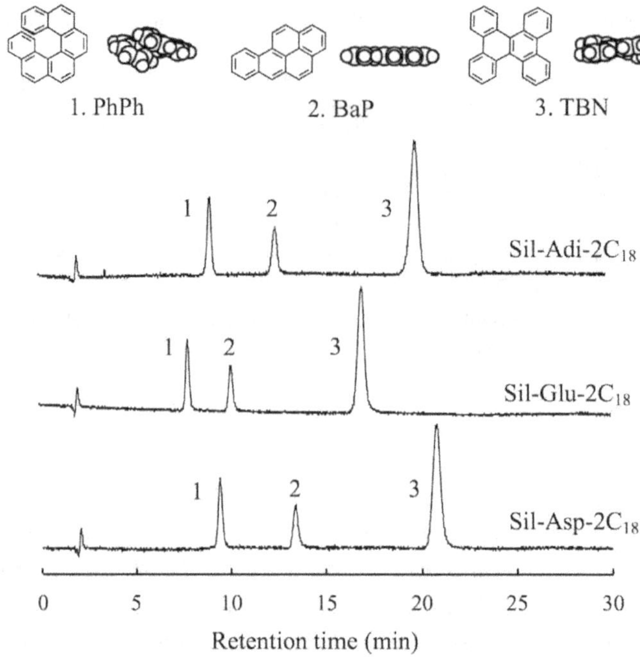

Figure 4. Chromatograms of phenanthro[3,4-c]phenanthrene (PhPh), benzo[a]pyrene (BaP) and 1,2:3,4:5,6:7,8-tetrabenzonaphthalene (TBN). Mobile phase: methanol-water (9:1), flow rate: 1.0 mL·min^{-1}, column temperature: 20 °C.

Table 2. Retention and separation factors (k and α) with C_{18}-amino acid derivative columns.

Analyte		**Sil-Asp-2C$_{18}$**		**Sil-Glu-2C$_{18}$**		**Sil-Adi-2C$_{18}$**		**ODS**	
		k	α	k	α	k	α	k	α
cis-Stilbene		1.01	-	0.99	-	1.03	-	1.74	-
trans-Stilbene		1.50	1.48	1.39	1.40	1.43	1.38	1.84	1.06
o-Terphenyl		1.32	-	0.88	-	1.23	-	2.27	-
m-Terphenyl		2.45	1.86	1.88	2.14	2.12	1.72	3.22	1.42
p-Terphenyl		3.14	2.39	2.16	2.46	2.50	2.03	3.26	1.43
Triphenylene		5.02	3.81	2.38	2.72	3.67	2.99	3.51	1.54
Triphenylene		5.02	-	2.38	-	3.67	-	3.51	-
Benz[a]anthracene		5.66	1.13	2.56	1.07	4.05	1.10	3.73	1.06
Chrysene		5.84	1.16	2.58	1.08	4.11	1.12	3.69	1.05
Naphthacene		7.90	1.57	3.11	1.31	5.08	1.38	4.26	1.21

Mobile phase: methanol-water (9:1), flow rate: 1.0 mL·min^{-1}, column temperature: 20 °C. The α values were calculated using the k of the first analyte and others respectively, in each group.

To further investigate the C_{18}-amino acid derivative columns, temperature dependencies of the selectivities for triphenylene/*o*-terphenyl ($\alpha_{triph/o\text{-}terp}$) and naphthacene/triphenylene ($\alpha_{naph/triph}$) were evaluated. Figure 5 reveals that a linear plot was observed for ODS. In contrast, bending points were observed in the plots of C_{18}-amino acid derivative columns. This suggests that a steric structure change related to the molecular interaction occurred on the stationary phase. Moreover, the higher selectivities of the C_{18}-amino acid derivative columns at temperatures below 25 °C are related to temperature-dependent molecular ordering. The ordering of the molecules on the surface of the stationary phase indicates that the organic phases are locally densely bonded. This is one of the reasons for the unique selectivity of the C_{18}-amino acid derivative columns. The aligned carbonyl groups may also contribute to the molecular shape selectivity of the C_{18}-amino acid derivative columns. The number of methylene groups between the carbonyl groups of the C_{18}-amino acid derivatives is odd except for in the Glu derivative. Because of this, it is presumed that the configuration of carbonyl groups in the Glu derivative is obviously different from that of the other amino acid derivatives. The alignment of the interaction sites on Sil-Glu-2C_{18} therefore must differs from those of the other amino acid derivatives, when it orientates on the surface of silica, and promotes recognition of the twisted structures of terphenyl isomers. For the same reason, the alignments of the interaction sites in Sil-Asp-2C_{18} and Sil-Adi-2C_{18} are better for recognition of planar and linear molecules than that in Sil-Glu-2C_{18}.

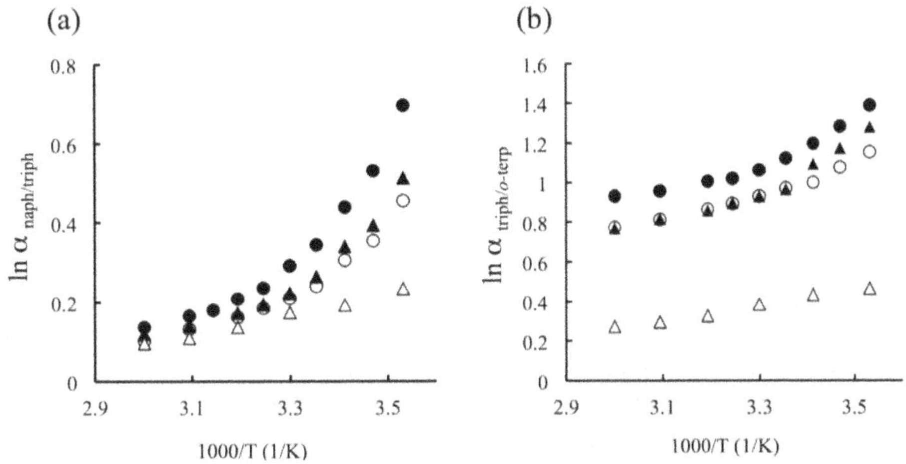

Figure 5. Van't Hoff plots for separation factor of (**a**) $\alpha_{naph/triph}$ and (**b**) $\alpha_{triph/o\text{-}ter}$ with amino acid derivatives bonded silica. Mobile phase: methanol-water (9:1), flow rate: 1.0 mL·min^{-1}. Black circle: Sil-Asp-2C_{18}, hollow circle: Sil-Glu-2C_{18}, black triangle: Sil-Adi-2C_{18}, and hollow triangle: octadecyl silica (ODS).

3.4. *Hydrophilic Interaction Chromatography*

The applicability of an amino acid derivative-bonded stationary phase to HILIC mode was evaluated using Sil-Asp-2C_1. The typical HILIC mode retention order was observed for phenol, resorcinol, and phloroglucinol with Sil-Asp-2C_1: higher *k* value was achieved by increasing the number of hydroxyl groups of the analyte (Table S1) [25]. Usually, a mixture of water and a polar organic solvent such as acetonitrile is used for the mobile phase in HILIC. The retention mechanism of HILIC involves liquid-liquid partitioning between the mobile phase and absorbed water on the surface of the stationary phase [26]. Thus, by increasing the content of water in the mobile phase, the retention in HILIC will decrease. Figure 6 shows the relationship between *k* and the proportion of acetonitrile in the mobile phase. Decreasing the content of acetonitrile in the mobile phase decreased the *k* values of all selected polar molecules, which is consistent with the usual retention behavior in HILIC. This is the typical HILIC mode retention behavior [25,27]. In addition, larger *k* values were obtained by molecules that show lower pKa values (e.g., pKa = 4.2 and 4.2 for adenine and cytosine [28]) among a sample set (e.g., pKa = 10.5 for thymine [28]). In the HILIC, in addition to the partitioning, ion exchange between positively charged molecules and negative charged silanol also affected the retentivity. This effect may be revealed in this study.

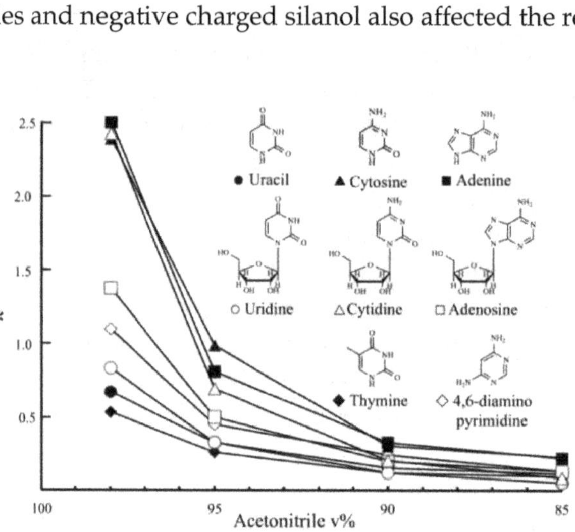

Figure 6. Effect of acetonitrile content ratio on the retention factors of polar molecules with Sil-Asp-2C_1. Mobile phase: acetonitrile and 10 mM ammonium acetate mixture, flow rate: 1.0 mL·min^{-1}, column temperature: 10 °C.

These results suggest that weak interaction site accumulation based on amino acids has the potential for application in molecular design for HILIC.

3.5. Separation of Tocopherol Isomers

Practical applications of the C_{18}-amino acid derivative columns were evaluated using the tocopherol isomers (α, β, γ, and δ isomers). Tocopherols are hydrophobic vitamins that have many beneficial bioeffects such as antioxidant and anti-inflammatory properties. For example, the importance of the role of tocopherols in retinal precursor cell differentiation is being increasingly studied [29]. Figure 7 depicts chromatograms of tocopherols obtained using the C_{18}-amino acid derivative columns and ODS. Separation of the β and γ isomers was not achieved using ODS, but was realized using the stationary phases with bonded octadecylated amino acid derivatives. In particular, Sil-Asp-2C_{18} successfully separated all the isomers in the reversed-phase mode.

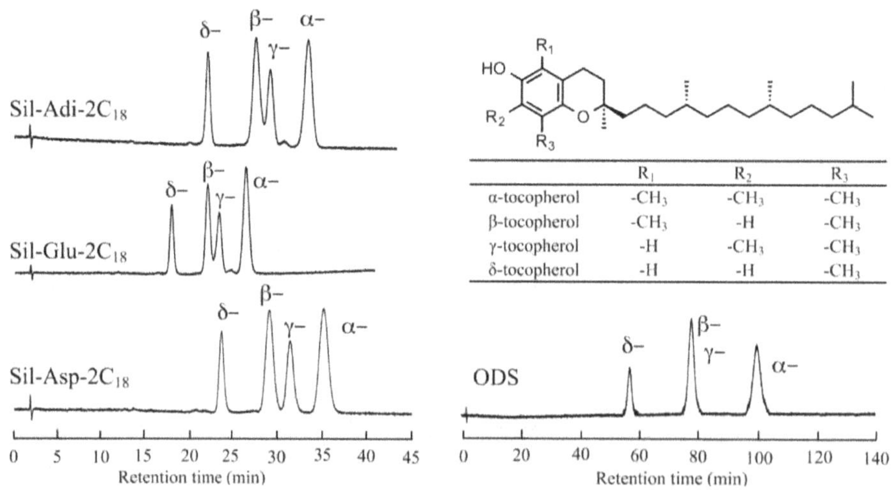

Figure 7. Chromatograms for tocopherol isomers with Sil-Asp-2C_{18}, Sil-Glu-2C_{18}, Sil-Adi-2C_{18} and ODS. Mobile phase: methanol-water (9:1), flow rate: 1.0 mL·min^{-1}, column temperature: 10 °C.

4. Conclusions

Organic stationary phases based on Asp, Glu, and Adi with aligned carbonyl groups were synthesized and bonded on silica particles. These stationary phases showed higher molecular shape selectivities for PAHs than ODS. The highest selectivity for *p*-/*o*-terphenyl and naphthacene/*o*-terphenyl were obtained by Sil-Glu-2C_{18} and Sil-Asp-2C_{18}, respectively. Sil-Asp-2C_{18}, Sil-Glu-2C_{18}, and Sil-Adi-2C_{18} differ in the number of methylene groups between carbonyl groups in the original amino acids; therefore, these results indicate that the alignment of the carbonyl groups as interaction sites affects the molecular shape selectivity of these organic phases. We also evaluated the use of Sil-Asp-2C_1 as a HILIC-mode stationary phase. Like common HILIC stationary phases, Sil-Asp-2C_1 achieved higher k values with increasing polarity. Effective separation of tocopherol isomers was realized with the C_{18}-amino acid derivative columns in reversed-phase mode, including the separation of β and γ isomers, which could not be obtained by ODS.

Acknowledgments: This work was partially supported by Grant-in-Aid for Scientific Research and the Sasakawa Scientific Research Grant.

Author Contributions: H.I. and M.T. conceived and designed the experiments; H.N. and T.L. performed the experiments, and analyzed the data. H.N. wrote the paper and H.I. and M.T. contributed to preparation and revisions of the paper.

Conflicts of Interest: The authors declare no conflict of interest.

References

1. Sun, M.; Feng, J.; Chen, W.; Li, L.; Duan, H.; Luo, C. Improvement of the chromatographic separation performance of an imidazolium ionic liquid functionalized silica column by in situ anion-exchange with dodecyl sulfonate and dodecylbenzene sulfonate anions. *J. Sep. Sci.* **2014**, *37*, 1283–1288. [CrossRef] [PubMed]
2. Qiu, H.; Mallik, A.K.; Sawada, T.; Takafuji, M.; Ihara, H. New surface-confined ionic liquid stationary phases with enhanced chromatographic selectivity and stability by co-immobilization of polymerizable anion and cation pairs. *Chem. Commun.* **2012**, *48*, 1299–1301. [CrossRef] [PubMed]
3. Zhang, M.; Mai, W.; Zhao, L.; Guo, Y.; Qiu, H. A polar-embedded C30 stationary phase: Preparation and evaluation. *J. Chromatogr. A* **2015**, *1388*, 133–140. [CrossRef]
4. O'Sullivan, G.; Scully, N.M.; Glennon, J.D. Polar-Embedded and Polar-Endcapped Stationary Phases for LC. *Anal. Lett.* **2010**, *43*, 1609–1629. [CrossRef]
5. Noguchi, H.; Charoenraks, T.; Takafuji, M.; Ihara, H. Effects of substitution groups of glutamide-derived molecular gels on molecular shape recognition. *J. Chromatogr. A* **2015**, *1392*, 56–62. [CrossRef] [PubMed]
6. Mallik, A.K.; Qiu, H.; Takafuji, M.; Ihara, H. High molecular-shape selectivity by molecular gel-forming compounds: Bioactive and shape-constrained isomers through the integration and orientation of weak interaction sites. *Chem. Commun.* **2011**, *47*, 10341–10343. [CrossRef] [PubMed]
7. Rahman, M.M.; Takafuji, M.; Ansarian, H.R.; Ihara, H. Molecular Shape Selectivity through Multiple Carbonyl-π Interactions with Noncrystalline Solid Phase for RP-HPLC. *Anal. Chem.* **2005**, *77*, 6671–6681. [CrossRef] [PubMed]
8. Ray, S.; Takafuji, M.; Ihara, H. Chromatographic evaluation of a newly designed peptide-silica stationary phase in reverse phase liquid chromatography and hydrophilic interaction liquid chromatography: Mixed mode behavior. *J. Chromatogr. A* **2012**, *1266*, 43–52. [CrossRef] [PubMed]
9. Ohyama, K.; Inoue, Y.; Kishikawa, N.; Kuroda, N. Preparation and characterization of surfactin-modified silica stationary phase for reversed-phase and hydrophilic interaction liquid chromatography. *J. Chromatogr. A* **2014**, *1371*, 257–260. [CrossRef] [PubMed]
10. Sander, L.C.; Wise, S.A. Synthesis and characterization of polymeric C18 stationary phases for liquid chromatography. *Anal. Chem.* **1984**, *56*, 504–510. [CrossRef]

11. Jinno, K.; Nagoshi, T.; Tanaka, N.; Okamoto, M.; Fetzer, J.C.; Biggs, W.R. Eluation behaviour of planar and non-planar polycyclic aromatic hydrocarbons on various chemically bonded stationary phases in liquid chromatography. *J. Chromatogr. A* **1987**, *392*, 75–82. [CrossRef]
12. Silva, C.R.; Airoldi, C.; Collins, K.E.; Collins, C.H. Preparation and characterization of a new C_{18} urea phase based on titanized silica. *J. Chromatogr. A* **2005**, *1087*, 29–37. [CrossRef] [PubMed]
13. Layne, J. Characterization and comparison of the chromatographic performance of conventional, polar-embedded, and polar-endcapped reversed-phase liquid chromatography stationary phases. *J. Chromatogr. A* **2002**, *957*, 149–164. [CrossRef]
14. Kučerová, G.; Kalíková, K.; Procházková, H.; Popr, M.; Jindřich, J.; Coufal, P.; Tesařová, E. Chromatographic Characterization of a New Cationic β-CD Based Stationary Phase Prepared by Dynamic Coating. *Chromatographia* **2016**, *79*, 529–536. [CrossRef]
15. Kakhki, R.M. Application of crown ethers as stationary phase in the chromatographic methods. *J. Incl. Phenom. Macrocycl. Chem.* **2013**, *75*, 11–22. [CrossRef]
16. Noguchi, H.; Takafuji, M.; Maurizot, V.; Huc, I. Chiral separation by a terminal chirality triggered *P*-helical quinoline oligoamide foldamer. *J. Chromatogr. A* **2016**, *1437*, 88–94. [CrossRef] [PubMed]
17. Zhang, C.; Ma, R.; Wang, H.; Sakai, R.; Satho, T.; Kakuchi, T.; Liu, L.; Okamoto, Y. Influence of Helical Structure on Chiral Recognition of Poly(phenylacetylene)s Bearing Phenylcarbamate Residues of L-Phenylglycinol and Amide Linage as Pendants. *Chirality* **2015**, *27*, 500–506. [CrossRef] [PubMed]
18. Fukumoto, T.; Ihara, H.; Sakaki, S.; Shosenji, H.; Hirayama, C. Chromatographic separation of geometrical isomers using highly oriented polymer-immobilized silica gels. *J. Chromatogr. A* **1994**, *672*, 237–241. [CrossRef]
19. Goto, Y.; Nakashima, K.; Mitsuishi, K.; Takafuji, M.; Sakaki, S.; Ihara, H. Selectivity enhancement of diastereomer separation in RPLC using crystalline-organic phase-bonded silica. *Chromatographia* **2002**, *56*, 19–23. [CrossRef]
20. Fowler, L.S.; Thomas, L.H.; Ellis, D.; Sutherland, A. A one-pot, reductive amination/6-endo-trig cyclisation for the stereoselective synthesis of 6-substituted-4-oxopipecolic acids. *Chem. Commun.* **2011**, *47*, 6569–6571. [CrossRef] [PubMed]
21. Berendsen, G.E.; Pikaart, K.A.; Galan, L. Preparation of various bonded phases for HPLC using monochlorosilanes. *J. Chromatogr.* **1980**, *3*, 1437–1464. [CrossRef]
22. Sander, L.C.; Pursch, M.; Wise, S.A. Shape Selectivity for Constrained Solutes in Reversed-Phase Liquid Chromatography. *Anal. Chem.* **1999**, *71*, 4821–4830. [CrossRef] [PubMed]
23. Jinno, K.; Kawasaki, K. Correlation between the Retention Data of Polycyclic Aromatic Hydrocarbons. *Chromatograhia* **1983**, *17*, 445–449. [CrossRef]
24. Wise, S.A.; Bonnett, W.J.; Guenther, F.R.; May, W.E. A Relationship between Reversed-Phase C_{18} Liquid Chromatographic Retention and the Shape of Polycyclic Aromatic Hydrocarbons. *J. Chromatogr. Sci.* **1981**, *19*, 457–465. [CrossRef]
25. Hemström, P.; Irgum, K. Hydrophilic interaction chromatography. *J. Sep. Sci.* **2006**, *29*, 1784–1821. [CrossRef] [PubMed]
26. Kahsay, G.; Song, H.; Schepdael, A.V.; Cabooter, D.; Adams, E. Hydrophilic interaction chromatography (HILIC) in the analysis of antibiotics. *J. Pharm. Biomed. Anal.* **2014**, *87*, 142–154. [CrossRef] [PubMed]
27. Wu, J.Y.; Bicker, W.; Linder, W. Separation properties of novel and commercial polar stationary phases in hydrophilic interaction and reversed-phase liquid chromatography mode. *J. Sep. Sci.* **2008**, *31*, 1492–1503. [CrossRef] [PubMed]
28. Verdolino, V.; Cammi, R.; Munk, B.H.; Schlegel, B. Calculation of pK_a Values of Nucleobases and the Guanine Oxidation Products Guanidinohydantoin and Spiroiminodihydantoin using Density Functional Theory and a Polarizable Continuum Model. *J. Phys. Chem. B* **2008**, *112*, 16860–16873. [CrossRef] [PubMed]
29. Deng, S.; Hou, G.; Xue, Z.; Zhang, L.; Zhou, Y.; Liu, C.; Liu, Y.; Li, Z. Vitamin E isomer δ-tocopherol enhances the efficiency of neural stem cell differentiation via L-type calcium channel. *Neurosci. Lett.* **2015**, *585*, 166–170. [CrossRef] [PubMed]

15

Simultaneous High Performance Liquid Chromatography Assay of Pentoxifylline, Bupivacaine HCl, Levocetirizine HCl, Tranilast and Fluticasone Propionate in Humco™ Sanare Advanced Scar Base

Troy Purvis [1,2]

[1] Humco Compounding, 313 East 12th Street, Suite 230, Austin, TX 78701, USA; tpurvis@humco.com; Tel.: +1-512-391-2330

[2] Humco, 7400 Alumax Road, Texarkana, TX 75501, USA

Academic Editor: Frank L. Dorman

Abstract: This article details the elements used in the verification method for the simultaneous high performance liquid chromatography (HPLC) assay of Pentoxifylline, Bupivacaine HCl, Levocetirizine HCl, Tranilast, and Fluticasone Propionate in Humco™ Sanare Advanced scar base. The method was proven to be linear over 50%–150% of the nominal concentration of the standard. The method was proven to be accurate over 50%–150%, with 98%–102% recovery of the actives from spiked placeboes over that range. The method exhibited specificity to the analytes listed, and it was shown to be precise, yielding acceptable results for system reproducibility and method repeatability. The method, as written, is considered to have been verified.

Keywords: pentoxifylline; bupivacaine HCl; levocetirizine HCl; tranilast; fluticasone propionate; high performance liquid chromatography (HPLC); Humco™ Sanare scar base

---

1. Introduction

Compounded formulations in silicone oil based creams containing Pentoxifylline, Bupivacaine HCl, Levocetirizine HCl, Tranilast, and Fluticasone Propionate are applied to scars to facilitate further healing of scar tissue while also minimizing the appearance of those scars. This particular active formulation facilitates the healing of scar tissue by containing a drug (Tranilast [1–3]), which reduces collagen formation in fibroblasts and inhibits the action of neurofibroma cells in keloid and hypertrophic scars. An antihistamine (Levocetirizine HCl [4,5]) drug is also incorporated, which prevents the release of endogenous inflammatory response factors and increases blood flow at the topical site of action. This formulation also contains anti-inflammatory drugs (Pentoxifylline [6] and Fluticasone Propionate [7]) along with a topical pain management drug (Bupivacaine HCl [8]). Additionally, Pentoxifylline has been shown to decrease collagen production while increasing the activity of collagenase in the dermis, thereby promoting the turnover of dermal cells [9,10]. Together these actives have been shown to be particularly effective at stimulating skin renewal on scar tissue, while reducing inflammation that might occur in healing scars.

Humco™ Sanare scar base is an anhydrous topical compounding base, containing silicone oils and polyethylene glycol (PEG) substituted silicones along with PEG substituted fixed oils which form a non-washable barrier that is uniquely suitable for application as scar therapy and for reducing symptoms of other skin conditions [11]. Additionally, Sanare contains an anti-inflammatory botanical agent, Carapa Guaianensis seed oil, a natural ingredient that promotes skin renewal and encourages healing of scar tissue [12–14]. Humco™ Sanare scar base has a smooth elegant yet non-greasy texture and is appropriate for application of compounded medications to areas that have begun to heal from burns, ulcers, abrasions, and other dermal injuries.

This report details the method verification requirements and corresponding acceptance criteria for the analytical method used to assay Pentoxifylline, Bupivacaine HCl, Levocetirizine HCl, Tranilast, and Fluticasone Propionate compounded in Humco™ Sanare scar base. Following verification, the method is suitable for analyzing samples compounded in-house and samples received from other pharmacies for analytical testing [15–17]. The ingredients in the formulation are listed in Table 1.

Table 1. Compounded formulation in Humco™ Sanare scar base.

Ingredient	%*w*/*w* in Formulation	Ingredient	%*w*/*w* in Formulation
Pentoxifylline	2%	Levocetirizine HCl	2%
Bupivacaine HCl	1%	Fluticasone Propionate	1%
Tranilast	2%	Humco™ Sanare Advanced Scar Base	quantity sufficient (q.s.)

2. Experimental Section

A reverse phase high performance liquid chromatography (HPLC) method was developed which uses a phosphate solution and acetonitrile to create a gradient to separate the components contained in the formulation. The details of the method, including HPLC instrument conditions, mobile phase preparation, and preparation of standards and samples are given. The method verification elements and acceptance criteria are also given.

2.1. Materials and Methods

2.1.1. Chromatographic Conditions

Column: Phenomenex® Gemini 150 × 4.6 mm C18 5 µm Part # 00F-4435-E0 or equivalent (Phenomenex, Torrance, CA, USA); Guard column: Phenomenex® SecurityGuard C18 Guard Column Part # KJ0-4282 (Phenomenex, Torrance, CA, USA); Column temperature: 25 °C; Mobile Phase A: 0.02 M Sodium phosphate dibasic, pH 3.0 with *o*-phosphoric acid; Mobile Phase B: Acetonitrile; Gradient profile: See Table 2.

Table 2. Mobile phase gradient profile.

Time (min)	%A	%B
0	85	15
5	85	15
42	25	75
45	85	15
52	85	15

Flow rate: 1.0 mL/min; Injection volume: 10 µL; Wavelength: 220 nm; Seal/needle rinse: 50/50 Acetonitrile/water; Run time: 52 min; Typical retention times: See Table 3.

Table 3. Typical retention times of actives.

Active	Approximate Retention Time (min)	Active	Approximate Retention Time (min)
Pentoxifylline	11.0–11.5	Tranilast	26.4–26.9
Bupivacaine HCl	13.7–14.3	Fluticasone Propionate	34.5–35.0
Levocetirizine HCl	19.7–20.4	-	-

2.1.2. Materials and Equipment

Pentoxifylline: United States Pharmacopeia or reference standard grade (Spectrum Chemical, Gardena, CA, USA); Bupivacaine HCl: United States Pharmacopeia or reference standard grade (Spectrum Chemical, Gardena, CA, USA); Levocetirizine HCl: United States Pharmacopeia or reference standard grade (Fisher Scientific, Pittsburgh, PA, USA); Tranilast: United States Pharmacopeia or reference standard grade (Fisher Scientific, Pittsburgh, PA, USA); Fluticasone Propionate: United States Pharmacopeia or reference standard grade (Fisher Scientific, Pittsburgh, PA, USA); Sodium phosphate dibasic: Reagent grade (Fisher Scientific, Pittsburgh, PA, USA); HPLC grade water: HPLC grade (Fisher Scientific, Pittsburgh, PA, USA); Acetonitrile: HPLC grade (Fisher Scientific, Pittsburgh, PA, USA); *o*-Phosphoric acid, 85%: HPLC grade (Fisher Scientific, Pittsburgh, PA, USA); Tetrahydrofuran: HPLC grade (Fisher Scientific, Pittsburgh, PA, USA); Humco™ Sanare scar base (Humco, Texarkana, TX, USA); Syringe filter: Whatman 25 mm GD/XF Syringe Filter with 0.45 µm PTFE Filter media Part #6974-2504 or equivalent (GE Sciences, Pittsburgh, PA, USA).

2.1.3. Mobile Phase A Preparation

A 1000-mL portion of purified water and 2.9 g sodium phosphate dibasic were combined and mixed well. The pH of the solution was adjusted to 3.0 $\pm$ 0.1 with *o*-phosphoric acid 85%.

2.1.4. Diluent (75/25 Water/Acetonitrile)

A 750-mL portion of HPLC grade water and 250 mL of acetonitrile were combined and mixed well. Volumes were scaled as necessary.

2.1.5. Standard Preparation

For the Stock Standard, actives were accurately weighed, to the nearest 0.1 mg. A quantity of 40 mg of Pentoxifylline, 20 mg of Bupivacaine HCl, 40 mg of Levocetirizine HCl, 40 mg Tranilast, and 20 mg of Fluticasone Propionate was transferred into a 100-mL volumetric flask. A 5-mL portion of THF was added along with 45 mL of acetonitrile to the flask. The solution was sonicated for about 5 min to dissolve the actives, and then the flask was diluted to volume with diluent. This was the Stock Standard solution.

For the Working Standard Solution, an aliquot of 5 mL of the Stock Standard solution was pipetted into a 10 mL volumetric flask. This solution was diluted to volume with diluent. This was the Working Standard solution.

2.1.6. Sample Preparation

For the Working Sample solution, about 2 g of the sample was weighed into a 200 mL volumetric flask. About 10 mL of THF was added to the flask followed by 90 mL of acetonitrile, and the sample was allowed to fully disperse with sonication. The flask was then diluted to volume with diluent. No further dilution was necessary. This was the Working Sample solution. Approximately, 3 mL of the sample was filtered using a 0.45 µm PTFE syringe filter into an appropriate HPLC vial for analysis.

2.2. Method Verification Elements

The following analytical method verification sections detail the documentation required to verify the performance characteristics of the procedure and ensure that it meets the requirements for the intended analytical applications. The acceptance criterion was the successful completion of each section. The verification included specificity, linearity, accuracy, and precision (system precision, or reproducibility, and method precision, or repeatability), and range [15–17].

2.2.1. Specificity

The specificity is the ability to assess unequivocally the analyte of interest in the presence of components that may be expected to be present, such as matrix components (preservatives or placebo peaks) or peaks in the blank. The Blank preparation and the Placebo preparation (Sanare scar base) were examined to ensure that no interference occurred at the retention time of any of the actives in the chromatograms.

2.2.2. Linearity

The linearity of an analytical procedure is its ability to elicit test results that are directly proportional to the concentration of the analyte in samples over a specified range. The analytical method must be shown to be linear over the range of 50%–150% of the nominal standard concentration, with the plot of concentration *vs.* analyte peak area for each analyte having a correlation coefficient (r^2) of $\geqslant 0.99$.

Limit of quantitation (LOQ) and limit of detection (LOD) for each of the actives should be determined by successive dilution of the Working Standard solution and applying the signal-to-noise ratio test to the resulting chromatograms. LOQ is the concentration at which the signal-to-noise ratio was about 10:1, while LOD is the concentration at which the signal-to-noise ratio was about 3:1.

2.2.3. Accuracy

The accuracy of an analytical procedure is the closeness of test results obtained by that procedure to the true value. The accuracy of this method should be verified by determining the recovery of a known amount of each analyte added to the sample matrix (a spiked placebo). The percent recovery of each analyte from the placebo spiked at 50%–150% of the nominal standard concentration should be determined to be 98%–102%. Additionally, the peak area % RSD (percent relative standard deviation) among sets of samples at each concentration must be $\leqslant 2.0\%$.

2.2.4. Precision

The precision of an analytical procedure is the degree of agreement among individual test results when the procedure is applied repeatedly to multiple samplings of a homogeneous sample. This is further broken down into system precision and method precision.

System Precision (Reproducibility)

The system precision or reproducibility evaluates the ability of the method to analyze a single preparation by injecting the sample six times. The peak area % RSD of each analyte among the six injections must be $\leqslant 2.0\%$.

Method Precision (Repeatability)

The method precision or repeatability evaluates the ability of the method to analyze multiple preparations of sample. This was determined by assaying three individual preparations injected in triplicate. The peak area % RSD of each analyte for the three individual preparations must be $\leqslant 2.0\%$.

2.2.5. Range

The range for an analytical procedure is established over the concentrations of the analytes where acceptable precision, accuracy, and linearity have been demonstrated. The range of the analytical method should be established by examining the precision, accuracy, and linearity studies.

3. Results and Discussion

3.1. System Suitability

System suitability of the method was proven using the parameters that are used by compendia to prove that the data generated are valid. The relative standard deviation of the peak area responses of Pentoxifylline, Bupivacaine HCl, Levocetirizine HCl, Tranilast, and Fluticasone Propionate for the first five consecutive injections and for all injections of the Working Standard solution was ⩽2.0%. The relative standard deviation of the retention times of Pentoxifylline, Bupivacaine HCl, Levocetirizine HCl, Tranilast, and Fluticasone Propionate for all Working Standard injections was ⩽0%. Theoretical plates for Pentoxifylline, Bupivacaine HCl, Levocetirizine HCl, Tranilast, and Fluticasone Propionate in the Working Standard solution were ⩾2000. The tailing factor for Pentoxifylline, Bupivacaine HCl, Levocetirizine HCl, Tranilast, and Fluticasone Propionate in the Working Standard solution was ⩽2.0. The resolution between each of the components of Pentoxifylline, Bupivacaine HCl, Levocetirizine HCl, Tranilast, and Fluticasone Propionate in the Working Standard solution was >1.5. No interference in the blank or placebo preparation (⩾0.3%) was observed at the retention time of Pentoxifylline, Bupivacaine HCl, Levocetirizine HCl, Tranilast, and Fluticasone Propionate.

The system suitability met all acceptance criteria, therefore, the system was suitable to analyze the samples for further method verification elements.

Table 4 gives the system suitability results and specifications obtained during the method verification.

Table 4. System Suitability of the high performance liquid chromatography (HPLC): n/a = not applicable. % RSD: percent relative standard deviation.

Active	Peak Area % RSD (n = 6)	Peak Area % RSD (overall)	Average Overall Retention Time % RSD	Average Theoretical Plates	Average Tailing	Average Resolution
Pentoxifylline	1%	1%	0%	44,833	1.1	n/a
Bupivacaine HCl	1%	1%	0%	70,171	1.4	12.1
Levocetirizine HCl	0%	1%	0%	258,990	1.1	36.9
Tranilast	0%	0%	0%	231,214	1.1	10.5
Fluticasone Propionate	0%	1%	0%	351,525	1.1	10.3
Specification	⩽2.0%	⩽2.0%	⩽2.0%	>2000	⩽2.0	>1.5

3.2. Specificity Results

3.2.1. Examining the Blank

The sample blank was assayed to verify that there are no significant peaks with similar retention times as Pentoxifylline, Bupivacaine HCl, Levocetirizine HCl, Tranilast, or Fluticasone Propionate. The Blank chromatogram exhibits no peaks (other than a small solvent front peak) beyond normal noise. There is a baseline ramp up to 39 min that is present, but this is a function of the gradient elution and not a true peak. No significant peaks (⩾0.3% of the analytes of interest), beyond the noise level were noted in the sample blank near the retention times of the analytes of interest.

3.2.2. Examining the Sample Matrix (Placebo)

The sample matrix without the active ingredient (also known as a placebo)—Humco™ Sanare scar base in this case—was assayed to verify that there are no significant peaks with similar retention times as Pentoxifylline, Bupivacaine HCl, Levocetirizine HCl, Tranilast, or Fluticasone Propionate. There were two an identified peaks (In Placebo 1 and In Placebo 2) in the Placebo chromatogram, representing components of the botanical extract contained in the product. These peaks do not interfere with analysis of the actives as they do not occur at the actives' retention times. As such, no significant peaks (⩾0.3% of the analytes of interest) were noted in the Placebo sample matrix near the retention time of the analytes of interest. Unidentified small peaks (retention time = 24.3 and 32.2 min) in the Standard and Sample chromatograms are process impurities of Tranilast that are present in the raw material.

Blank, Placebo, Working Standard, and Sample chromatograms are given in Figures 1–4.

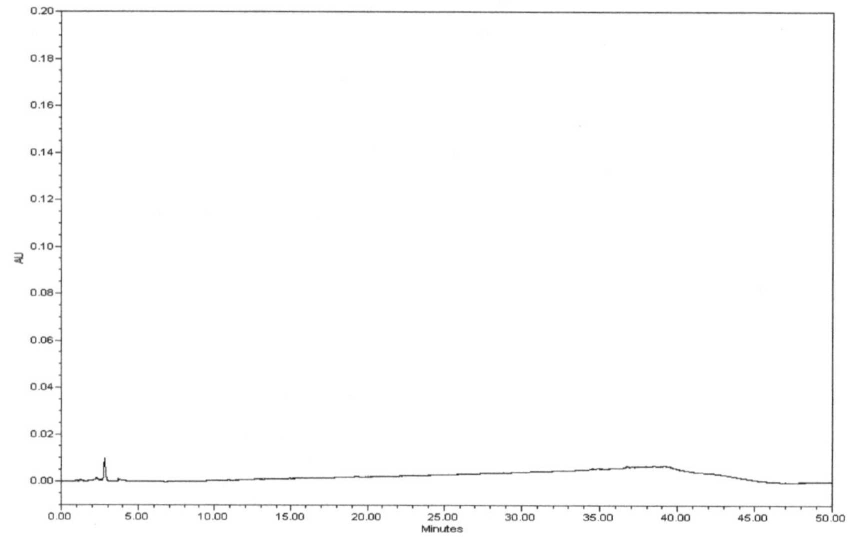

Figure 1. Blank chromatogram.

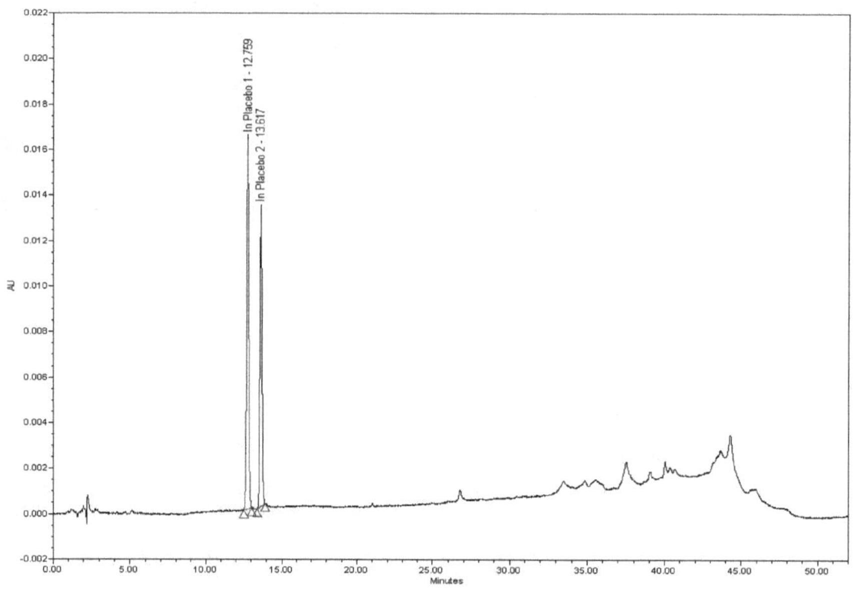

Figure 2. Placebo (Humco™ Sanare scar base) chromatogram.

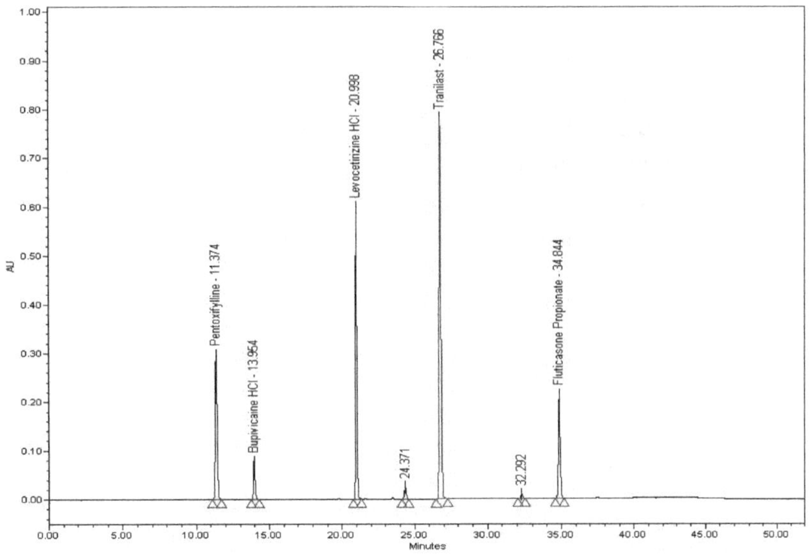

Figure 3. Working Standard solution chromatogram.

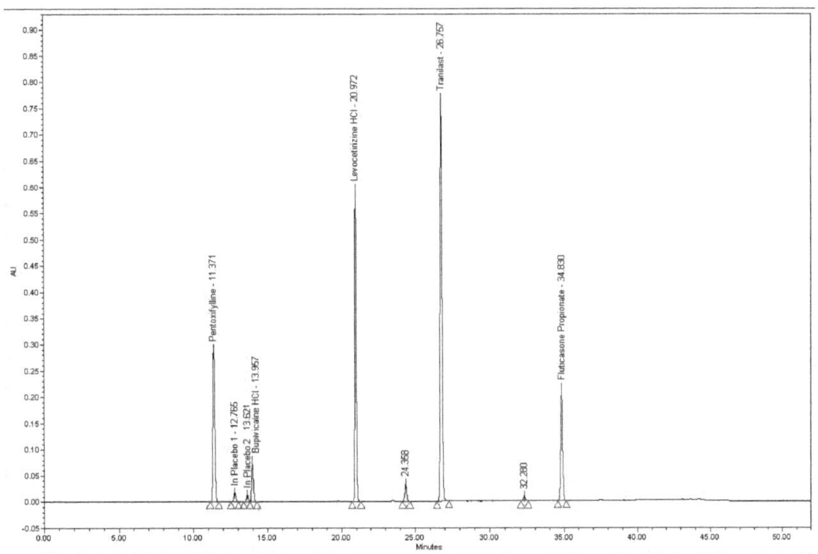

Figure 4. Sample preparation chromatogram.

3.2.3. Specificity Discussion

There was no interference in the blank or Placebo (Humco™ Sanare scar base) chromatograms at the retention times of Pentoxifylline, Bupivacaine HCl, Levocetirizine HCl, Tranilast, or Fluticasone Propionate. Therefore, the acceptance criteria for Specificity of Pentoxifylline, Bupivacaine HCl, Levocetirizine HCl, Tranilast, or Fluticasone Propionate are met.

3.3. Linearity Results

Linearity for Pentoxifylline, Bupivacaine HCl, Levocetirizine HCl, Tranilast, or Fluticasone Propionate was conducted over a range of 50%–150% of the nominal analytes in the prepared sample concentration. Five concentrations were tested within the range of 50%–150%.

3.3.1. Experiments

The Stock Standard solution of Pentoxifylline, Bupivacaine HCl, Levocetirizine HCl, Tranilast, or Fluticasone Propionate was prepared. The linearity was accomplished by making injections of varying volumes of the Working Standard solution. During calibration of the HPLC, the injection volumes are verified for accuracy over the stated range; therefore, this method for testing linearity of the components is acceptable. For example, for the 50% level, 5 μL of the Working Standard was injected; for the 80% level, 8 μL of the Working Standard was injected; for the 100% level, 10 μL of the Working Standard was injected; for the 120% level, 12 μL of the Working Standard was injected; and for the 150% level, 15 μL of the Working Standard was injected. Serial dilutions of the Working Standard were made to determine the approximate LOQ and LOD of the five actives.

Table 5 gives the Stock Standard weights for each compound and the effective concentrations (mg/mL) represented by the varied injection volumes for each of the linearity levels.

Table 5. Concentrations of each active used in the linearity and accuracy evaluations. (Conc.: concentration).

Compound	Standard Weight (mg)	50% Conc. (mg/mL)	80% Conc. (mg/mL)	100% Conc. (mg/mL)	120% Conc. (mg/mL)	150% Conc. (mg/mL)
Pentoxifylline	39.4	0.0985	0.1576	0.1970	0.2364	0.2955
Bupivacaine HCl	19.5	0.0488	0.0780	0.0975	0.1170	0.1463
Levocetirizine HCl	40.0	0.1000	0.1600	0.2000	0.2400	0.3000
Tranilast	41.2	0.1030	0.1648	0.2060	0.2472	0.3090
Fluticasone Propionate	20.5	0.0513	0.0820	0.1025	0.1230	0.1538

3.3.2. Linearity of Actives

Table 6 gives the linearity results for each active in the method verification, while Table 7 gives the LOQ and LOD concentrations of each active as determined in the method verification.

Table 6. The linearity level, theoretical and actual concentrations, and percent recovery as well as the r^2-value (correlation coefficient) for each active.

Active	Linearity %	Theoretical Conc. (mg/mL)	Actual Conc. (mg/mL)	% Recovery	r^2
Pentoxifylline	50	0.0985	0.0982	99.7%	0.9999
	80	0.1576	0.1582	100.3%	
	100	0.1970	0.1965	99.7%	
	120	0.2364	0.2370	100.2%	
	150	0.2955	0.2952	99.9%	
Bupivacaine HCl	50	0.0488	0.0488	100.0%	0.9999
	80	0.0780	0.0782	100.3%	
	100	0.0975	0.0971	99.6%	
	120	0.1170	0.1171	100.1%	
	150	0.1463	0.1463	100.0%	
Levocetirizine HCl	50	0.1000	0.0996	99.6%	0.9999
	80	0.1600	0.1605	100.3%	
	100	0.2000	0.1999	99.9%	
	120	0.2400	0.2403	100.1%	
	150	0.3000	0.2997	99.9%	
Tranilast	50	0.1030	0.1027	99.7%	0.9999
	80	0.1648	0.1653	100.3%	
	100	0.2060	0.2057	99.8%	
	120	0.2472	0.2476	100.1%	
	150	0.3090	0.3088	99.9%	
Fluticasone Propionate	50	0.0513	0.0511	99.6%	0.9999
	80	0.0820	0.0823	100.4%	
	100	0.1025	0.1024	99.9%	
	120	0.1230	0.1231	100.1%	
	150	0.1538	0.1536	99.9%	

Table 7. Limit of quantitation (LOQ) and limit of detection (LOD) concentrations for each active (determined by signal to noise ratio).

Active	LOQ (mg/mL)	LOD (mg/mL)
Pentoxifylline	0.00131	0.00044
Bupivacaine HCl	0.00975	0.00325
Levocetirizine HCl	0.00100	0.00054
Tranilast	0.00103	0.00034
Fluticasone Propionate	0.00513	0.00171

3.3.3. Linearity Discussion

The correlation coefficient (r^2) from the plotted area response *versus* concentration for Pentoxifylline is 0.9999, which is ⩾0.99; for Bupivacaine HCl is 0.9999, which is ⩾0.99; for Levocetirizine HCl is 0.9999, which is ⩾0.99; for Tranilast is 0.9999, which is ⩾0.99; and for Fluticasone Propionate is 0.9999, which is ⩾0.99. The data used for calculation of the linearity are represented in Table 6.

The average percent recovery for Pentoxifylline is 99.7%–100.3%, which is in the range 98%–102%; for Bupivacaine HCl is 99.6%–100.3%, which is in the range 98%–102%; for Levocetirizine HCl is 99.6%–100.3%, which is in the range 98%–102%; for Tranilast is 99.7%–100.3%, which is in the range 98%–102%; and for Fluticasone Propionate is 99.6%–100.4%, which is in the range 98%–102% of the amount prepared for the 50%–150% level.

All acceptance criteria for the linearity of Pentoxifylline, Bupivacaine HCl, Levocetirizine HCl, Tranilast, and Fluticasone Propionate are met.

3.4. *Accuracy Results*

The accuracy of the method was proven by using spiked placebo solutions that were prepared by spiking in the appropriate amount of the analytes of interest into the sample matrix and assayed using a standard. The spiked placebo preparation spiked with each of the analytes of interest (Pentoxifylline, Bupivacaine HCl, Levocetirizine HCl, Tranilast, and Fluticasone Propionate) over a range of 50%–150% of the nominal standard concentration. These solutions were assayed, and the data were compared with the amount prepared *versus* the amount recovered.

3.4.1. Accuracy of Actives

The concentrations of the spiked placebo solutions were the same as those represented in Table 5, and results of the accuracy evaluation are given in Table 8.

Table 8. Accuracy level, theoretical and actual concentrations of each active, percent recovery of each active, and the % RSD of triplicate injections at each accuracy level for each active.

Compound	Accuracy %	Theoretical Conc. (mg/mL)	Actual Conc. (mg/mL)	% Recovery	% RSD
Pentoxifylline	50	0.1035	0.1035	100.0%	0.3
	80	0.1656	0.1629	98.4%	0.2
	100	0.2070	0.2101	101.5%	0.4
	120	0.2484	0.2496	100.5%	0.2
	150	0.3105	0.3089	99.5%	0.7
Bupivacaine HCl	50	0.0498	0.0497	99.9%	0.2
	80	0.0796	0.0786	98.8%	0.2
	100	0.0995	0.1008	101.3%	0.1
	120	0.1194	0.1198	100.4%	0.2
	150	0.1493	0.1486	99.6%	0.6

Table 8. *Cont.*

Compound	Accuracy %	Theoretical Conc. (mg/mL)	Actual Conc. (mg/mL)	% Recovery	% RSD
Levocetirizine HCl	50	0.0995	0.0991	99.6%	0.3
	80	0.1592	0.1588	99.7%	0.6
	100	0.1990	0.2006	100.8%	0.1
	120	0.2388	0.2383	99.8%	0.1
	150	0.2985	0.2981	99.9%	0.5
Tranilast	50	0.1030	0.1040	101.0%	0.2
	80	0.1648	0.1634	99.2%	0.4
	100	0.2060	0.2050	99.5%	0.2
	120	0.2472	0.2487	100.6%	0.0
	150	0.3090	0.3089	100.0%	0.5
Fluticasone Propionate	50	0.0510	0.0515	101.0%	0.3
	80	0.0816	0.0809	99.1%	0.5
	100	0.1020	0.1023	100.2%	0.2
	120	0.1224	0.1219	99.6%	0.1
	150	0.1530	0.1535	100.3%	0.5

3.4.2. Accuracy Discussion

The recovery for the Pentoxifylline, Bupivacaine HCl, Levocetirizine HCl, Tranilast, and Fluticasone Propionate was within the acceptance criteria of 98%–102%. The % RSD among the accuracy preparations was ≤2.0% RSD, meeting the acceptance criteria.

The accuracy of Pentoxifylline, Bupivacaine HCl, Levocetirizine HCl, Tranilast, and Fluticasone Propionate meets the acceptance criteria.

3.5. Precision Results

The system precision (reproducibility) and method precision (repeatability) were evaluated using preparations of the spiked placebo. The system precision evaluated the ability of the method to analyze a single sample preparation by injecting the sample six times. The method precision evaluated the ability of the method to analyze multiple preparations of sample. This was determined by assaying three individual preparations.

3.5.1. System Precision Results

System precision (reproducibility) results are given in Table 9.

Table 9. System precision results (% RSD) for six injections of a single sample preparation.

Active	Peak Area % RSD; (n = 6 Injections)
Pentoxifylline	0.8%
Bupivacaine HCl	0.6%
Levocetirizine HCl	0.2%
Tranilast	0.2%
Fluticasone Propionate	0.5%

3.5.2. System Precision Discussion

The % RSD for the Pentoxifylline, Bupivacaine HCl, Levocetirizine HCl, Tranilast, and Fluticasone Propionate peak areas for the six replicate injections is ≤2.0%. The system precision acceptance criteria are met.

3.5.3. Method Precision Results

Method Precision (Repeatability) results are given in Table 10.

Table 10. Method precision results (% RSD) for three individual preparations of a sample.

Active	Peak Area % RSD; (n = 3 Preparations)
Pentoxifylline	1.0%
Bupivacaine HCl	0.6%
Levocetirizine HCl	0.8%
Tranilast	0.8%
Fluticasone Propionate	1.0%

3.5.4. Method Precision Discussion

The % RSD for the Pentoxifylline, Bupivacaine HCl, Levocetirizine HCl, Tranilast, and Fluticasone Propionate peak areas for the three preparations is ⩽2.0%. The method precision acceptance criteria are met.

3.6. *Range Results and Discussion*

The results of the precision, accuracy, and linearity each pass the respective specifications over the 50%–150% nominal range for each analyte specified in this study. The range for the method is concluded to be 50%–150% the nominal standard concentration of each analyte in the study.

4. Conclusions

The method verification elements of linearity, accuracy, specificity, precision, and range [15–17] met each of the respective elements' acceptance criteria; therefore, the analytical method is considered to be verified for its intended purposes, as defined previously.

Acknowledgments: The author wishes to thank the Humco Holding Group Board of Directors for approving funding for the work performed in this study.

Conflicts of Interest: The author declares no conflict of interest.

References

1. Yamamoto, M.; Yamauchi, T.; Okano, K.; Takahashi, M.; Watabe, S.; Yamamoto, Y. Tranilast, an anti-allergic drug, down-regulates the growth of cultured neurofibroma cells derived from neurofibromatosis type. *Tohoku J. Exp. Med.* **2009**, *217*, 193–201. [CrossRef] [PubMed]
2. Tranilast Inhibits Cytokine-Induced Nuclear Factor κB Activation in Vascular Endothelial Cells. Available online: http://molpharm.aspetjournals.org/content/62/4/856.full (accessed on 1 December 2015).
3. Tranilast, a Selective Inhibitor of Collagen Synthesis in Human Skin Fibroblasts. Available online: http://jb.oxfordjournals.org/cgi/content/abstract/116/4/892 (accessed on 1 December 2015).
4. Grant, J.A.; Riethuisen, J.M.; Moulaert, B.; DeVos, C.; Gamalero, C.; Descalzi, D.; Folli, C.; Passalacqua, G.; Canonica, G.W. A double-blind, randomized, single-dose, crossover comparison of levocetirizine with ebastine, fexofenadine, loratadine, mizolastine, and placebo: Suppression of histamine-induced wheal-and-flare response during 24 hours in healthy male subjects. *Ann. Allergy Asthma Immunol.* **2002**, *88*, 190–197. [CrossRef]
5. Pasquali, M.; Baiardini, I.; Rogkakou, A.; Riccio, A.M.; Gamalero, C.; Descalzi, D.; Folli, C.; Passalacqua, G.; Canonica, G.W. Levocetirizine in persistent allergic rhinitis and asthma: Effects on symptoms, quality of life and inflammatory parameters. *Clin. Exp. Allergy* **2006**, *36*, 1161–1167. [CrossRef] [PubMed]
6. Trental, Pentoxil (pentoxifylline) dosing, indications, interactions, adverse effects, and more. Available online: http://reference.medscape.com/drug/trental-pentoxil-pentoxifylline-342179 (accessed on 3 December 2015).
7. Rhen, T.; Cidlowski, J.A. Anti-inflammatory action of glucocorticoids—New mechanisms for old drugs. *N. Engl. J. Med.* **2005**, *353*, 1711–1723. [CrossRef] [PubMed]
8. Bupivacaine Hydrochloride: The American Society of Health-System Pharmacists. Available online: http://www.drugs.com/monograph/bupivacaine-hydrochloride.html (accessed on 2 May 2015).

9. Pentoxifylline: A drug with wide spectrum applications in dermatology. Available online: http://escholarship.org/uc/item/6270c6vm (accessed on 30 November 2015).
10. Ward, A.; Clissold, S.P. Pentoxifylline. A review of its pharmacodynamic and pharmacokinetic properties, and its therapeutic efficacy. *Drugs* **1987**, *34*, 50–97. [CrossRef] [PubMed]
11. Zatz, J.L.; Steinberg, D.C. Topical protective and cosmetic products. In *Transdermal and Topical Drug Delivery Systems*; Ghosh, T.K., Pfister, W.R., Su, Y., Eds.; Interpharm Press Inc.: Inglewood, CO, USA, 1997; Chapter 17; p. 624.
12. Nayak, B.; Kanhai, J.; Milne, D.M.; Pereira, L.P.; Swanston, W.H. Experimental Evaluation of Ethanolic Extract of *Carapa guianensis* L. Leaf for Its Wound Healing Activity Using Three Wound Models. *Evid. Based Complement. Alternat. Med.* **2009**. [CrossRef] [PubMed]
13. Hammer, M.L.; Johns, E.A. Tapping an Amazonian plethora: Four medicinal plants of Marajó Island, Pará (Brazil). *J. Ethnopharmacol.* **1993**, *40*, 53–75. [CrossRef]
14. Nayak, B.; Kanhai, J.; Milne, D.M.; Swanston, W.H.; Mayers, S.; Eversley, M.; Rao, A.V. Investigation of the wound healing activity of *Carapa guianensis* L. (Meliaceae) bark extract in rats using excision, incision, and dead space wound models. *J. Med. Food* **2010**, *13*, 1141–1146. [CrossRef] [PubMed]
15. Validation of Compendial Procedures; United States Pharmacopoeia. Available online: http://www.ofnisystems.com/wp-content/uploads/2013/12/USP36_1225.pdf (accessed on 15 August 2015).
16. Validation of Analytical Procedures: Text and Methodology. Available online: http://www.ich.org/products/guidelines/quality/quality-single/article/validation-of-analytical-procedures-text-and-methodology.html (accessed on 1 August 2015).
17. *Guidance for Industry: Analytical Procedures and Methods Validation for Drugs and Biologics*; Food and Drug Administration Center for Drug Evaluation Research: Silver Spring, MD, USA, 2015.

16

Method for the Determination of Benzene Metabolite *t,t*-Muconic Acid in Urine by HPLC-UV with an Ion Exclusion Column

Rafaela de Paiva Gomes, Ananda Lima Sanson, Fabiana Aparecida Lobo, Robson José de Cássia Franco Afonso and Mauricio Xavier Coutrim *

Department of Chemistry, Institute of Exact and Biological Sciences, Federal University of Ouro Preto, Campus Morro do Cruzeiro, s/n, Bauxita, 35400-000 Ouro Preto, MG, Brazil; rafaela_p_gomes@hotmail.com (R.P.G.); ananda_lima@hotmail.com (A.L.S.); fabiana@iceb.ufop.br (F.A.L.); robsonafonso@iceb.ufop.br (R.J.C.F.A.)

* Correspondence: mcoutrim@iceb.ufop.br

Academic Editor: Frank L. Dorman

Abstract: The benzene metabolite *trans,trans*-muconic acid (*tt*-MA) is widely used as a biological indicator of exposure to this xenobiotic. An analytical method was developed and validated for the determination of urinary *tt*-MA using solid phase extraction onto a Strata SAX cartridge. The analysis was performed by HPLC with an Aminex HPX-87H ion exclusion column and a UV diode array detector. The sample preparation conditions were optimized using a 2^{4-1} factorial design. Comparison of the slopes of standard analytical curves prepared in aqueous solution and urine showed that the biological matrix suppressed the *tt*-MA signal by around 50%, so the analytical curves were prepared with *tt*-MA standard at low concentrations in pooled urine. The analytical curves in the range of 5–500 $\mu g \cdot L^{-1}$ showed determination coefficients values (R^2) > 0.99 for *tt*-MA standard in water and R^2 > 0.98 for *tt*-MA standards in pooled urine. The coefficients of variation obtained using seven replicates were lower than 3.6%, and recoveries of *tt*-MA from solutions containing 5, 25, 50 and 100 $\mu g \cdot L^{-1}$ of the analyte were in the range 85%–90%, demonstrating the satisfactory precision and accuracy of the method. The limits of detection and quantification were 0.11 and 0.36 $\mu g \cdot L^{-1}$, respectively. The benefits of this new method developed is the possibility of complete chromatographic peak separation for the determination of *tt*-MA at baseline, without matrix components' interference as normally found in the C18 column. This is the first time that this chromatographic column has been used for the analysis of *tt*-MA in urine.

Keywords: benzene; *trans,trans*-muconic acid; exposure biomarker; environmental exposure; Aminex HPX-87H ion exclusion column

---

1. Introduction

Despite substantial progress over the last 20 years in understanding the risks posed by benzene, there is a continuing need for the development of sensitive techniques able to determine the compound at low levels, in order to assist in the implementation of control strategies. Various methods have been reported for the evaluation of the contamination of humans by benzene. These normally measure benzene in its original form and have been applied to exhaled air [1–4], blood [4–7] and urine [6,8–10]. However, various difficulties have been described, related to problems in sample preparation (for air samples), the small fraction of benzene that remains unaltered (in urine) and invasive sample collection (of blood). An alternative to the evaluation of exposure to benzene by directly measuring its concentration in air is to determine the concentrations of its metabolites in urine or blood that involve the use of biological exposure indicators (BEIs). Two of the metabolites of benzene that can be used as a BEI are *trans,trans*-muconic acid (*tt*-MA) and S-phenyl mercapturic acid (SPMA), which are biomarkers that are more sensitive and specific, compared to urinary phenol [11]. This exposure biomarker has therefore been adopted in Brazilian legislation for the monitoring of occupational exposure to benzene [12]. The widespread use of *tt*-MA as a BEI is due to the good correlation obtained between the level of the compound in urine and the concentrations of benzene in air and blood, even after exposure to low concentrations of benzene.

Various techniques have been developed for the determination of *tt*-MA in urine. The most widely-used sample preparation procedure employs solid phase extraction (SPE) with ion exchange resins [13–18]. This procedure reduces the quantities of solvents required and provides better recoveries, as well as clearer extracts, compared to the non-extraction procedure [19,20].

Only a few studies have employed gas chromatography for the determination of urinary *tt*-MA [21–25]. On the other hand, there has been widespread use of high performance liquid chromatography (HPLC) with reverse phase columns and ultraviolet and visible spectrophotometry (UV-VIS) or mass spectrometry (MS) detection [26–42]. One way to separate anions by chromatography is to use cation exchange columns where the ions are separated by ion exclusion. In this type of chromatography, the analytical conditions allow that part of the mobile phase to remain immobilized in the stationary phase, and the separation is due to the differences in the partitioning of the molecular solute between the solvent from the mobile phase and the solvent immobilized on the resin [43]. However, no analytical method has been described in the literature for the determination of urinary *tt*-MA as a BEI for benzene using HPLC with an Aminex HPX-87H column, although this analyte and other volatile organic acids have been determined in various other sample matrices using this type of column [44–46].

The objective of this work was to develop a new technique for the measurement of *tt*-MA in urine, using SPE (SAX cartridges) and HPLC with an Aminex HPX-87H column and UV detection.

2. Experimental Section

2.1. Instruments and Accessories

The extraction of *tt*-MA employed Strata SAX cartridges containing 500 mg of trimethylaminopropylsilane (Phenomenex®, Torrance, CA, USA), a manifold (Phenomenex®) and a vacuum pump (Model 131, Prismatec Ind Com, Itu, SP, Brazil). Analysis was performed with an HPLC (Model 20A, Shimadzu, Columbia, MD, USA) equipped with a diode array. The column used was an Aminex HPX-87H (300 × 7.8 mm, Bio-Rad Laboratories, Hercules, CA, USA), recommended for the separation of carboxylic acids, installed in a temperature-controlled oven (Model TC300, GBC Scientific Equipment Pty Ltd., Hamsphire, IL, USA), and the chromatograms were recorded at 264 nm using LabSolutions System software, version 5.53 SP3 (Shimadzu, Laval, Quebec, PQ, Canada). Additional items of equipment were micropipettes of various volumes, an analytical balance (Model AUY 220, Shimadzu) and a membrane filtration system (PTFE, 0.22-µm pore size, Phenomenex®).

2.2. Reagents, Solutions and Samples

The reagents used were methanol, acetic acid, acetonitrile and sulfuric acid, all chromatography grade (JT Baker®, Center Valley, PA, USA), *trans,trans*-muconic acid (analytical grade, 99% min. purity, Sigma-Aldrich®, St. Louis, MO, USA) and pH 8.0 tris-phosphate buffer (Sigma-Aldrich®). The ultra-pure water (conductivity of 0.065 µS) was obtained from a LAB-UPW system (TKA, Niederelbert, Germany).

A standard solution of *tt*-MA (10 $mg \cdot L^{-1}$) was prepared in water and diluted to obtain standards at lower concentrations. Ten different urine samples were obtained from non-smoker volunteers (5 males and 5 females) and were combined to produce a pooled urine sample.

2.3. Optimization of the Procedure

A 2^{4-1} factorial design was used to define the sample preparation conditions. The chromatographic conditions were optimized in a univariate manner.

2.3.1. Preparation of Samples

The sample cleanup procedure was based on the method described previously by Ducos *et al.* (1990) [27] and is illustrated in the flow diagram of Figure 1.

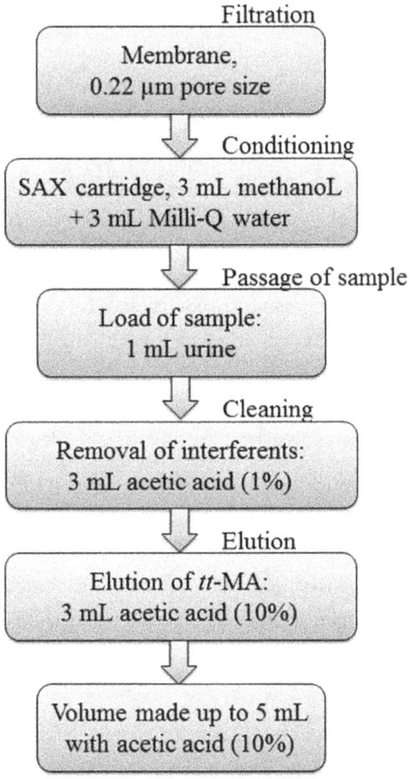

Figure 1. Flow diagram illustrating the sample cleanup procedure. *tt*-MA, *trans,trans*-muconic acid.

The sample was filtered through a 0.22 μm membrane, after which the *tt*-MA was extracted using a SAX cartridge. The extraction was performed under vacuum using a manifold with a capacity for 12 cartridges. The entire procedure was performed in triplicate.

2.3.2. Optimization of the Sample Preparation Procedure

A factorial design was used to evaluate the need for the buffering of the samples, as well as to adjust other factors related to sample transfer, cleanup and elution. An appropriate electronic spreadsheet was used for the preparation and interpretation of the experimental designs [47]. Four variables were investigated (sample buffering, removal of interferents (washing), final volume and sample volume), at three levels (Table 1).

Table 1. Variables and levels used in the 2^{4-1} fractional factorial design.

Variable	**Level**		
	−1	**0**	**1**
Solution buffering	No buffering	pH 8.0	pH 10.0
Removal of interferents (washing)	No addition of water	1.5 mL of water	3.0 mL of water
Final volume	3.0 mL	4.0 mL	5.0 mL
Sample volume	1.0 mL	2.0 mL	3.0 mL

The parameters shown in Table 1 were used to generate 11 experiments, as shown in Table 2. The experiments were performed in random order. Throughout the optimization procedure, the response was based on the analyte peak area.

Table 2. Investigation of the experimental variables using a 2^{4-1} fractional factorial design with three replicates at the central point (CP).

Tests \ Variable	Buffering of the Solution	Washing	Final Volume	Sample Volume
Test 1	no buffering (−1)	no water addition (−1)	3.0 mL (−1)	1.0 mL (−1)
Test 2	pH 10.0 (1)	no water addition (−1)	3.0 mL (−1)	3.0 mL (1)
Test 3	no buffering (−1)	3.0 mL of water (1)	3.0 mL (−1)	3.0 mL (1)
Test 4	pH 10.0 (1)	3.0 mL of water (1)	3.0 mL (−1)	1.0 mL (−1)
Test 5	no buffering (−1)	no water addition (−1)	5.0 mL (1)	3.0 mL (1)
Test 6	pH 10.0 (1)	no water addition (−1)	5.0 mL (1)	1.0 mL (−1)
Test 7	no buffering (−1)	3.0 mL of water (1)	5.0 mL (1)	1.0 mL (−1)
Test 8	pH 10.0 (1)	3.0 mL of water (1)	5.0 mL (1)	3.0 mL (1)
CP 1	pH 8.0 (0)	1.5 mL of water (0)	4.0 mL (0)	2.0 mL (0)
CP 2	pH 8.0 (0)	1.5 mL of water (0)	4.0 mL (0)	2.0 mL (0)
CP 3	pH 8.0 (0)	1.5 mL of water (0)	4.0 mL (0)	2.0 mL (0)

Note: Level (1): highest value; Level (−1): lowest value; Level (0): central point (CP).

2.3.3. Chromatographic Conditions

Development of the new analytical method was initially based on the procedure described by Ducos *et al.* (1990) [27], with modifications in terms of the types of chromatographic precolumn and main column employed, as well as the mobile phase and the wavelength used to detect *tt*-MA. In this step of the work, the optimization of the variables was performed in a univariate manner. Nine chromatographic mobile phase compositions were tested, together with two different flow rates and the presence or absence of the precolumn.

2.4. *Validation of the Method*

The analytical method was validated considering the following parameters: selectivity, linearity, working range, limits of detection and quantification, precision and accuracy.

A comparison was made of analytical curves constructed using solutions prepared in water and in the sample matrix. In the case of the sample matrix, the curves employed the results obtained for standard solutions prepared by adding aliquots of *tt*-MA solutions to volumes of the pooled urine sample composed of the samples obtained from the ten non-smoker volunteers. The *tt*-MA solutions were prepared by fortification of the pooled urine with an intermediate standard solution obtained by dilution of the 10 mg·L^{-1} solution of *tt*-MA in water.

The standards prepared in the sample matrix were subsequently submitted to the same treatment employed for the samples: filtration through a 0.22 μm membrane, solid phase extraction, and injection onto the chromatographic column (as described in Section 2.3.1).

2.4.1. Selectivity

The selectivity of the method was evaluated by comparing the analyte retention times obtained for the samples and the standard solutions and by comparing the absorption spectra of the chromatographic peaks obtained for urine samples fortified with *tt*-MA standard with those for standard solutions of *tt*-MA prepared in water.

2.4.2. Matrix Interferences

Interferences caused by the presence of the biological matrix were assessed by comparing the coefficients of determination obtained for analytical curves constructed using aqueous standard solutions and standard solutions prepared using the pooled urine sample. The SPE extraction was performed under the optimized conditions.

2.4.3. Analytical Curve Adjustment and Working Range

The analytical curve adjustment and working range of the method were evaluated using the determination coefficients (R^2) obtained for the analytical curves constructed using the standards prepared in the pooled urine, in the concentration range 5–500 μg·L^{-1}.

2.4.4. Limits of Detection and Quantification

The limits of detection (LOD) and quantification (LOQ) of the method were determined using the average area of the baseline noise of several chromatograms obtained for the analysis of the pooled urine sample. The calculated LOD and LOQ concentration values were equivalent to 3- and 10-times the average area of the baseline noise, respectively.

2.4.5. Precision and Accuracy

The precision of the technique was measured using the coefficient of variation (CV) value obtained for repetitions of the entire procedure, from sample preparation up to injection onto the chromatographic column. Values of up to 20% are normally considered acceptable for analytical methods applied to environmental samples with complex matrices [48,49].

Accuracy was determined using the percentage recovery of *tt*-MA added to urine at concentrations of 5.0, 50 and 500 μg·L^{-1}, with nine replicates [48].

3. Results and Discussion

3.1. Optimization of Sample Preparation

Urine is a complex matrix that includes protein, salt, hormones, drugs and a variety of other compounds. Therefore, it is recommended that a pretreatment might be required in order to obtain a satisfactory chromatography performance. Qualitative tests of protein precipitation were performed using four systems: (1) pure sample; (2) sample with 0.01 mol·L^{-1} sulfuric acid; (3) sample with 0.01 mol·L^{-1} sulfuric acid and acetonitrile (65:35); and (4) 0.01 mol·L^{-1} sulfuric acid, used as a reference. All of the systems were kept at a temperature of 55 °C for 24 h, for the observation of turbidity. However, no cloudiness was observed in any of the cases, and then it was adopted only passing the sample through a 0.22 μm membrane.

A 2^{4-1} fractional factorial design with triplicate central points was used to optimize the extraction process, considering the following variables: sample buffering, remove of interferents, final volume and the volume of sample used in the extraction process. In the normal probability plot (Figure 2), point C1, corresponding to sample buffering, showed a large negative effect. Some of the previous studies reported in the literature used extraction without buffering [27,30,36,38,41], while in other work concerning the exposure of humans to benzene, better analytical results were achieved by buffering of samples [29,50,51].

The sample volume (C4 in Figure 2) showed the greatest positive effect, indicating that the best extraction conditions were obtained using 3 mL of urine, which provided a greater quantity of urinary *tt*-MA.

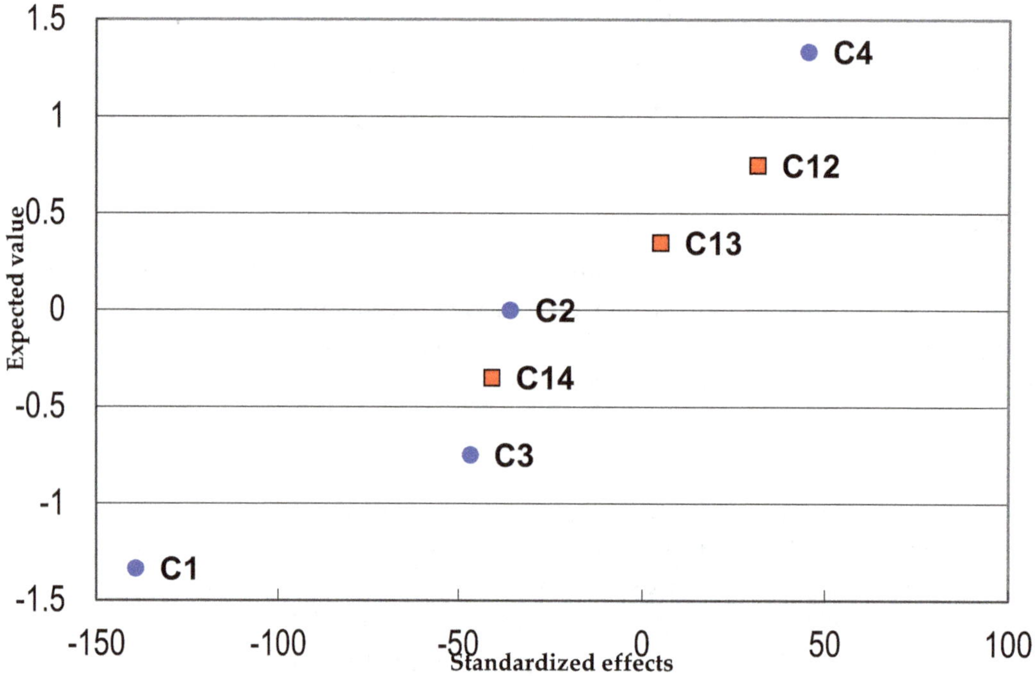

Figure 2. Normal probability plot showing the effects of sample buffering (C1), removal of interferents (C2), final volume (C3) and sample volume (C4).

In the fractional factorial design, the coefficient of variation of the tests was obtained by the analysis of the triplicate in the central point. The coefficient of variation of the signal obtained under the conditions was 1.95%. The highest *tt*-MA peak areas were obtained for Test 5, followed by Test 3 (Table 3).

Table 3. Optimization using the 2^{4-1} fractional factorial design. X_1: buffering of sample; X_2: removal of interferents; X_3: final volume; X_4: sample volume; y: area of the *tt*-MA peak.

Test	X_1	X_2	X_3	X_4	y
1	−1	−1	−1	−1	106,519
2	1	−1	−1	1	38,378
3	−1	1	−1	1	116,419
4	1	1	−1	−1	33,805
5	−1	−1	1	1	124,593
6	1	−1	1	−1	14,086
7	−1	1	1	−1	44,460
8	1	1	1	1	14,293
CP1	0	0	0	0	38,802
CP2	0	0	0	0	37,527
CP3	0	0	0	0	37,513
CP (mean)					37,947
CP (standard deviation)					740.2
CP (RSD, %)					1.95

Comparing the levels of four variables tested in the two best conditions, 3 and 5, we can see that variables that interfered more with the system of extraction, sample buffering (C1) and sample volume (C4) have the same levels in these two tests, confirming that the other variables (C2 and C3) do not interfere with the system.

The optimized sample preparation conditions (Table 4) were used in all of the subsequent experiments.

Table 4. Optimized sample preparation conditions.

Parameter	Condition
Buffering	None
Sample volume	3.0 mL
Sample washing	5.0 mL of acetic acid (1%)
Final volume	5.0 mL of acetic acid (10%)

3.2. Evaluation of Analytical Conditions

As shown in the results of the factorial design, the absence of buffering therefore provided the best analytical conditions. The first mobile phase used for the chromatographic separation of *tt*-MA was 0.01 mol·L^{-1} sulfuric acid. Other mobile phases tested were sulfuric acid at a concentration of 0.02 mol·L^{-1}, as well as sulfuric acid solutions with the addition of an organic modifier (acetonitrile) at different ratios (65:35, 75:25, 85:15, 90:10 and 95:5). However, none of these mobile phases showed better performance than sulfuric acid at a concentration of 0.01 mol·L^{-1}. Typical chromatograms obtained for the separation of *tt*-MA using this mobile phase at a flow rate of 0.6 mL·min^{-1} are shown in Figure 3.

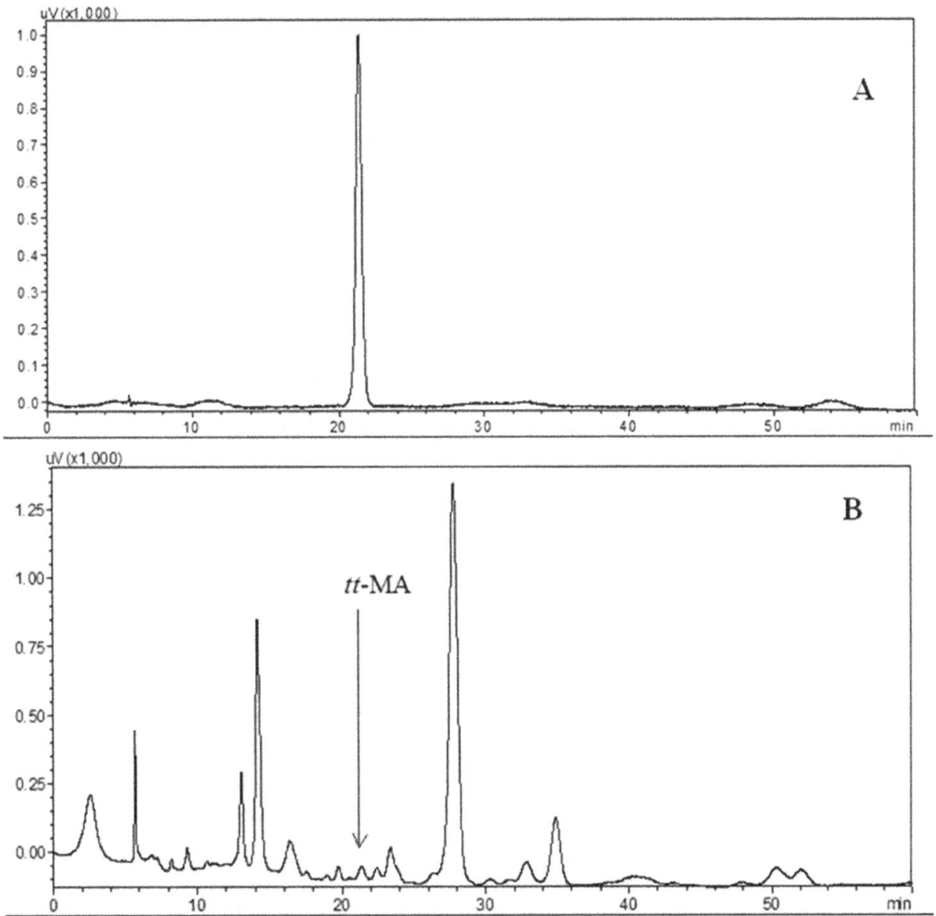

Figure 3. Chromatograms obtained for the separation of *tt*-MA using the Aminex HPX-87H (300 mm × 7.8 mm) ion exchange column with a mobile phase of 0.01 mol·L^{-1} sulfuric acid at a flow rate of 0.6 mL·min^{-1}, detection by diode array detector (DAD) at 264 nm. Chromatogram A: standard solution of *tt*-MA (500 µg· L^{-1}); Chromatogram B: *in natura* urine sample.

3.3. Validation of the Method

After optimization of the chromatographic conditions (summarized in Table 5) and the sample preparation procedure, the method was validated to ensure that it complied with established norms [48,49,52].

Table 5. Optimized chromatographic conditions employed for the validation of the method for the determination of *trans,trans*-muconic acid.

Chromatographic Conditions	Description
Column	Aminex HPX-87H (300 mm × 7.8 mm) ion exclusion
Mobile phase flow rate	0.6 mL·min^{-1}
Injection volume	20 µL (automatic injection system)
Column temperature	55 °C
Run time	60 min

3.3.1. Selectivity

Since this work employed high performance liquid chromatography, the evaluation of the selectivity of the method employed the analyte retention times, comparing the retention times obtained for the *in natura* sample, standard solutions and fortified samples. A comparison was also made of the absorption diode array detector (DAD) spectra of the species eluted at the retention time of *tt*-MA in the sample chromatograms and the spectra obtained for the *tt*-MA standard solutions.

The presence of *tt*-MA peaks in the sample chromatograms was confirmed using four different techniques, depending on the level of uncertainty in the evaluation: (1) by comparing the retention time of the *tt*-MA peak in the sample chromatogram with the retention time of the peak obtained for a *tt*-MA standard, under the same conditions; (2) by the increase in the area of the peak resulting from the addition of *tt*-MA standard to the sample; (3) by comparison of the absorption spectrum of the *tt*-MA peak obtained for a standard solution with that of the peak obtained for the sample; and (4) according to the degree of purity of the eluted peak, as determined using the LabSolutions software, version 5.53 SP3 (Shimatzu, Laval, PQ, Canada). In the last case, the evaluation is made of the absorption spectra at the retention times corresponding to the upslope, top and downslope regions of the chromatographic peak, comparing them in order to identify any possibility of co-elution of other substances. The results of these different procedures applied to the chromatographic peaks showed that the method was selective to *tt*-MA.

3.3.2. Interferences due to the Biological Matrix

The analytical curve obtained using standards prepared in aqueous solution showed a slope (angular coefficient) that was greater than the slope of the curve for standards prepared in the pooled urine from the ten volunteers (all of whom were non-smokers with no history of occupational exposure to benzene). These results (Figure 4) indicated that the presence of the sample matrix depressed the analyte extraction efficiency by around 50% (Equation (1)), resulting in a decrease in sensitivity. Despite the existence of matrix interferences, this is not a problem, since all of the analytical curves used for the quantification of *tt*-MA in urine samples were constructed using standards prepared in the pooled urine.

$$\text{Matrix effect (\%)} = \frac{X_1}{X_2} \times 100 \tag{1}$$

where:

X_1: *tt*-MA peak area for the analytical curve prepared using aqueous standards;
X_2: *tt*-MA peak area for the analytical curve prepared using pooled urine.

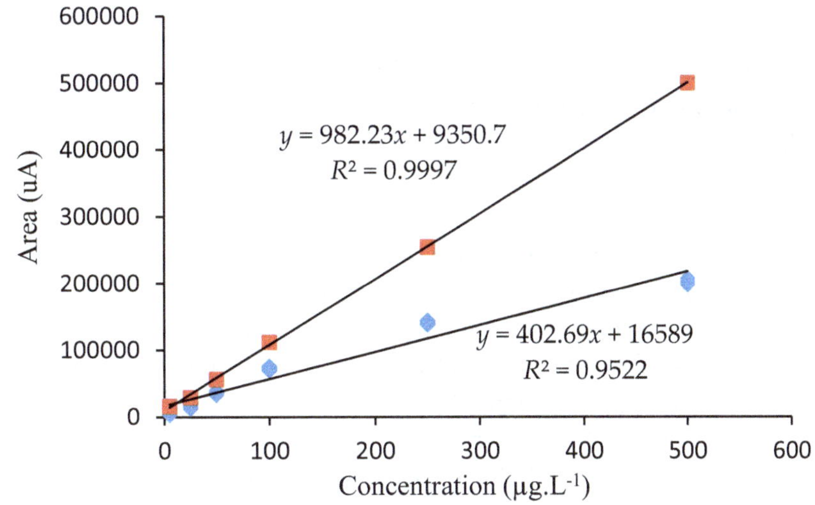

Figure 4. Analytical curves obtained for the determination of *trans,trans*-muconic acid in urine. The square red and blue points correspond to the values obtained for standards prepared in aqueous solution and pooled urine, respectively.

3.3.3. Analytical Curve Adjustment and Working Range

A coefficient of determination (R^2) value exceeding 0.99 was obtained for linear fitting of the analytical curve prepared using aqueous *tt*-MA standards in the concentration range 5–500 $\mu g{\cdot}L^{-1}$. An R^2 value of 0.98 was obtained for linear fitting of the curve prepared using standards prepared in the pooled urine at concentrations in the range 5–250 $\mu g{\cdot}L^{-1}$, while a quadratic fit with an R^2 value exceeding 0.99 was obtained for standards prepared in the pooled urine at concentrations in the range 5–500 $\mu g{\cdot}L^{-1}$ (Figure 5). These results were in accordance with established norms and were considered satisfactory [48,52].

The concentrations used to determine the working range were based on *tt*-MA concentrations reported in the literature [42,50]. The results showed that linear fits (determination coefficient, $R^2 = 0.9909$) could be applied to analytical curves prepared using concentrations of *tt*-MA in urine up to 100 $\mu g{\cdot}L^{-1}$. At higher concentrations (up to 500 $\mu g{\cdot}L^{-1}$), a quadratic fit (determination coefficient, $R^2 = 0.9982$) could be used, or the samples could be diluted after the extraction step, in order to use the linear model.

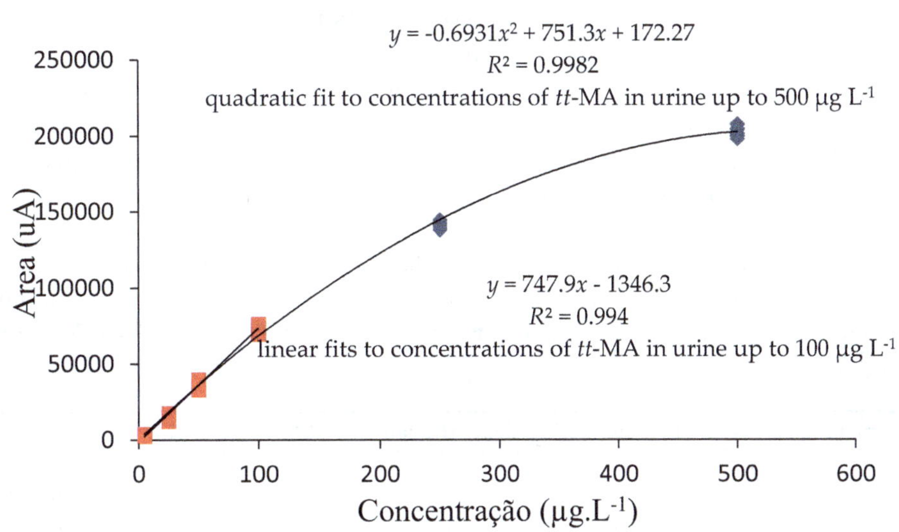

Figure 5. Analytical curves constructed using standards (nine replicates) prepared in pooled urine.

3.3.4. Limits of Detection and Quantification

The limits of detection (LOD) and quantification (LOQ) of the method were calculated using the signal-to-noise ratio, as described previously. The values obtained were 0.11 $\mu g \cdot L^{-1}$ (LOD) and 0.36 $\mu g \cdot L^{-1}$ (LOQ).

3.3.5. Precision and Accuracy

The precision of the method was determined from the intra-experiment repeatability, considering the *tt*-MA concentrations obtained for one sample extract that was analyzed seven times on the same day. No evaluation was made of inter-experiment repeatability (analyses on different days). The results were calculated as coefficient of variation (CV) values (Table 6). The highest value obtained was 1.47%, demonstrating that the method provided good precision. The results also showed that the precision was independent of the amount of *tt*-MA present in the sample. Although the intermediate precision was not determined, the results indicated that there was little variation for the measurements made on different days or by different analysts.

Table 6. Coefficients of variation (CV) for the determination of *tt*-MA in the pooled urine sample and in pooled urine enriched with *tt*-MA at concentrations of 5.0, 50.0 and 500.0 $\mu g \cdot L^{-1}$, using nine replicates ($n = 9$).

Urine Sample	Average Peak Area (μA)	CV (%)
Pool of 10 samples	96,461	1.22
Pool with addition of 5.0 $\mu g \cdot L^{-1}$	99,892	0.53
Pool with addition of 50.0 $\mu g \cdot L^{-1}$	133,092	1.47
Pool with addition of 500 $\mu g \cdot L^{-1}$	299,546	1.16

The accuracy of the method was determined using the recovery of *tt*-MA standards added to the pooled urine at concentrations of 5.0, 50.0 and 500.0 $\mu g \cdot L^{-1}$. The concentration of *tt*-MA in the unfortified pooled urine was 88.6 $\mu g \cdot L^{-1}$, calculated using the linear equation describing the analytical curve constructed using aqueous standards.

The concentrations of *tt*-MA quantified in the fortified urine samples using the quadratic equation describing the analytical curve constructed using standards prepared in the pooled urine represented recoveries close to 100% (Table 7), with the exception of the sample fortified with 5.0 $\mu g \cdot L^{-1}$ of *tt*-MA, for which a recovery of 87.20% was obtained. The lower recovery found for an added concentration of 5 $\mu g \cdot L^{-1}$ can be explained by the fact that the concentration of *tt*-MA already present in the urine (88.6 $\mu g \cdot L^{-1}$) was much higher than the amount added. Ideally, the additions of *tt*-MA standard should be made to samples of urine containing lower levels of the analyte. However, it is impossible to obtain *a priori* knowledge of the concentrations of *tt*-MA in pooled urine samples. Despite the fact that the samples used here were obtained from volunteers for whom low concentrations of *tt*-MA were expected, metabolic differences between individuals contribute to uncertainty.

Table 7. Recoveries obtained for quantification using the analytical curve constructed using standards prepared in the pooled urine from ten non-smoker volunteers. Amounts of *tt*-MA standard equivalent to concentrations of 5.0, 50.0 and 500.0 $\mu g \cdot L^{-1}$ were added to pooled urine with a *tt*-MA content of 88.6 $\mu g \cdot L^{-1}$ ($n = 9$).

Concentration of *tt*-MA Added ($\mu g \cdot L^{-1}$)	Concentration of *tt*-MA Found ($\mu g \cdot L^{-1}$)	Recovery (%)
5.0	4.4	87.20
50.0	50.9	101.84
500.0	510.6	102.12

Although the quantification of urinary *tt*-MA using analytical curves constructed employing standards prepared in the pooled urine made the method more labor-intensive, the accuracy was much improved. In the case of analytical methods involving complex matrices, such as urine, recovery values in the range 80%–120% are acceptable [49,52]. The results obtained here could therefore be considered satisfactory. Although the analysis time was higher in these proposed conditions than using reverse phase, the results obtained here could therefore be considered satisfactory because the separation of the different compounds, particularly *tt*-muconic acid, was well performed by a combination of different mechanisms: hydrophobicity, size exclusion and ion exclusion. Thus, the chromatographic resolutions were much better, allowing a separation of *tt*-muconic acid at baseline without interference.

4. Conclusions

The method, not yet described in the literature, was optimized and validated for the determination of urinary *tt*-MA, using solid phase extraction with Strata SAX cartridges and chromatographic separation with an Aminex HPX 87H column. This method presented low limits of detection and quantification (0.11 and 0.36 $\mu g \cdot L^{-1}$, respectively) and better resolution compared to other techniques described in the literature.

Acknowledgments: The authors thank Federal University of Ouro Preto (UFOP), National Counsel of Technological and Scientific Development (CNPq)—financing 47983/2013-3 and Research Support Foundation of the State of Minas Gerais (FAPEMIG)—financing APQ 00918-13.

Author Contributions: Rafaela de Paiva Gomes participated in the research as an undergraduate student, and his research project was led by Mauricio Xavier Coutrim. The other authors, Ananda Lima Sanson, Fabiana Aparecida Lobo and Robson Jose de Cassia Franco Afonso, were collaborators in the development of the project at various stages of research.

Conflicts of Interest: The authors declare no conflict of interest.

References

1. Protano, C.; Andreoli, R.; Manini, P.; Guidotti, M.; Vitali, M. A tobacco-related carcinogen: Assessing the impact of smoking behaviours of cohabitants on benzene exposure in children. *Tob. Control* **2012**, *21*, 325–329. [CrossRef] [PubMed]
2. Guarieiro, L.L.N.; Vasconcellos, P.C.; Solci, M.C. Air Pollutants from the burning of fossil fuels and biofuels: A brief review. *Rev. Virtual Quim.* **2011**, *3*, 434–445. [CrossRef]
3. Manini, P.; de Palma, G.; Andreoli, R.; Poli, D.; Mozzoni, P.; Folesani, G.; Mutti, A.; Apostoli, P. Environmental and biological monitoring of benzene exposure in a cohort of Italian taxi drivers. *Toxicol. Lett.* **2006**, *167*, 142–151. [CrossRef] [PubMed]
4. Kivistö, H.; Pekari, K.; Peltonen, K.; Svinhufvud, J.; Veidebaum, T.; Sorsa, M.; Aitio, A. Biological monitoring of exposure to benzene in the production of benzene and in a cokery. *Sci. Total Environ.* **1997**, *199*, 49–63. [CrossRef]
5. Choi, Y.; Shin, D.; Park, S.; Chung, Y.; Kim, M. Biological monitoring of benzene in residents living near petrochemical industrial areas in Korea. *J. Occup. Health* **2000**, *42*, 31–37. [CrossRef]
6. Kirkeleit, J.; Riise, T.; Bråtveit, M.; Pekari, K.; Mikkola, J.; Moen, B.E. Biological monitoring of benzene exposure during maintenance work in crude oil cargo tanks. *Chem.-Biol. Interact.* **2006**, *164*, 60–67. [CrossRef] [PubMed]
7. Brugnone, F.; Perbellini, L.; Maranelli, G.; Romeo, L.; Guglielmi, G.; Lombardini, F. Reference values for blood benzene in the occupationally unexposed general population. *Int. Arch. Occup. Environ. Health* **1992**, *64*, 179–184. [CrossRef]
8. Aprea, C.; Sciarra, G.; Bozzi, N.; Pagliantini, M.; Perico, A.; Bavazzano, P.; Leandri, A.; Carrieri, M.; Scapellato, M.; Bettinelli, M.; *et al.* Reference Values of Urinary *trans,trans*-muconic Acid: Italian Multicentric Study. *Arch. Environ. Contam. Toxicol.* **2008**, *55*, 329–340. [CrossRef] [PubMed]
9. Fustinoni, S.; Campo, L.; Mercadante, R.; Manini, P. Methodological issues in the biological monitoring of urinary benzene and *S*-phenylmercapturic acid at low exposure levels. *J. Chromatogr. B* **2010**, *878*, 2534–2540. [CrossRef] [PubMed]

10. Waidyanatha, S.; Rothman, N.; Fustinoni, S.; Smith, M.T.; Hayes, R.B.; Bechtold, W.; Dosemeci, M.; Guilan, L.; Yin, S.; Rappaport, S.M. Urinary benzene as a biomarker of exposure among occupationally exposed and unexposed subjects. *Carcinogenesis* **2001**, *22*, 279–286. [CrossRef] [PubMed]
11. Coutrim, M.X.; de Carvalho, L.R.F.; Arcuri, A.S.A. Evaluation of the analytical methods to determinate benzene metabolites as potencial biomarkers for determining human exposure to benzene in air. *Quím. Nova* **2000**, *23*, 653–663. [CrossRef]
12. Ministry of Labor and Employment of Brazil. Portaria No. 34, 20 December 2001, Attachment: Protocolo para a Utilização de Indicador Biológico da Exposição Ocupacional ao Benzeno. Available online: http://www.capecanaveral4045.com/legislacao/port_34_2001_benzeno_indicador_biologico.html (accessed on 12 September 2015).
13. Menezes, M.; Balbão, M.S.; Siqueira, M.E.P.B.; Martins, I. Influence of tobacco smoking on urinary excretion of *trans,trans*-muconic acid. *Braz. J. Pharm. Sci.* **2008**, *44*, 459–464.
14. Roma-Torres, J.; Teixeira, J.P.; Silva, S.; Laffon, B.; Cunha, L.; Mendez, J.; Mayan, O. Evaluation of genotoxicity in a group of workers from a petroleum refinery aromatics plant. *Mutat. Res.* **2006**, *604*, 19–27. [CrossRef] [PubMed]
15. Violante, F.; Sanguinetti, G.; Barbieri, A.; Accorsi, A.; Mattioli, S.; Cesari, R.; Fimognari, C.; Hrelia, P. Lack of correlation between environmental or biological indicators of benzene exposure at parts per billion levels and micronuclei induction. *Environ. Res.* **2003**, *91*, 135–142. [CrossRef]
16. Kouniali, A.; Cicolella, A.; Gonzalez-Flesca, N.; Dujardin, R.; Gehanno, J.-F.; Bois, F.Y. Environmental benzene exposure assessment for parent-child pairs in Rouen. *France. Sci. Total Environ.* **2003**, *308*, 73–82. [CrossRef]
17. Cocco, P.; Tocco, M.G.; Ibba, A.; Scano, L.; Ennas, M.G.; Flore, C.; Randaccio, F. *trans,trans*-Muconic acid excretion in relation to environmental exposure to benzene. *Int. Arch. Occup. Environ. Health* **2003**, *76*, 456–460. [CrossRef] [PubMed]
18. Marrubini, G.; Coccini, T.; Manzo, L. Direct analysis of urinary *trans,trans*-muconic acid by coupled column liquid chromatography and spectrophotometric ultraviolet detection: Method applicability to human urine. *J. Chromatogr. B* **2001**, *758*, 295–303. [CrossRef]
19. Waidyanatha, S.; Rothman, N.; Li, G.; Smith, M.T.; Yin, S.; Rappaport, S.M. Rapid determination of six urinary benzene metabolites in occupationally exposed and unexposed subjects. *Anal. Biochem.* **2004**, *327*, 184–199. [CrossRef] [PubMed]
20. Kim, S.; Vermeulen, R.; Waidyanatha, S.; Johnson, B.A.; Lan, Q.; Rothman, N.; Smith, M.T.; Zhang, L.; Li, G.; Shen, M.; *et al.* Using urinary biomarkers to elucidate dose-related patterns of human benzene metabolism. *Carcinogenesis* **2006**, *27*, 772–781. [CrossRef] [PubMed]
21. Mudiam, M.K.R.; Chauhan, A.; Singh, K.P.; Gupta, S.K.; Jain, R.; Ch, R.; Murthy, R.C. Determination of *t,t*-muconic acid in urine samples using a molecular imprinted polymer combined with simultaneous ethyl chloroformate derivatization and pre-concentration by dispersive liquid-liquid microextraction. *Anal. Bioanal. Chem.* **2013**, *405*, 341–349. [CrossRef] [PubMed]
22. Sørensen, M.; Skov, H.; Autrup, H.; Hertel, O.; Loft, S. Urban benzene exposure and oxidative DNA damage: Influence of genetic polymorphisms in metabolism genes. *Sci. Total Environ.* **2003**, *309*, 69–80. [CrossRef]
23. Wiwanitkit, V.; Suwansaksri, J.; Soogarun, S. A note on urine *trans, trans* muconic acid level among a sample of Thai police: Implication for an occupational health issue. *Yale J. Biol. Med.* **2003**, *76*, 103–108. [PubMed]
24. Ruppert, T.; Scherer, G.; Tricker, A.R.; Rauscher, D.; Adlkofer, F. Determination of urinary *trans,trans*-muconic acid by gas chromatography-mass spectrometry. *J. Chromatogr. B* **1995**, *666*, 71–76. [CrossRef]
25. Franqui, L.S.; Vieira, A.C.; Maia, P.P.; Figueiredo, E.C. Molecularly imprinted solid-phase extraction of urinary *trans,trans*-muconic acid and analysis by gas chromatography-mass spectrometry. *Quím. Nova* **2012**, *35*, 1577–1581. [CrossRef]
26. Johnson, E.S.; Langård, S.; Lin, Y.-S. A critique of benzene exposure in the general population. *Sci. Total Environ.* **2007**, *374*, 183–198. [CrossRef] [PubMed]
27. Ducos, P.; Gaudin, R.; Robert, A.; Francin, J.; Maire, C. Improvement in HPLC analysis of urinary *trans,trans*-muconic acid, a promising substitute for phenol in the assessment of benzene exposure. *Int. Arch. Occup. Environ. Health* **1990**, *62*, 529–534. [CrossRef] [PubMed]
28. Negri, S.; Bono, R.; Maestri, L.; Ghittori, S.; Imbriani, M. High-pressure liquid chromatographic-mass spectrometric determination of sorbic acid in urine: Verification of formation of *trans,trans*-muconic acid. *Chem.-Biol. Interact.* **2005**, *153*, 243–246. [CrossRef] [PubMed]

29. Schroijen, C.; Baeyens, W.; Schoeters, G.; den Hond, E.; Koppen, G.; Bruckers, L.; Nelen, V.; van de Mieroop, E.; Bilau, M.; Covaci, A.; *et al.* Internal exposure to pollutants measured in blood and urine of Flemish adolescents in function of area of residence. *Chemosphere* **2008**, *71*, 1317–1325. [CrossRef] [PubMed]
30. Serena, P.; Tapparo, A.; Bombi, G. Direct determination of *t,t*-muconic acid in human urine by two-dimensional liquid chromatography. *Analyst* **2000**, *125*, 689–692. [CrossRef] [PubMed]
31. Melikian, A.A.; Connor, R.; Prahalad, A.; Hu, P.; Li, H.; Kagan, M.; Thompson, S. Determination of the urinary benzene metabolites *S*-phenylmercapturic acid and *trans,trans*-muconic acid by liquid chromatography-tandem mass spectrometry. *Carcinogenesis* **1999**, *20*, 719–726. [CrossRef] [PubMed]
32. Senzolo, C.; Frignani, S.; Pavoni, B. Environmental and biological monitoring of occupational exposure to organic micropollutants in gasoline. *Chemosphere* **2001**, *44*, 67–82. [CrossRef]
33. Barbieri, A.; Sabatini, L.; Accorsi, A.; Roda, A.; Violante, F. Simultaneous determination of *t,t*-muconic, *S*-phenylmercapturic and *S*-benzylmercapturic acids in urine by a rapid and sensitive liquid chromatography/electrospray tandem mass spectrometry method. *Rapid Commun. Mass Spectrom.* **2004**, *18*, 1983–1988. [CrossRef] [PubMed]
34. Manini, P.; Andreoli, R.; Niessen, W.M.A. Liquid chromatography-mass spectrometry in occupational toxicology: A novel approach to the study of biotransformation of industrial chemicals. *J. Chromatogr. A* **2004**, *1058*, 21–37. [CrossRef]
35. Fustinoni, S.; Buratti, M.; Campo, L.; Colombi, A.; Consonni, D.; Pesatori, A.C.; Bonzini, M.; Farmer, P.; Garte, S.; Valerio, F.; *et al.* Urinary *t,t*-muconic acid, *S*-phenylmercapturic acid and benzene as biomarkers of low benzene exposure. *Chem.-Biol. Interact.* **2005**, *153*, 253–256. [CrossRef] [PubMed]
36. Lee, B.; Ong, H.; Ong, Y.; Ong, C. A sensitive liquid chromatographic method for the spectrophotometric determination of urinary *trans,trans*-muconic acid. *J. Chromatogr. B* **2005**, *818*, 277–283. [CrossRef] [PubMed]
37. Navasumrit, P.; Chanvaivit, S.; Intarasunanont, P.; Arayasiri, M.; Lauhareungpanya, N.; Parnlob, V.; Settachan, D.; Ruchirawat, M. Environmental and occupational exposure to benzene in Thailand. *Chem.-Biol. Interact.* **2005**, *153*, 75–83. [CrossRef] [PubMed]
38. Carrieri, M.; Bonfiglio, E.; Scapellato, M.L.; Maccà, I.; Tranfo, G.; Faranda, P.; Paci, E.; Bartolucci, G.B. Comparison of exposure assessment methods in occupational exposure to benzene in gasoline filling-station attendants. *Toxicol. Lett.* **2006**, *162*, 146–152. [CrossRef] [PubMed]
39. Scherer, G.; Engl, J.; Urban, M.; Gilch, G.; Janket, D.; Riedel, K. Relationship between machine-derived smoke yields and biomarkers in cigarette smokers in Germany. *Regul. Toxicol. Pharm.* **2007**, *47*, 171–183. [CrossRef] [PubMed]
40. Ray, M.; Roychoudhury, S.; Mukherjee, S.; Lahiri, T. Occupational benzene exposure from vehicular sources in India and its effect on hematology, lymphocyte subsets and platelet P-selectin expression. *Toxicol. Ind. Health* **2007**, *23*, 167–175. [CrossRef] [PubMed]
41. Chakroun, R.; Faidi, F.; Hedhili, A.; Charbaji, K.; Nouaigui, H.; Laiba, M.B. Inhalant Abuse Detection and Evaluation in Young Tunisians. *J. Forensic Sci.* **2008**, *53*, 232–237. [CrossRef] [PubMed]
42. Weaver, V.; Buckley, T.; Groopman, J.D. Lack of specificity of *trans,trans*-muconic acid as a benzene biomarker after ingestion of sorbic acid-preserved foods. *Cancer Epidemiol. Biomarkers Prev.* **2000**, *9*, 749–755. [PubMed]
43. Fritz, J.S. Review: Principles and applications of ion-exclusion chromatography. *J. Chromatogr.* **1991**, *546*, 111–118. [CrossRef]
44. Kelebek, H.; Selli, S. Determination of volatile, phenolic, organic acid and sugar components in a Turkish cv. Dortyol (*Citrus sinensis* L. Osbeck) orange juice. *J. Agric. Food Chem.* **2011**, *91*, 1855–1862. [CrossRef] [PubMed]
45. Eyeghe-Bickong, H.; Alexandersson, E.O.; Gouws, L.; Young, P.; Vivier, M.A. Optimisation of an HPLC method for the simultaneous quantification of the major sugars and organic acids in grapevine berries. *J. Chromatogr. B* **2012**, *885*, 43–49. [CrossRef] [PubMed]
46. Jeon, J.; Brigham, C.J.; Kim, Y.; Kim, H.; Yi, D.; Kim, H.; Rha, C.; Sinskey, A.; Yang, Y. Biosynthesis of poly(3-hydroxybutyrate-*co*-3-hydroxyhexanoate) (P(HB-*co*-HHx)) from butyrate using engineered Ralstonia eutropha. *Appl. Microbiol. Biotech.* **2014**, *98*, 5461–5469. [CrossRef] [PubMed]
47. Teófilo, R.F.; Ferreira, M.M.C. Chemometrics II: Spreadsheets for experimental design calculations, a tutorial. *Quím. Nova* **2006**, *29*, 338–350. [CrossRef]

48. Brazil, National Institute of Metrology (INMETRO). Orientação Sobre Validação de Métodos Analíticos, 3rd Review, February 2010. Available online: http://www.inmetro.gov.br/sidoq/arquivos/cgcre/doq/doq-cgcre-8_03.pdf (accessed on 14 September 2015).
49. Ribani, M.; Bottoli, C.B.G.; Collins, C.H.; Jardim, I.C.S.F.; Melo, L.F.C. Validation for chromatographic and electrophoretic methods. *Quím. Nova* **2004**, *27*, 771–780. [CrossRef]
50. Martins, I.; Siqueira, M.E.P.B. Determination of *trans,trans*-muconic acid in urine: Validation of a high performance liquid chromatographic method. *Braz. J. Pharmac. Sci.* **2002**, *38*, 205–213.
51. Tranfo, G.; Paci, E.; Sisto, R.; Pigini, D. Validation of an HPLC/MS/MS method with isotopic dilution for quantitative determination of *trans,trans*-muconic acid in urine samples of workers exposed to low benzene concentrations. *J. Chromatogr. B* **2008**, *867*, 26–31. [CrossRef] [PubMed]
52. Brazilian National Health Surveillance Agency (ANVISA). Resolution RE No. 899, 29 May 2003, Attachment: Guia para Validação de Métodos Analíticos e Bioanalíticos, pp. 56–59. Available online: http://pesquisa.in.gov.br/imprensa/jsp/visualiza/index.jsp?data=02/06/2003&jornal=1&pagina=56&totalArquivos=176 (accessed on 14 September 2015).

17

Low-Temperature Mobile Phase for Peptide Trapping at Elevated Separation Temperature Prior to Nano RP-HPLC-MS/MS

Matthias Schöbinger, Oskar-James Klein and Goran Mitulović *

Clinical Institute of Laboratory Medicine, Medical University of Vienna, Währinger Gürtel 18-20, A-1090 Wien, Austria; matthias.schoebinger@chello.at (M.S.); ojklein@hotmail.com (O.-J.K.)
* Correspondence: goran.mitulovic@meduniwien.ac.at

Academic Editor: Michael J. Dunphy

Abstract: Nano HPLC-MS/MS separation and detection of peptides for proteomic analysis is usually performed upon tryptic digest of proteins and peptide pre-concentration on trap columns. Pre-concentration on trap columns is needed for sample wash (removal of salts and impurities), sample focusing prior to separation, and volume reduction. Usually, trap columns are mounted on selection valves close to the separation column in order to keep the void volume low and to enable injection of large sample amounts onto nano-separation column. Since separation columns are operated at elevated temperature of ⩾45 °C and they are mounted on the same valve as the trap column (in the column oven); loading samples at elevated temperature will result with significant loss of analytes. A method for loading samples on a trap column at 60 °C was developed and optimized. No sample loss was observed when the optimized method was used for analysis of standards and of complex biological samples.

Keywords: peptides; temperature; trapping; mobile phase; proteomics

1. Introduction

Nano HPLC and mass spectrometry (MS) are analytical methods of choice for separation and detection of peptides generated through enzymatic digestion of proteins. Samples' complexity depends on its origin and the enzymatic method employed for proteins' digestion; however, the number of peptides generated largely surpasses the capacity of separation columns applied [1–6], which needs to be optimized.

Separation of peptides for proteomics analysis is performed using nano HPLC columns with inner diameters (ID) of 50 μm to 100 μm. These small ID will contribute to increased sensitivity of the separation and detection method, which is electrospray ionization—mass spectrometry (ESI-MS) but will also limit the amount of sample in terms of volume that can be injected [7] on a nano-separation column. The use of trap columns is a method, which can provide help. By applying trap columns, injections of volumes larger than 5 μL are possible without risking the damage of the separation column or peak broadening due to the enforced peak displacement on the column.

The effect of temperature on separation performance of reversed phase columns for peptides has been published in several publications [8–16]. There are several studies addressing the use of elevated column temperature but only a limited number of experiments have investigated the effect of low temperature on peptide trapping. The most remarkable study for using cooling and heating phases of the entire separation column or only a column segment was performed by Eghbali *et al.* [8]. The authors described the use of selective cooling of the predefined section of the separation column, which enabled peak focusing at the end of the column. Authors correctly noted that peaks are broadening during the trapping procedure and that the lower temperature of the column enables zone focusing. Very detailed information is provided on how temperature changes affect the trapping efficiency and retention time behavior of both proteins and peptides on the separation column. It was effectively shown that peptides and proteins can be trapped on the stationary phase even under gradient conditions, and being eluted with significantly better peak shape in comparison to heated column only. However, the method described is significantly different that the approach described in this manuscript. Here, columns, the trapping column and the separation column are operated at elevated temperature but the loading mobile phase is cooled and the cooling effect applies only to the trapping column and does not directly influence the separation performance. Interested readers are encouraged to read this paper and compare the effects described with those described in the current manuscript.

Modern nano-chromatographic systems for separations of peptides are currently operated using UHPLC conditions, which implies the use of particles with small diameters packed in long separation columns. Although columns packed with sub-3 μm and sub-2 μm particles are getting more ubiquitous, the use of separation columns packed with 3 μm-particles (conventional) are more broadly spread. The use of columns packed with sub-2 μm particles induces a high backpressure, and separation at elevated column temperature is often employed [17–23] in order to reduce the backpressure and improve column performance. Separation columns packed with conventional particles can also be applied for UHPLC separations, without loss of separation performance when higher flow rates and higher operational temperatures are applied.

As for the current study, a conventionally-packed separation column with 3 μm particles was applied using conventional nano-separation approach.

In modern chromatographic systems, trap columns are always mounted on the same switching valve as the separation column, Figure 1. This approach is needed in order to minimize the void volume and enable fast sample transfer from the trap column to the separation column and avoid additional separation on the trap column. These settings require trap columns to be operated in the column oven along with the separation column, using the same operating temperature. In cases where separation column is operated at elevated temperature, this set-up can lead to insufficient sample trapping and subsequent sample loss. Mounting the column outside of the column oven is not a reasonable solution in this case due to long connecting tubing and subsequently increased void volume. Separate temperature control for the trap column and for the separation column would be an option for improved sample trapping at elevated temperatures. However, this solution is difficult to implement and operate; therefore, other options to maintain optimal sample loading must be considered.

The most convenient option is the use of precooled mobile phases for sample loading, which would enable the lowering of the trapping column's temperature and enhancing the trapping efficiency. Here, we describe the use of standard loading phase in our laboratory, aqueous trifluoroacetic acid (TFA), for sample loading at different operating temperatures of the column oven. The temperature of the elution mobile phase (nano flow) was not altered, enabling it to reach the trap column and the separation column at the temperature of the column oven during the elution.

All experiments were carried out using an RSLC nano HPLC system (ThermoFisher, Germering, Germany) with the C18 trapping and separation column having an inner diameter (ID) of 300 μm and 75 μm, respectively.

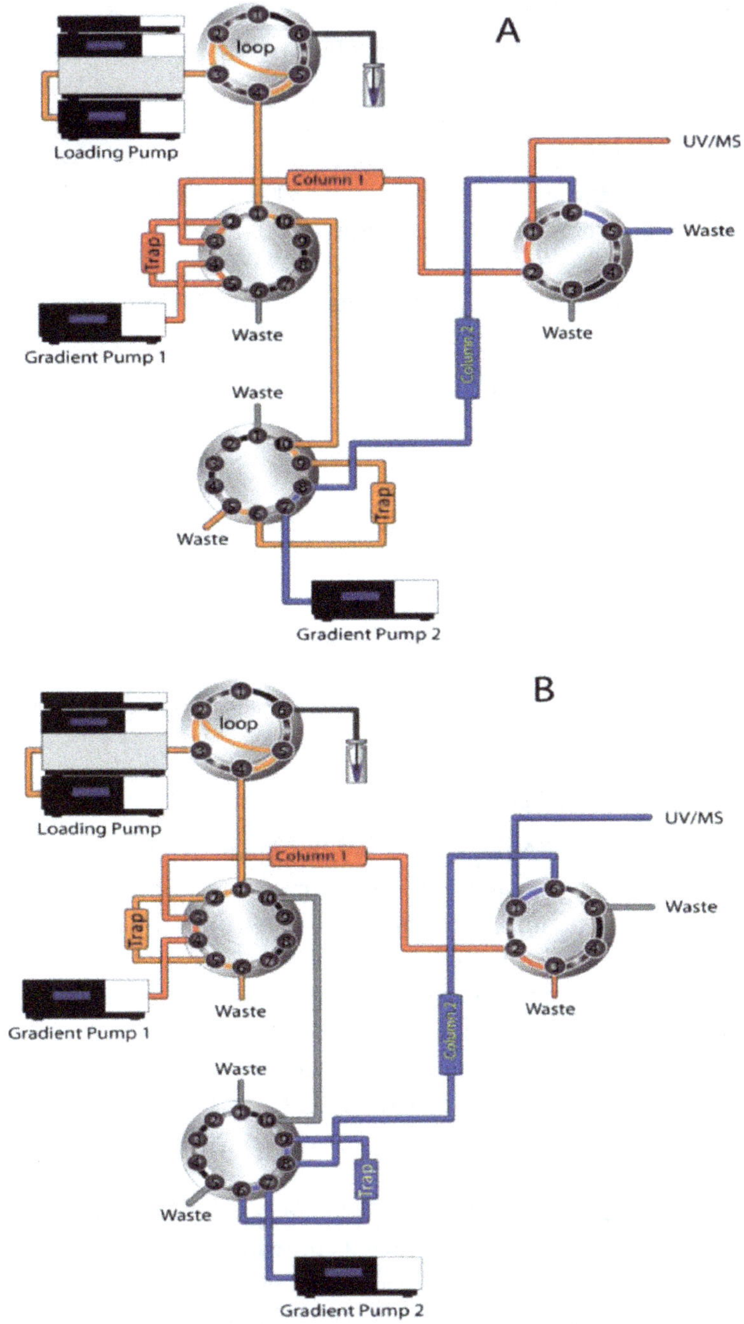

Figure 1. Plumbing scheme showing the (dual) nano HPLC separation system with (**A**) trap column and nano column 1 operated in eluting loading position while trap column and nano column 2 are equilibrated and prepared for the following sample injection; and (**B**) trap column 1 and nano column 1 are being equilibrated while the analytes are being eluted from trap column 2 and nano column 2. The trap column was operated in a back-flush mode ensuring minimal peak broadening and preventing peptide separation before the transfer on the nano separation column.

2. Experimental Section

2.1. Chemicals

HPLC Grade water was obtained from an in-house Millipore water purification system (Millipore, Wien, Austria), acetonitrile (AcN), methanol (MeOH), both as Chromasolv, dithiothreitol, iodoacetamide, trifluoroacetic acid, formic acid, and bovine serum albumin (BSA) was purchased from Sigma-Aldrich (Sigma-Aldrich, Wien, Austria). Trypsin (sequencing grade) was purchased from Promega GmbH (Wien, Austria). Trifluoroethanol (TFE) was purchased from Merck-Millipore (Wien, Austria).

2.2. LC/MS Instrumentation

All experiments were performed using an Ultimate 3000 RSLC nano HPLC system, PepMap C18 trapping column (TC, 5 μm particle size, 300 Å pore size, 300 μm ID × 5 mm length), and the PepMap C18 nano separation column (3 μm particle size, 100 Å pore size, 75 μm ID × 250 mm length), all purchased from ThermoFisher Scientific, Vienna, Austria. Both columns were operated in a column oven at:

1. ambient temperature (approx. 25 °C);
2. 45 °C, and;
3. at 60 °C.

Loading mobile phase was always aqueous 0.1% TFA, which was pumped through the trap column at 30 μL/min using:

1. ambient temperature (approx. 25 °C), and;
2. Cooled to 5 °C.

Mixing following mobile phases composed reversed phase separation gradient:

A: 5% AcN, 0.1% FA,
B: 50% AcN, 30% MeOH, 10% TFE, 0.1% FA at 300 nL/min.

The separation gradient for BSA peptides was generated as follows: 3% B for 10 min rising to 35% B at 40 min, 35% B at 70 min, kept at 90% B from 51 min to 55 min. Finally, the separation column was equilibrated at 3% B from 56 min to 60 min. The trap column was switched online with the separation column at 8 min and switched back to the load position at 50 min.

The separation gradient for peptides from the biological sample was generated as follows: 3% B for 15 min rising to 30% B at 70 min, rising to 60% B at 90 min, keeping at 90% B from 91 min to 100 min. Finally, the separation column was equilibrated using 3% B from 101 min to 125 min. The trap column was switched online with the separation column at 10 min and switched back to load position at 115 min.

The trap column was operated in a back-flush mode, meaning that the sample was eluted using the nano-flow, pumped in opposite direction of the loading flow. This elution method ensures low void volume, prevents peak broadening, and prevents unwanted separation on the trap column.

All peptides were detected using UV detection at 214 nm and electrospray ionization with a maXis Impact q-ToF mass spectrometer (Bruker, Bremen, Germany) equipped with a captive spray source. Ionization was performed at 1600 V, MS/MS using top ten of all ions was performed for all species with two or more positive charges while single charged ions were excluded.

2.3. Samples

Tryptically-digested BSA was used as a standard sample to investigate the influence of the trap column temperature on separation efficiency. For the complex sample, proteins obtained from human urine were tryptically digested upon protein precipitation using slightly modified Wessel-Fluegge method [24,25].

2.4. Data Analysis

All HPLC–UV data was analyzed using Chromeleon 6.8 (ThermoFisher, Germering, Germany), Raw MS data was transformed into Mascot *mgf* files using Data Analysis 4.2 (Bruker, Bremen, Germany). Database search of MS/MS data was performed using ProteinScape 3.1 (Bruker,

Bremen, Germany) and Mascot 2.5.1. (Matrix Science, London, UK), SwissProt protein databases (www.uniprot.org) at the most current version was used for MS/MS data search. Trypsin was selected as digesting enzyme with two miscleavages allowed, carbamidomethylation on C and oxidation on M were selected as fixed and variable modifications, respectively.

3. Results and Discussion

The relation between the retention factor k and the absolute temperature in a chromatographic system is described by van't Hoff equation [26]:

$$\ln(k) = -\frac{\Delta H^0}{\Delta RT} + \frac{\Delta S^0}{R} + \ln(\varphi) \quad (1)$$

Here, ΔH^0 is the enthalpy of the retention and ΔS^0 is the entropy of the retention at the given temperature T, R is the universal gas constant, and ϕ the phase ration. Obviously, the ln(k) is inversely proportional to the operating temperature of the separation column and retention of analytes is influenced by operating temperature of the column and the mobile phase. According to the Equation (1), retention time will be higher for separations performed at lower temperatures, which is also true for the trapping efficiency of a column. Thus, an analyte will be trapped more efficiently on a trap column when loaded at lower temperature. Since the trapping column was operated in a column oven, it was not possible to provide efficient solution for separate cooling and heating of the trapping column within the existing column oven. Therefore, the decision was made to use the precooled loading mobile phase in order to enhance the trapping efficiency of tryptic peptides.

Trapping column was constantly flushed with the loading mobile phase, but its temperature was not assessed separately. It can also be assumed that a certain increase of the temperature of the precooled loading mobile phase occurs during the loading process within the heated column oven. The temperature difference of the loading mobile phase before and after the trapping column was, however, not measured.

Separation column's operational temperature has significant impact on the thermodynamic process of peptide binding and elution influencing retention time and peak shape. In case of nano-separations, as used for proteomics analysis, it can be assumed that the temperature of the column oven is actually the operational temperature of both the trap and separation column due to their small dimensions and thin outer walls, which enables fast heating and fast thermal equilibration.

Tryptically-digested BSA is routinely used in our laboratory to test the performance of the separation and detection system. Figure 2A shows the separation of the BSA tryptic digest performed with both the trap column and separation column operated at 60 °C without a precooled loading mobile phase. In Figure 2B, the same separation was performed using a precooled loading mobile phase (5 °C) and otherwise identical conditions.

Reproducible separation of complex biological samples is of great importance and it implies reproducible trapping of peptides generated upon enzymatic digestion. Figure 3 shows the reproducibility of the BSA separation using cooled loading mobile phase at 5 °C. Reproducibility of the retention times is shown in Figure 4, ranging from 0.05% for the first eluting peptide and 0.48% for the last eluting peptide for sample loading with the precooled mobile phase. The retention time stability and reproducibility is, therefore comparable with the separation performed without cooling of the mobile, which ranges from 0.2% for the first eluted peptide at 23.1 min and 0.46% for the latest eluting peptide at 55.5 min. Temperature stability of all system's components is very important for reproducible trapping and separation, as shown in Figure 3 for first two injections. Although the trap column was flushed with the precooled loading mobile phase, certain time was needed for a complete equilibration of the separation system.

Obviously, some peptides were lost during the loading process with the non-cooled loading mobile phase as compared to analysis with the precooled solvent. Table 1 shows summarized results for peptide separations of the standard sample and of the biological sample. For BSA samples, the highest number of identified peptides, which implies the highest sequence coverage, was achieved when a precooled mobile phase at 5 °C was used for sample loading and the separation was performed at 60 °C. The use of the precooled loading solvent resulted with improved trapping of peptides on the trap column and higher separation temperature provided improved peptide separation on the nano column, which, again, results with better ionization and higher detection and identification rates.

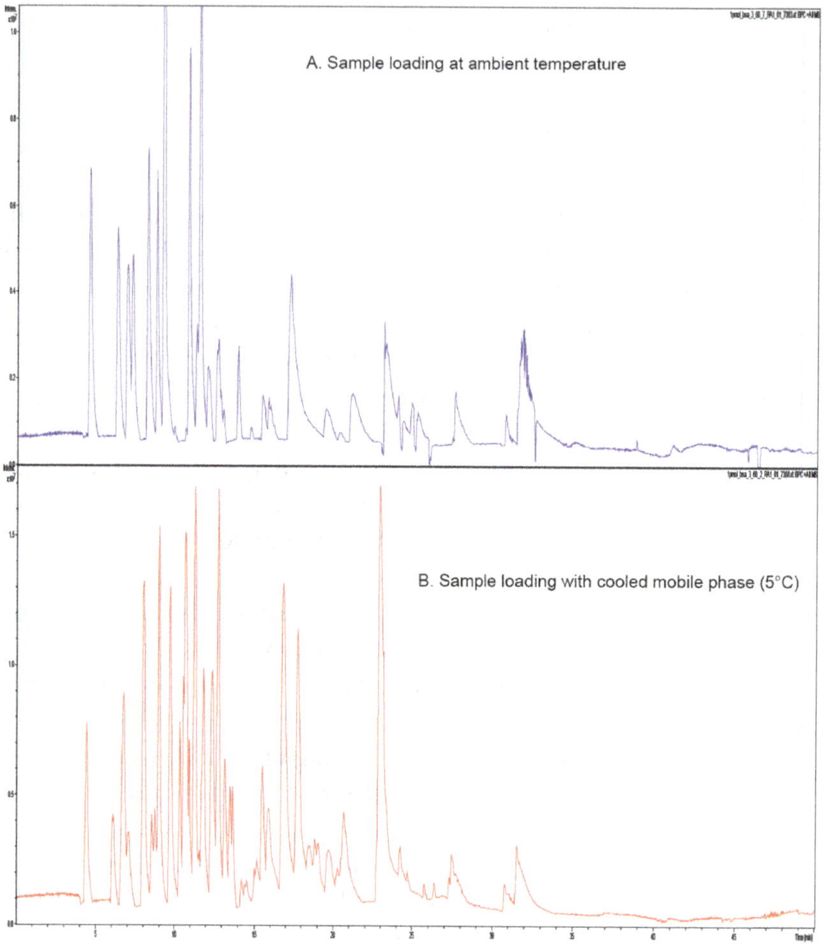

Figure 2. (**A**) MS-Base Peak Chromatogram showing the separation of BSA tryptic peptides on a nano separation column and trap column in operated in a column oven at 60 °C. Loading mobile phase was not precooled and it was loaded at ambient temperature; and (**B**) MS-Base Peak Chromatogram showing the separation of BSA tryptic peptides on a nano separation column and trap column operated in a column oven at 60 °C. Loading mobile phase was precooled for sample loading.

Table 1. Separation of BSA tryptic digest using different loading conditions resulted with identification of higher number of peptides when precooled loading mobile phase was used. The use of higher separation temperature (60 °C *vs.* 45 °C) also contributed to higher sequence coverage. For the biological sample, more proteins and peptides were identified when sample loading was performed using a precooled loading mobile phase.

T (C; Loading MP/Column Oven)	Sequence Coverage (%)	Proteins #	Peptides #
BSA (5 °C/45 °C)	56.3	1	49
BSA (5 °C/60 °C)	66.9	1	60
BSA (RT/60 °C)	48.1	1	39
Urine (5 °C/60 °C)	n.a.	344	1415
Urine (RT/60 °C)	n.a.	182	442

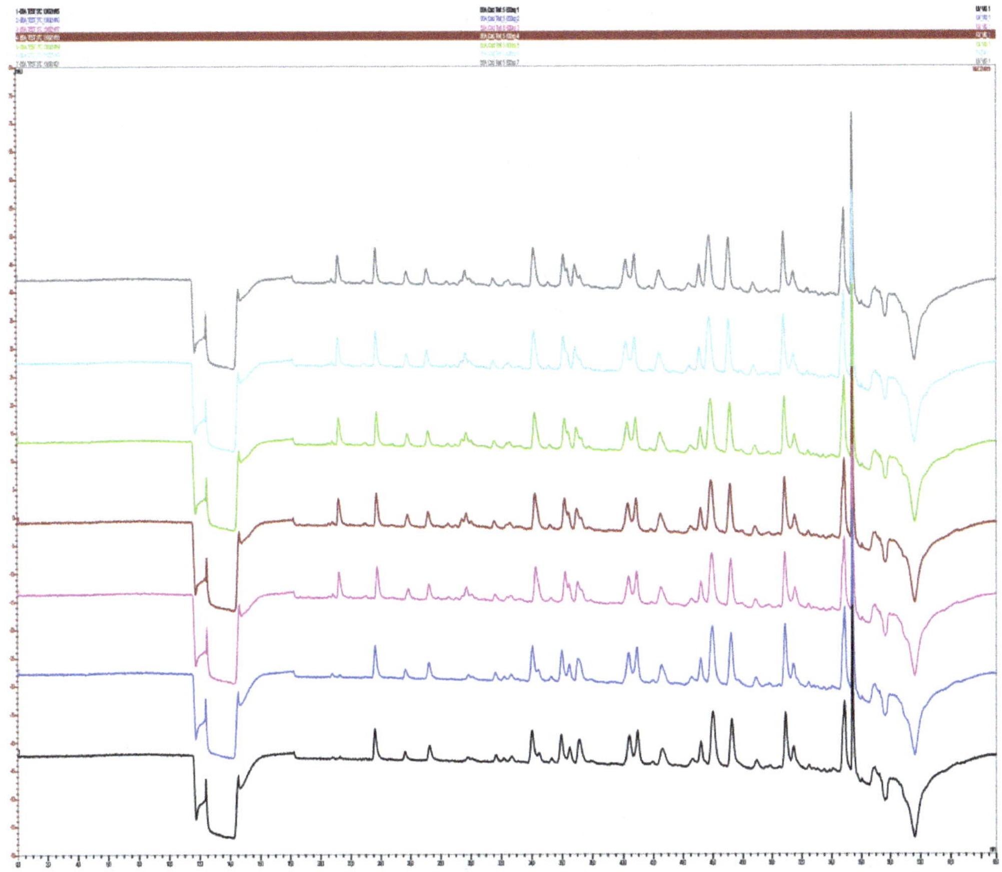

Figure 3. Reproducibility of the separation of BSA tryptic peptides with both separation column and trap column in a column oven at 60 °C and the precooled loading mobile phase to 5 °C. Seven consecutive injections of 1 pMol digested BSA are shown. Two runs (from the bottom) show that the separation system needed a certain time to equilibrate.

Enzymatic digestion of biological samples generates a large amount of peptides with a broad array of different physicochemical properties. The injected sample volume and loading time depend on the sample's matrix, applied flow rates, and the trap column's dimension. In our laboratory, an injection and separation protocol was established for sample loading and subsequent wash of the autosampler and transfer lines, which was also used for current measurements of peptides from urinary peptides [25]. The amount of urinary proteins differs strongly from person to person and published results report numbers between a few hundred and several thousands of identified proteins [27–34].

Upon digestion of urinary proteins, 20 µL of digested sample was diluted with 30 µL of 0.1% aqueous TFA and 40 µL were injected on the trap column using two loading conditions, based on results obtained from tests with BSA:

(A) Heated trap column and separation column operated at 60 °C, using loading mobile phase at ambient temperature.
(B) Heated trap column and separation column operated at 60 °C, using precooled loading mobile phase at 5 °C.

Figure 5 shows the MS-BPC for separations of tryptic urinary peptides upon sample loading at different temperature of the loading solvent. Using loading solvent at ambient temperature resulted with significant sample loss (Figure 5A) in comparison to the use of the precooled loading phase (Figure 5B). This result is confirmed by the number of identified peptides and proteins for both experiments as shown in Table 1. Obviously, there is a significant loss of information, which is higher when compared to analysis of BSA only. As opposed to the digest of a single protein, enzymatic digest of a complex sample results with significantly higher number of peptides, competing for a binding site on the stationary phase. Higher operating temperature weakens the bonds between the analyte (peptides) and promotes elution rather than retention. Therefore, peptides with lower binding affinity face competition from other peptides with higher binding affinity and are additionally hindered in biding to the stationary phase by high temperature. According to Equation (1), these peptides will be eluted significantly earlier from the column as compared to loading and binding at lower temperature. For urinary peptides analyzed in this experiment, it was possible to detect almost twice as many proteins using the precooled loading mobile phase as compared to the non-cooled phase.

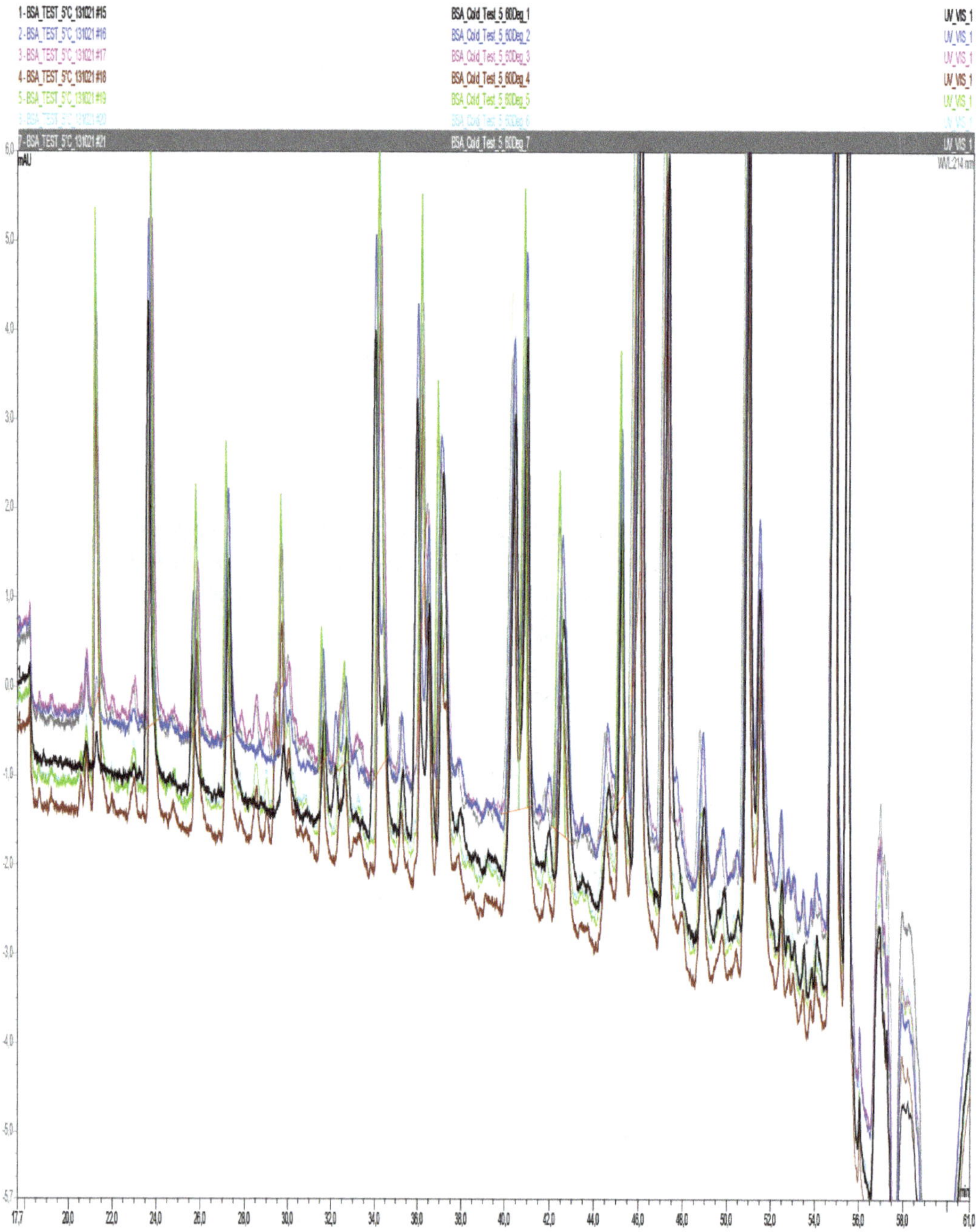

Figure 4. Zoom in of seven consecutive, overlapped, chromatograms showing separation of BSA tryptic peptides. The retention time's reproducibility ranged from 0.05% for the first peak eluted at 21.2 min and 0.48% for the peak eluted at 55.3 min.

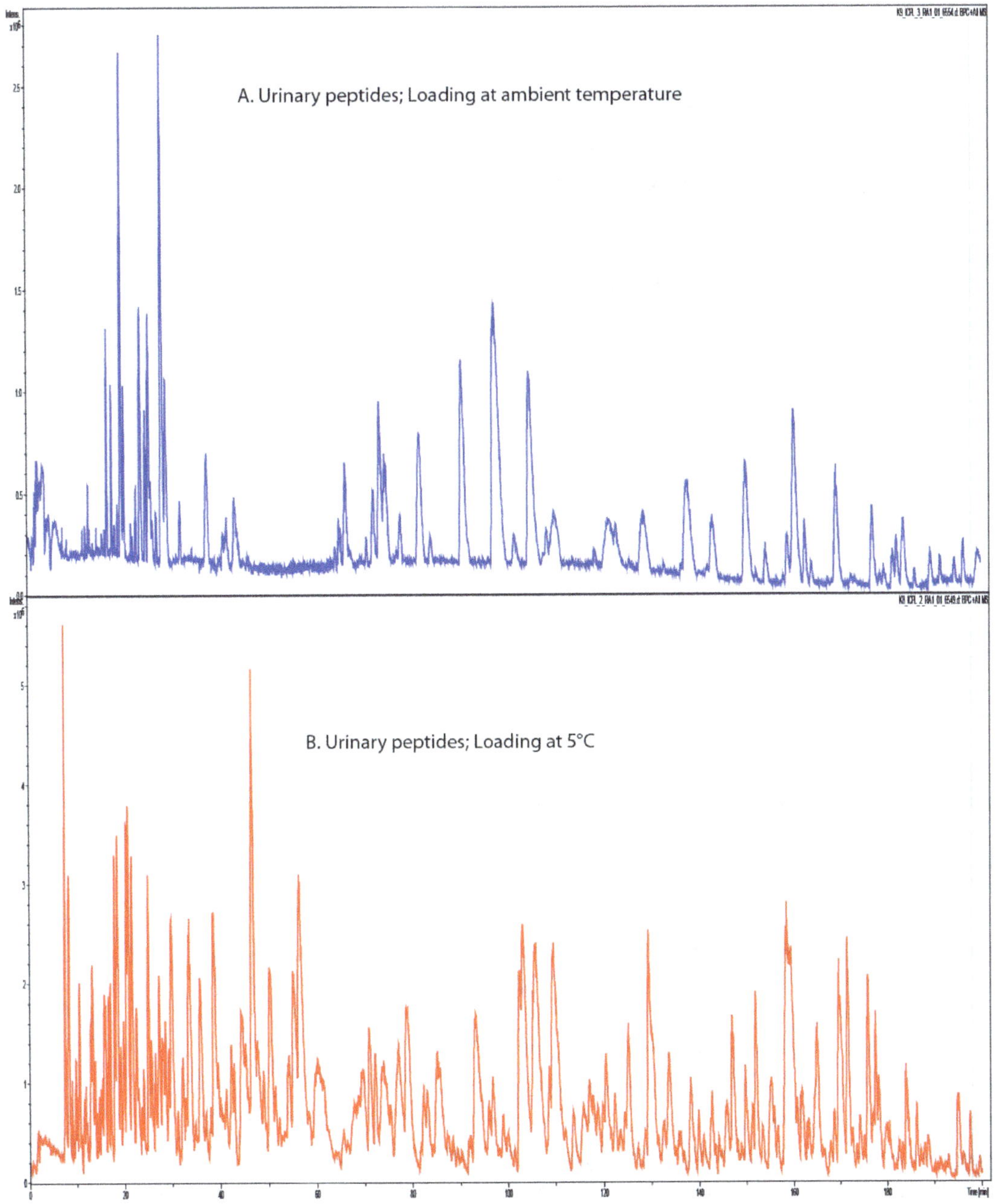

Figure 5. MS-Base Peak Chromatogram showing separation of tryptic peptides obtained from urinary proteins. (**A**) Trap column and the nano separation column were operated at 60 °C loading mobile phase was not precooled; and (**B**) tryptic peptides from urinary proteins were separated using identical gradient as in (**A**) but with precooled loading mobile phase.

In terms of reproducibility of protein identifications, two sets of samples were injected in triplicate using precooled mobile phase and loading mobile phase at ambient temperature. Figure 6A shows the number of overlapping identifications for the precooled mobile phase and Figure 6B for the loading mobile phase at ambient temperature. As it can be seen, the amount of proteins identified with precooled mobile phase is significantly higher and the number of overlapping identifications is higher as compared to identifications when loading solvent was used at ambient temperature. Some proteins have been identified only in one of the injections, this effect has been regularly observed for proteomics analysis and it is due to instability of electrospray ionization of peptides in complex samples.

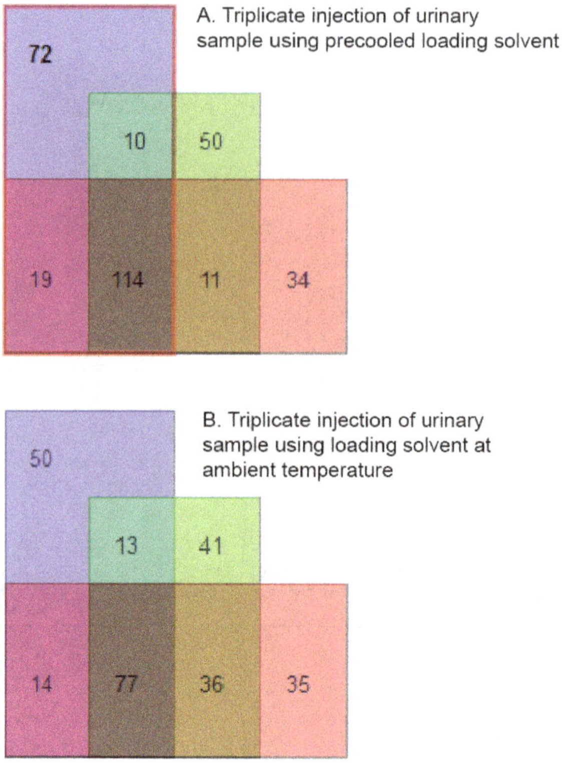

Figure 6. Reproducibility of protein identifications based on triplicate injections of tryptic urinary peptides using precooled (**A**) loading mobile phase and (**B**) mobile phase at ambient temperature.

4. Conclusions

The use of precooled loading mobile phase enabled improved peptide trapping on C18 trap columns operated in a column oven at 60 °C and prevented samples losses during the loading step. Due to the small volume of the trap column, it was possible to achieve fast cooling of the stationary phase despite operating temperature of 60 °C.

Further, it could be shown that the reproducibility of peptides' retention times was not affected due to the use of the precooled loading mobile phase and it remained stable and high.

We were able to show that a normal laboratory cooling bath set to 5 °C can be safely used for routinely cooling the loading mobile phase.

Positive results achieved during the tests encouraged us to apply these conditions as a standard protocol for of all peptide separations in our laboratory.

Acknowledgments: Acknowledgments: This work was supported by the FP7 project #282506, Prot-HISPRA (www.prot-hispra.eu).

Author Contributions: Author Contributions: Matthias Schöbinger and Oskar-James Klein performed hands-on experiments; Goran Mitulović designed the experiments and analyzed the data.

Conflicts of Interest: Conflicts of Interest: The authors declare no conflict of interest.

References

1. Mitulović, G.; Smoluch, M.; Chervet, J.-P.; Steinmacher, I.; Kungl, A.; Mechtler, K. An improved method for tracking and reducing the void volume in nano hplc-ms with micro trapping columns. *Anal. Bioanal. Chem.* **2003**, *376*, 946–951. [PubMed]
2. Mitulovic, G.; Stingl, C.; Smoluch, M.; Swart, R.; Chervet, J.P.; Steinmacher, I.; Gerner, C.; Mechtler, K. Automated, on-line two-dimensional nano liquid chromatography tandem mass spectrometry for rapid analysis of complex protein digests. *Proteomics* **2004**, *4*, 2545–2557. [CrossRef] [PubMed]

3. Mitulović, G. New HPLC techniques for proteomics analysis: A short overview of latest developments. *J. Liquid Chromatogr. Relat. Technol.* **2014**, *38*, 390–403.
4. Eeltink, S.; Dolman, S.; Swart, R.; Ursem, M.; Schoenmakers, P.J. Optimizing the peak capacity per unit time in one-dimensional and off-line two-dimensional liquid chromatography for the separation of complex peptide samples. *J. Chromatogr. A* **2009**, *1216*, 7368–7374. [PubMed]
5. Eeltink, S.; Dolman, S.; Vivo-Truyols, G.; Schoenmakers, P.; Swart, R.; Ursem, M.; Desmet, G. Selection of column dimensions and gradient conditions to maximize the peak-production rate in comprehensive off-line two-dimensional liquid chromatography using monolithic columns. *Anal. Chem.* **2010**, *82*, 7015–7020. [PubMed]
6. Detobel, F.; Broeckhoven, K.; Wellens, J.; Wouters, B.; Swart, R.; Ursem, M.; Desmet, G.; Eeltink, S. Parameters affecting the separation of intact proteins in gradient-elution reversed-phase chromatography using poly(styrene-co-divinylbenzene) monolithic capillary columns. *J. Chromatogr. A* **2010**, *1217*, 3085–3090. [PubMed]
7. Chervet, J.P.; Ursem, M.; Salzmann, J.P. Instrumental requirements for nanoscale liquid chromatography. *Anal. Chem.* **1996**, *68*, 1507–1512. [PubMed]
8. Eghbali, H.; Sandra, K.; Tienpont, B.; Eeltink, S.; Sandra, P.; Desmet, G. Exploring the possibilities of cryogenic cooling in liquid chromatography for biological applications: A proof of principle. *Anal. Chem.* **2012**, *84*, 2031–2037. [CrossRef] [PubMed]
9. Melander, W.; Campbell, D.E.; Horváth, C. Enthalpy-entropy compensation in reversed-phase chromatography. *J. Chromatogr. A* **1978**, *158*, 215–225.
10. Melander, W.R.; Chen, B.-K.; Horvàth, C. Mobile phase effects in reversed-phase chromatography: I. Concomitant dependence of retention on column temperature eluent composition. *J. Chromatogr. A* **1979**, *185*, 99–109.
11. Antia, D.F.; Horváth, C. High-performance liquid chromatography at elevated temperatures: Examination of conditions for the rapid separation of large molecules. *J. Chromatogr. A* **1988**, *435*, 1–15. [CrossRef]
12. Fekete, S.; Horváth, K.; Guillarme, D. Influence of pressure and temperature on molar volume and retention properties of peptides in ultra-high pressure liquid chromatography. *J. Chromatogr. A* **2013**, *1311*, 65–71. [CrossRef] [PubMed]
13. Ilisz, I.; Pataj, Z.; Gecse, Z.; Szakonyi, Z.; Fülöp, F.; Lindner, W.; Péter, A. Unusual temperature-induced retention behavior of constrained β-amino acid enantiomers on the zwitterionic chiral stationary phases zwix(+) and zwix(−). *Chirality* **2014**, *26*, 385–393. [PubMed]
14. Maes, K.; van Liefferinge, J.; Viaene, J.; van Schoors, J.; van Wanseele, Y.; Béchade, G.; Chambers, E.E.; Morren, H.; Michotte, Y.; vander Heyden, Y.; *et al.* Improved sensitivity of the nano ultra-high performance liquid chromatography-tandem mass spectrometric analysis of low-concentrated neuropeptides by reducing aspecific adsorption and optimizing the injection solvent. *J. Chromatogr. A* **2014**, *1360*, 217–228. [PubMed]
15. Cao, P.; Müller, T.K.H.; Ketterer, B.; Ewert, S.; Theodosiou, E.; Thomas, O.R.T.; Franzreb, M. Integrated system for temperature-controlled fast protein liquid chromatography. II. Optimized adsorbents and "single column continuous operation". *J. Chromatogr. A* **2015**, *1403*, 118–131. [CrossRef] [PubMed]
16. Groskreutz, S.R.; Horner, A.R.; Weber, S.G. Temperature-based on-column solute focusing in capillary liquid chromatography reduces peak broadening from pre-column dispersion and volume overload when used alone or with solvent-based focusing. *J. Chromatogr. A* **2015**, *1405*, 133–139. [CrossRef] [PubMed]
17. Zou, G.; Xiong, K.; Jiang, C.; Li, H.; Wang, Y.; Zhang, S.; Qian, Y. Magnetic Fe_3O_4 nanodisc synthesis on a large scale via a surfactant-assisted process. *Nanotechnology* **2005**, *16*, 1584–1588.
18. Han, G.; Ye, M.; Liu, H.; Song, C.; Sun, D.; Wu, Y.; Jiang, X.; Chen, R.; Wang, C.; Wang, L.; *et al.* Phosphoproteome analysis of human liver tissue by long-gradient nanoflow lc coupled with multiple stage ms analysis. *Electrophoresis* **2010**, *31*, 1080–1089. [PubMed]
19. Eeltink, S.; Dolman, S.; Detobel, F.; Swart, R.; Ursem, M.; Schoenmakers, P.J. High-efficiency liquid chromatography-mass spectrometry separations with 50 mm, 250 mm, and 1 m long polymer-based monolithic capillary columns for the characterization of complex proteolytic digests. *J. Chromatogr. A* **2010**, *1217*, 6610–6615. [CrossRef] [PubMed]
20. Šesták, J.; Kahle, V. Constant pressure mode extended simple gradient liquid chromatography system for micro and nanocolumns. *J. Chromatogr. A* **2014**, *1350*, 68–71. [CrossRef] [PubMed]

21. Neue, U.D.; Mazzeo, J.R. A theoretical study of the optimization of gradients at elevated temperature. *J. Sep. Sci.* **2001**, *24*, 921–929. [CrossRef]
22. Hartonen, K.; Riekkola, M.-L. Liquid chromatography at elevated temperatures with pure water as the mobile phase. *TrAC Trends Anal. Chem.* **2008**, *27*, 1–14. [CrossRef]
23. Farias, S.E.; Kline, K.G.; Klepacki, J.; Wu, C.C. Quantitative improvements in peptide recovery at elevated chromatographic temperatures from microcapillary liquid chromatography-mass spectrometry analyses of brain using selected reaction monitoring. *Anal. Chem.* **2010**, *82*, 3435–3440. [PubMed]
24. Wessel, D.; Flügge, U.I. A method for the quantitative recovery of protein in dilute solution in the presence of detergents and lipids. *Anal. Biochem.* **1984**, *138*, 141–143. [CrossRef]
25. Mitulović, G.; Stingl, C.; Steinmacher, I.; Hudecz, O.; Hutchins, J.R.A.; Peters, J.-M.; Mechtler, K. Preventing Carryover of Peptides and Proteins in Nano LC-MS Separations. *Anal. Chem.* **2009**, *81*, 5955–5960. [PubMed]
26. Neue, U. *HPLC Columns. Theory, Technology and Practice*; Wiley—VCH Publishers: New York, NY, USA, 1997; p. 43.
27. Trnka, P.; Ivanova, L.; Hiatt, M.J.; Matsell, D.G. Urinary biomarkers in obstructive nephropathy. *Clin. J. Am. Soc. Nephrol.* **2012**, *7*, 1567–1575. [PubMed]
28. Alberice, J.V.; Amaral, A.F.S.; Armitage, E.G.; Lorente, J.A.; Algaba, F.; Carrilho, E.; Márquez, M.; García, A.; Malats, N.; Barbas, C. Searching for urine biomarkers of bladder cancer recurrence using a liquid chromatography-mass spectrometry and capillary electrophoresis-mass spectrometry metabolomics approach. *J. Chromatogr. A* **2013**, *1318*, 163–170. [PubMed]
29. Dear, J.W.; Street, J.M.; Bailey, M.A. Urinary exosomes: A reservoir for biomarker discovery and potential mediators of intrarenal signalling. *Proteomics* **2013**, *13*, 1572–1580. [CrossRef] [PubMed]
30. Becker, L.; Kheirandish-Gozal, L.; Peris, E.; Schoenfelt, K.Q.; Gozal, D. Contextualised urinary biomarker analysis facilitates diagnosis of paediatric obstructive sleep apnoea. *Sleep Med.* **2014**, *15*, 541–549. [PubMed]
31. Beasley-Green, A.; Burris, N.M.; Bunk, D.M.; Phinney, K.W. Multiplexed LC-MS/MS Assay for Urine Albumin. *J. Proteome Res.* **2014**, *13*, 3930–3939. [PubMed]
32. Wu, Q.; Wu, D.; Guan, Y. Hybrid titania-zirconia nanoparticles coated adsorbent for highly selective capture of nucleosides from human urine in physiological condition. *Anal. Chem.* **2014**, *21*, 10122–10130. [CrossRef] [PubMed]
33. Smith, C.; Batruch, I.; Bauca, J.; Kosanam, H.; Ridley, J.; Bernardini, M.; Leung, F.; Diamandis, E.; Kulasingam, V. Deciphering the peptidome of urine from ovarian cancer patients and healthy controls. *Clin. Proteom.* **2014**, *11*. [CrossRef] [PubMed]
34. Yu, Y.; Suh, M.-J.; Sikorski, P.; Kwon, K.; Nelson, K.E.; Pieper, R. Urine sample preparation in 96-well filter plates for quantitative clinical proteomics. *Anal. Chem.* **2014**, *86*, 5470–5477. [PubMed]

18

Analyses of Indole Compounds in Sugar Cane (*Saccharum officinarum* L.) Juice by High Performance Liquid Chromatography and Liquid Chromatography-Mass Spectrometry after Solid-Phase Extraction

Jean Wan Hong Yong [1], Liya Ge [2], Wei San Wong [3], Zhen Ma [4] and Swee Ngin Tan [4,*]

1 ARC Centre for Mine Site Restoration, School of Biological Sciences, University of Western Australia, 35 Stirling Highway, Crawley, Perth 6009, Western Australia, Australia; john.yong@uwa.edu.au
2 Residues and Resource Reclamation Centre, Nanyang Environment and Water Research Institute, Nanyang Technological University, 1 Cleantech Loop, Singapore 637141, Singapore; lyge@ntu.edu.sg
3 Advanced Environmental Biotechnology Centre, Nanyang Environment and Water Research Institute, Nanyang Technological University, 1 Cleantech Loop, Singapore 637141, Singapore; weisan@ntu.edu.sg
4 Natural Sciences and Science Education Academic Group, Nanyang Technological University, 1 Nanyang Walk, Singapore 637616, Singapore; mazhen@ntu.edu.sg
* Correspondence: sweengin.tan@nie.edu.sg

Academic Editor: Juan F. García-Reyes

Abstract: Simultaneous quantitative analysis of 10 indole compounds, including indole-3-acetic acid (IAA, one of the most important naturally occurring auxins) and some of its metabolites, by high performance liquid chromatography (HPLC) and liquid chromatography-mass spectrometry (LC-MS) after solid-phase extraction (SPE) was reported for the first time. The analysis was carried out using a reverse phase HPLC gradient elution, with an aqueous mobile phase (containing 0.1% formic acid) modified by methanol. Furthermore, a novel SPE procedure was developed for the pre-concentration and purification of indole compounds using C_{18} SPE cartridges. The combination of SPE, HPLC, and LC-MS was applied to screen for the indole compounds present in sugar cane (*Saccharum officinarum* L.) juice, a refreshing beverage with various health benefits. Finally, four indole compounds were successfully detected and quantified in sugar cane juice by HPLC, which were further unequivocally confirmed by LC-MS/MS experiments operating in the multiple reaction monitoring (MRM) mode.

Keywords: indole compounds; high performance liquid chromatography; liquid chromatography-tandem mass spectrometry; solid-phase extraction; sugar cane juice

---

1. Introduction

Sugar cane (*Saccharum officinarum* L.), a widely distributed crop in the world, has many economic dimensions, especially for the sugar industry, in many tropical and subtropical countries [1]. In many countries where sugar cane is cultivated, general population like to drink sugar cane juice as a delicious, refreshing, and popular beverage. Past research works have revealed several interesting biochemical properties of sugar cane, e.g., certain phenolic analytes and flavonoids in sugar cane were considered to have certain antioxidant activity [1,2]. In addition, sugar cane juice contains sugar, vitamins, inorganic minerals and plant growth regulators (i.e., phytohormones). Therefore, sugar cane juice has also been used in the plant tissue culture industry [3], though not as widely used as coconut water [4,5]. For the tissue culture work performed in our group (unpublished data), sugar cane juice supplementation is able to support tissue culture growth. Thus, this work aims to provide some evidence on the phytohormone (e.g., auxin) composition of the sugar cane juice, as such work has not been previously reported. Characterization of phytohormone composition of sugar cane could promote its wider applications in the tissue culture industry.

Indole, an aromatic heterocyclic organic compound, is the potent basic pharmacodynamic nucleus that has been reported to possess a wide variety of biological and clinical properties [6]. All indole compounds have a bicyclic structure, consisting of a six-membered benzene ring fused to a five-membered nitrogen-containing pyrrole ring (Figure 1). The indole structure can be found in many organic compounds like tryptophan, tryptophan-containing proteins, alkaloids, pigments, etc. Some earlier studies reported that plant tissues contain many indole compounds [7], which included the most important phytohormone, i.e., auxin. Auxins play various important functions in plant growth and development, such as cell division, elongation, embryo formation, and they also function as 'chemical signals' among cells, tissues and organs. Indole-3-acetic acid (IAA) is the main active endogenous auxin in most plants [8]. IAA is bio-synthesized from the tryptophan or indole, via indole-3-pyruvic acid (IPA), indole-3-acetamide, indole-3-acetonitrile, etc. [8–12]. In addition, some other indole compounds have been reported to have beneficial effects in biomedical research involving human health: relief from insomnia, enhanced anticonvulsive activity of antiepileptic drugs, some antioxidant activity, as starting materials for some anti-inflammatory drugs, etc. [13–21]. For example, IPA is a potential drug used for treating anxiety and insomnia [13–15].

Considering the low concentrations of indole compounds in plant tissues, different sample preparation methods—mainly including liquid–liquid extraction (LLE) and solid phase extraction (SPE)—have been employed for preconcentration and purification of indole compounds [22–24]. Compared with LLE, SPE is a simpler, yet more effective and versatile method for different sample matrices.

Analyses of indoles or phytohormones (e.g., anxins) by gas chromatography (GC) and gas chromatography-mass spectrometry (GC-MS) had been previously reported [23–27]. Also, there were some reports on the analyses of indoles and/or phytohormones by capillary electrophoresis (CE) [28,29]. Compared with the other chromatographic procedures, high performance liquid chromatography (HPLC), and liquid chromatography-mass spectrometry (LC-MS) have several advantages: high efficiency coupled with high sample capacity; rapid speed of analysis; simplicity of sample recovery; ability to analyze non-derivatized samples, etc. Although HPLC and LC-MS have been successfully used to analyze some indoles or phytohormones [30–35], few of these studies attempted to simultaneously analyze and detect different kinds of indole compounds within a complex plant matrix using HPLC or LC-MS.

Current work proposes a simple and sensitive HPLC method for analyzing IAA, IPA, 5 indole carboxylic acid isomers, 3-acetylindole, and indole-3-acetamide in a single analysis. The developed methodology was used to screen for the presence of endogenous indole compounds in sugar cane juice after application of a new SPE method for pre-concentration and purification of indole compounds. The presence of four indole compounds, which included IAA, was further confirmed by LC-MS/MS assay based on its characteristic fragmentation pattern. With HPLC and LC-MS based methods, IAA, IPA, 3-acetylindole, and indole-2-carboxylic acid were successfully identified and quantified.

2. Materials and Methods

2.1. Reagents and Materials

All indole standards were purchased from Sigma-Aldrich (Steinheim, Germany) except IAA, which was obtained from PhytoTechnology Laboratories (Shawnee Mission, KS, USA). Figure 1 shows the chemical structures of the 10 indole compounds. All the standards were dissolved in methanol with the concentration ranging from 250 to 500 μM and stored at temperature 4 °C. Methanol and acetonitrile (HPLC grade) were purchased from Sigma-Aldrich (St. Louis, MO, USA). Formic acid (analytical reagent grade) was purchased from Merck (Darmstadt, Germany). Filter paper (12.5 cm, No. 542, Whatman, Maidstone, Kent, England) was used to filter the sugar cane juice before SPE. Ultrapure water (Milford, MA, USA) was used throughout the analysis. The pH of the sample solution was adjusted to 3 by adding formic acid and monitored using a pH meter (CORNING 440, Corning Glass Works, Corning, NY, USA).

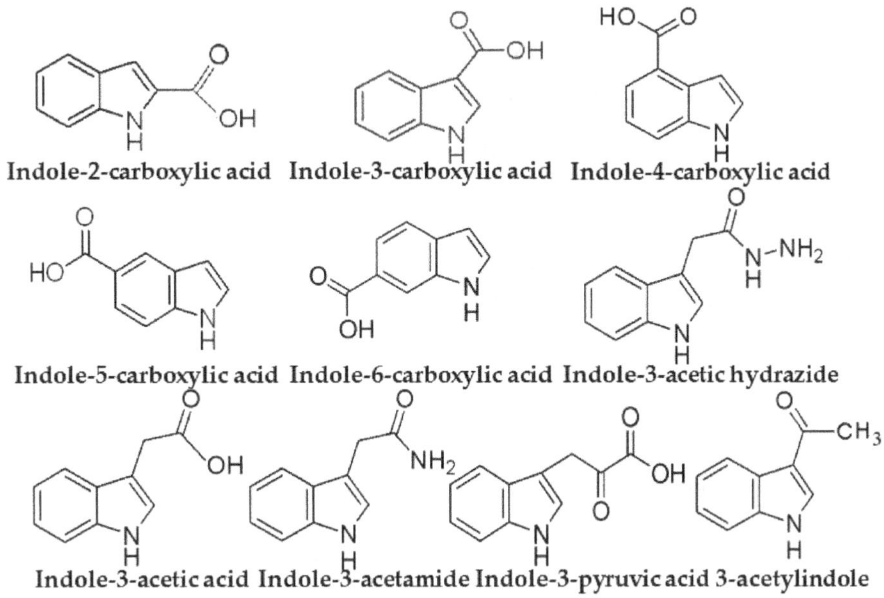

Figure 1. Chemical structures of the 10 indole compounds.

2.2. HPLC Instrumental Set-Up and Procedure

The analysis of indole compounds was performed using a high performance liquid chromatography system (Waters 2695 Separations Module, Waters, Milford MA, USA) linked simultaneously to a PDA system (PDA 2996 detector, Waters). Typically, the PDA system was operated at a detection wavelength of 280 nm. Data were processed by the accompanying system software (Millennium32 Software, Data Handling System for Windows, version 4.0).

The initial HPLC running condition was methanol–formic acid buffer (20:80, v/v). Solvent (A) consisted of 0.1% formic acid (w/v); and solvent (B) consisted of methanol. The flow rate was 0.3 mL·min^{-1} throughout the whole separation. Separations were performed using a C_{18} reverse-phase column (Zorbax SB-C_{18} 100 Å, 3.5 μm, 2.1 mm × 150 mm, Agilent Technologies, Palo Alto, CA, USA), at 30 °C. The injection volume of standards and purified sugar cane juice samples were 5 μL and 10 μL, respectively.

2.3. LC-MS/MS Conditions

LC (Model 1100 Series, Agilent Technologies) linked with LC/MSD Trap XCT (Agilent Technologies, Palo Alto, CA, USA) by an electrospray ionization (ESI) interface was used in scan mode for standards. The same C_{18} reverse-phase column and solvents as HPLC were employed with an injection volume of 5 μL. The column thermostat was set at 40 °C.

ESI-MS analysis was performed in the negative mode, and the ion trap was scanned at m/z 50–400 in full scan mode. The maximum accumulation time for the ion trap was set at 200 ms and the target count was set at 100,000. The actual accumulation time was controlled by ion charge control (ICC), which was used to prevent ion saturation in the ion trap. The nebulizer gas pressure, drying gas flow rate, drying gas temperature, and capillary voltage for the ESI source were set at 30 psi, 8 L·min^{-1}, 350 °C, and 3500 V, respectively. Other instrument parameters were optimized for generating the highest signal intensities.

The multiple reaction monitoring (MRM) modes were used to monitor the transitions from the precursor ions to the most abundant product ions. MRM were performed in the smart mode with helium as collision gas. Table 1 showed the MRM transition results of different indole compounds. The data were processed using LC/MSD Chemstation software.

Table 1. MRM transition result of 10 indole compounds.

Indole Compounds	MRM Transition (*m/z*)
Indole-2-carboxylic acid	160→115
Indole-3-carboxylic acid	160→115
Indole-4-carboxylic acid	160→115
Indole-5-carboxylic acid	160→115
Indole-6-carboxylic acid	160→115
Indole-3-acetic acid (IAA)	173→129
Indole-3-acetic hydrazide	188→115
Indole-3-acetamide (IAM)	173→129
Indole-3-pyruvic acid (IPA)	202→157
3-acetylindole	158→115

2.4. Retention of Indole Compounds on SPE Cartridges

The SPE experiments were conducted using C_{18} SPE cartridges (J.T. Baker, Phillipsburg, NJ, USA; 500 mg, 3 mL). In the preliminary experiments, standard solutions containing 10 indole compounds (2 μmol for each standard) dissolved in 100 mL acidified ultra-pure water (pH adjusted to 3 by formic acid) were extracted using SPE, which was previously conditioned with 2 mL methanol, 2 mL methanol-water solution (50:50, *v*/*v*), 2 mL methanol-water solution (30:70, *v*/*v*), and 2 mL acidified water (pH adjusted to 3 by formic acid) to obtain the optimum extraction conditions. Typically, the cartridge was washed with 5 mL acidified water (pH adjusted to 3 by formic acid), and then eluted with 5 mL methanol-water solution (80:20, *v*/*v*). The recovery of each indole standard was measured by HPLC method.

2.5. Sample Preparation

Sugar cane juice was crushed from fresh green sugar cane stems (Malaysia), and the analysis was performed within one week after harvest, since the quality of juice would change upon delayed extraction and storage [36]. Before further pre-concentration and purification, the pH of 500 mL sugar cane juice was adjusted to 3 by adding formic acid and filtered through filter paper (Whatman, 12.5 cm, No. 542) to remove the suspended matters. The optimum novel SPE procedure using C_{18} SPE cartridges (Section 2.4) was applied to pre-concentrate and purify the putative indole compounds. The SPE eluate was lyophilized to dryness, and the dried residue was reconstituted in 1 mL methanol. The collected samples were finally analyzed by HPLC and LC-MS.

2.6. Recovery Study of SPE Procedure

The extraction recoveries of 10 indole compounds using the C_{18} cartridges under optimum SPE conditions were also investigated. Unlabeled 10 indole standards (0.2 μmol each) were added to 50 mL sugar cane juice, and extracted according to the same sample preparation procedure. The spiked extracted sample was then subsequently analyzed using HPLC measurement.

3. Results and Discussion

3.1. Development HPLC-PDA Method

Initially, different stationary phases (C_{18} and C_8) and mobile phases (methanol and acetonitrile) were compared. As the column with C_{18} provided a better peak shape in most cases, it was used for further study. Even though methanol and acetonitrile provided similar results, methanol was selected for the further study due to the lower cost. Different gradient elution programs were tested to find the optimum separation conditions for the 10 indole compounds. With reference to the separation efficiency and a reasonable analytical time, the following gradient elution approach was eventually used as the preferred method: the column was initialized isocratically with methanol-formic acid buffer (20:80, v/v) for 5 min, then a linear gradient to methanol-formic acid buffer (30:70, v/v) in 15 min, which was later maintained isocratically for 15 min. Under the optimum separation conditions, all 10 compounds were successfully separated within 35 min. After each analysis, the column was washed with 95:5 (v/v) methanol-formic acid buffer for 5 min, and then formic acid buffer-methanol 90:10 (v/v) for ca. 30 min was used for re-equilibrated after this procedure.

Considering the difference UV absorbances of different indole compounds, 280 nm was chosen as the compromised optimum detecting wavelength with the programmable PDA system. Figure 2 showed the well resolved HPLC chromatogram of the 10 indole standard mixture under optimum conditions within 35 min. The HPLC–PDA method also allowed for the identification of the 10 indole compounds by comparing their retention times and the concomitant UV spectra with those of the respective standards.

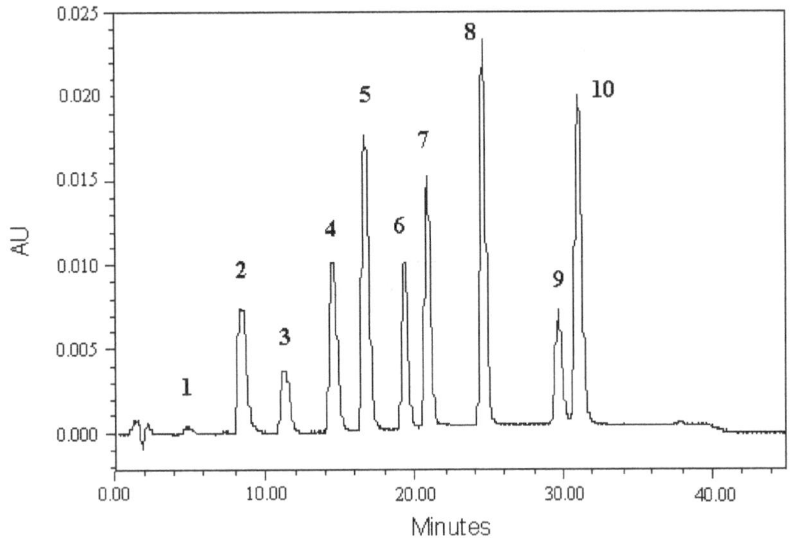

Figure 2. HPLC chromatogram of standard mixture of indole compounds (50 μM for each compound) at 280 nm. Peak identities: 1, indole-3-acetic hydrazide; 2, acetamide; 3, indole-4-carboxylic acid; 4, indole-5-carboxylic acid; 5, indole-3-carboxylic acid; 6, IAA; 7, 3-acetylindole; 8, indole-6-carboxylic acid; 9, indole-3-pyruvic acid; 10, indole-2-carboxylic acid. Note: Injection volume was 5 μL. HPLC running conditions were described in Sections 2.2 and 3.1.

3.2. Validation of the Method

Based on the optimum conditions, validation of the method was investigated by the relative standard deviations (RSDs) of retention time and peak area, linearity, and limits of quantification (LOQs). The reproducibility of the retention time and peak area of 10 indole compounds under optimum HPLC conditions was investigated by doing repeated injections (n = 6) of a mixture of the standards at a concentration of 50 μM. The RSDs of the retention time for all analytes were in the range of 0.08%–0.24%; the RSDs of peak area were in the range of 0.2%–6.0%. A linear correlation was found between the concentration and the peak area for all analytes in the range of 12.5–200 μM except indole-3-acetic hydrazide (25–200 μM); typically, R^2 values were in the range of 0.989–0.999, which was good enough for analytical purpose. Based on a signal-to-noise ratio of 10, the LOQs were in the range of 0.08–1.72 μM. Three independent injections were carried out for each calibration point. All the response characteristics of the indole standards using HPLC analysis were summarized in Table 2.

Table 2. Response characteristics of indole compound standards using HPLC.

Indole Compounds	Retention Times [a] Mean (min)	Retention Times [a] RSD (%)	Peak Area [a] Mean (μAu·s)	Peak Area [a] RSD (%)	Equation of Calibration Curve [b]	R^2	Linear Range (μM)	LOQ (μM) [c]
Indole-3-acetic hydrazide	4.827	0.10	11,362	6.0	$y = 31.8x + 9717.1$	0.998	25–200	1.09
Indole-3-acetamide	8.432	0.08	290,551	0.4	$y = 6231.3x - 9126.4$	0.999	12.5–200	0.08
Indole-4-carboxylic acid	11.331	0.15	135,491	1.5	$y = 2920.1x - 13671$	0.989	12.5–200	1.64
Indole-5-carboxylic acid	14.574	0.17	333,576	0.7	$y = 7086.2x - 14788.3$	0.999	12.5–200	1.50
Indole-3-carboxylic acid	16.723	0.18	545,307	0.6	$y = 11269.9x - 13352.6$	0.999	12.5–200	1.34
Indole-3-acetic acid	19.358	0.19	284,600	0.3	$y = 5999.5x - 13144.3$	0.999	12.5–200	1.15
3-acetylindole	20.867	0.18	398,848	1.0	$y = 8465.8x - 13598.7$	0.999	12.5–200	1.49
Indole-6-carboxylic acid	24.659	0.24	589,492	1.7	$y = 12959x - 17287.8$	0.992	12.5–200	1.48
Indole-3-pyruvic acid	29.637	0.22	208,408	2.8	$y = 5459.3x - 61613.2$	0.996	12.5–200	1.72
Indole-2-carboxylic acid	31.044	0.21	615,529	0.2	$y = 13249.1x - 35713.8$	0.999	12.5–200	0.11

[a] The data were measured with repeated injections ($n = 6$) of a mixture of indole standards at a concentration of 50 μM each; [b] In the calibration equation, x represents concentration of the analyte (μM) and y represents the peak area (μAu· s); [c] The LOQ were estimated based on S/N = 10.

Intra- and inter-day variations were chosen to further determine the precision of the HPLC method. The intra-day variation was determined by analyzing in triplicate the same standard mixtures dissolved in methanol solution for six times within one day. While for the inter-day variability test, the solution was examined in triplicate for three consecutive days and repeated one week later. The RSDs of the retention time (t_R) and peak area (p_a) were taken as the measure of precision (Table 3). These results are within the acceptable criteria for precision.

Table 3. Precision of 10 indole compounds for retention time (t_R) and peak area (p_a) expressed as RSDs (%).

Indole Compounds	Intra-Day Variability [a] RSD (t_R [b] %)	Intra-Day Variability [a] RSD (p_a [c] %)	Inter-Day Variability [d] RSD (t_R %)	Inter-Day Variability [d] RSD (p_a %)
Indole-3-acetic hydrazide	0.10	6.0	0.12	7.1
Indole-3-acetamide	0.08	0.4	0.09	1.3
Indole-4-carboxylic acid	0.15	1.5	0.16	1.5
Indole-5-carboxylic acid	0.17	0.7	0.19	2.1
Indole-3-carboxylic acid	0.18	0.6	0.19	1.6
Indole-3-acetic acid	0.19	0.3	0.20	0.9
3-acetylindole	0.18	1.0	0.20	1.2
Indole-6-carboxylic acid	0.24	1.7	0.25	1.9
Indole-3-pyruvic acid	0.22	2.8	0.23	1.4
Indole-2-carboxylic acid	0.21	0.2	0.23	0.8

[a] The intra-day variation was determined by analyzing in triplicate the same standard mixtures (50 μM for each indole compound) dissolved in methanol solution for six times within one day; [b] The retention time; [c] The peak area; [d] The inter-day variation was investigated by analyzing in triplicate the same standards mixtures (50 μM for each indole compound) for three consecutive days and repeated one week later.

3.3. Optimization of the SPE Conditions

To develop a new method for measuring indole compounds, the C_{18} SPE cartridge was chosen since it provided good recovery performance in our previous phytohormone study which included a few indole compounds [37]. In the earlier work, the SPE was optimized to simultaneously pre-concentrate different classes of phytohormones in a single assay. The new SPE method developed was found to be useful for the analysis of the 10 indole compounds.

During the initial optimization process, ultrapure water was used without any pH adjustment in the conditioning and washing steps. Later, different compositions of methanol-water solution were used as the eluent solution (90:10, 80:20, 70:30, 60:40, 50:50, v/v). As shown in Figure 3a, 80:20 (v/v) methanol-water solution was finally selected as the compromised solution based on the recoveries of the 10 indole compounds. Typically, recoveries of the 10 indoles were in the range of 45%–91% under the optimized condition (80:20 (*v*/*v*) methanol-water). The effect of washing water solution pH on the recoveries of the 10 indole compounds was also investigated. The different pH values (2.5, 3, 3.5) of washing water solution were adjusted by formic acid. As shown in Figure 3b, when using ultrapure water (pH adjusted to 3 by formic acid) for conditioning and washing the cartridges, the recoveries of all indoles were from 68% to 98%, which were considered to be optimal extraction recovery rates.

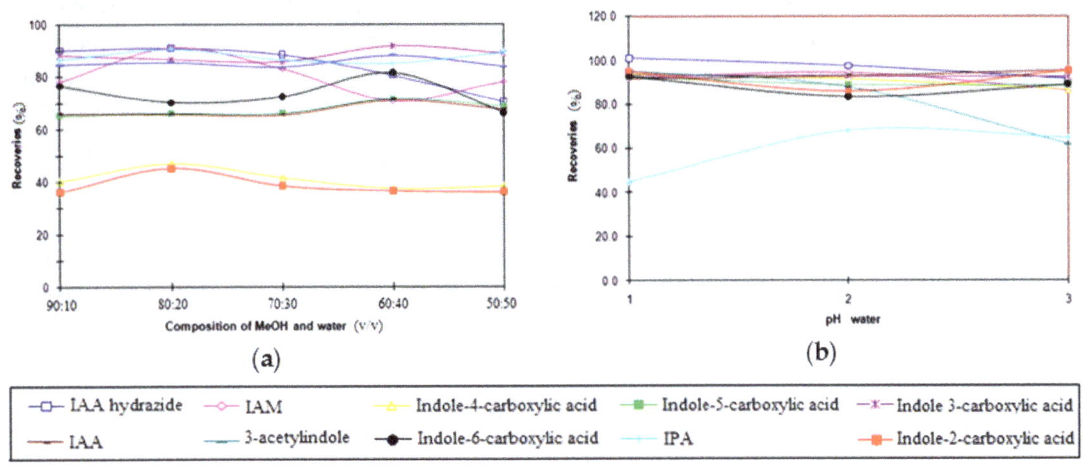

Figure 3. The effects of different methanol and water composition (**a**) in eluting solution, different pH values of water (**b**) used in conditioning and washing steps which were adjusted by formic acid, on the recoveries of indole compounds using C_{18} SPE columns.

The optimized C_{18} SPE method was able to pre-concentrate *ca.* indole compounds 500-fold, and this approach was later used to pre-concentrate and purify the putative indoles present naturally in sugar cane juice, prior to HPLC and LC-MS analyses. The schematic pre-concentration protocol based on the recoveries of indole standards in the washing and eluting solutions was presented in Figure 4.

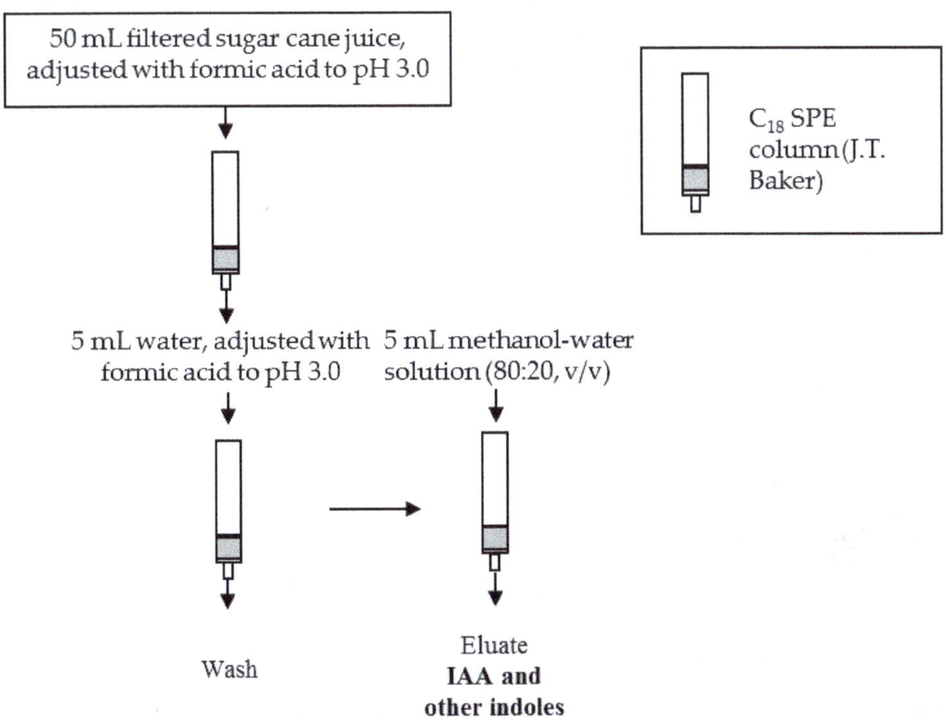

Figure 4. Pre-concentration procedure for indoles in sugar cane juice by SPE C_{18} cartridge.

3.4. Analysis of Indole Compounds in Sugar Cane Juice Using HPLC

Under the optimum operating conditions, HPLC method was applied to determinate the various indole compounds present in the sugar cane juice. Typical chromatogram for the sample after SPE was shown in Figure 5a. On the bases of retention times as well as the results obtained by using the standard addition method (Figure 5b), the presence of IAA, 3-acetylindole, IPA, and indole-2-carboxylic acid in the sugar cane juice was successfully identified. The approximate concentrations of IAA, 3-acetylindole, IPA, and indole-2-carboxylic acid in the sugar cane juice, disregarding the loss during the extraction and purification steps, were shown in Table 4. Based on absolute concentration, the major indole compound in sugar cane juice is IAA, the most physiologically active auxin in plants [38]. Furthermore, indole-2-carboxylic acid was successfully identified among five positional isomers (indole-2-carboxylic acid, indole-3-carboxylic acid, indole-4-carboxylic acid, indole-5-carboxylic acid, and indole-6-carboxylic acid). Figure 5a showed possible interferences from the plant matrix during the analytical process. The identity of the analytes of interest for other future application(s), if detected near to the interfering matrix peak(s), could be independently confirmed by LC-MS/MS experiments.

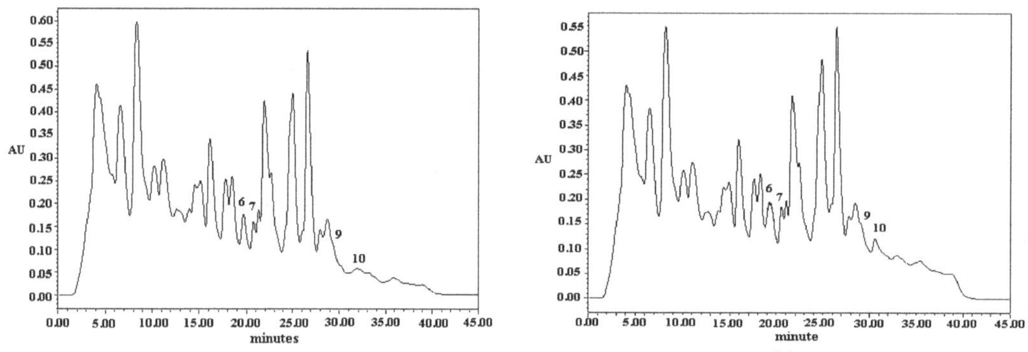

(a) (b)

Figure 5. HPLC chromatograms of (**a**) sugar cane juice extracts; (**b**) spiked sugar cane juice extracts by IAA, 3-acetylindole, IPA and indole-2-carboxylic acid at 280 nm. Note: Injection volume was 10 µL. Sugar cane juice extracts were purified earlier using C_{18} SPE column (Section 2.5). HPLC running conditions and peak identities were the same as described in Figure 2.

Table 4. Estimated concentrations of 10 indoles in sugar cane juice, recoveries of 10 indoles for SPE procedure and estimated original concentrations of 10 indoles in sugar cane juice.

Indole Compounds	Estimated Concentrations (µM)	Recovery (%)	Estimated Original Concentrations (µM)
Indole-3-acetic hydrazide	n.d.	92%	Below 1.18
Indole-3-acetamide	n.d.	80%	Below 0.10
Indole-4-carboxylic acid	n.d.	79%	Below 2.08
Indole-5-carboxylic acid	n.d.	65%	Below 2.31
Indole-3-carboxylic acid	n.d.	78%	Below 1.72
Indole-3-acetic acid	1.05 ± 0.23	68%	1.54 ± 0.34
3-acetylindole	0.57 ± 0.11	62%	0.92 ± 0.18
Indole-6-carboxylic acid	n.d.	87%	Below 1.70
Indole-3-pyruvic acid	0.21 ± 0.08	61%	0.34 ± 0.13
Indole-2-carboxylic acid	0.34 ± 0.17	82%	0.41 ± 0.21

n.d.: not detected.

3.5. LC-ESI-MS/MS Measurement

The main aim of the LC-MS/MS investigation was to further confirm the identities of the putative indole compounds found in sugar cane juice using HPLC analysis. The optimized HPLC separation condition was adopted for LC-MS/MS experiment.

Due to the potential high interferences in our HPLC results as shown in Figure 5, LC-MS/MS experiments were carried out to re-confirm the results of HPLC measurement. Based on the findings of four putative indoles using HPLC, IAA, 3-acetylindole, IPA, and indole-2-carboxylic acid were detected without significant interferences from the plant matrix using LC-MS/MS with MRM mode (Figure 6). The estimated concentrations of IAA, 3-acetylindole, IPA, and indole-2-carboxylic acid in the sugar cane juice were 1.14 μM, 4.1×10^{-1} μM, 1.3×10^{-1} μM, and 2.7×10^{-1} μM, based on the LC-MS measurement, respectively, which were similar in the range with values obtained using the HPLC method. The slight variation of the results obtained by HPLC and LC-MS could be due to the impurity affecting the UV peaks, or partially due to the natural variation exhibited by different sugar cane batches. Thus, from the results obtained using the LC-MS/MS investigation, IAA and other three indole compounds were unequivocally found to be present in the sugar cane juice sample, based on retention times with LC analysis and their MRM mass transitions in the LC-MS measurement mode.

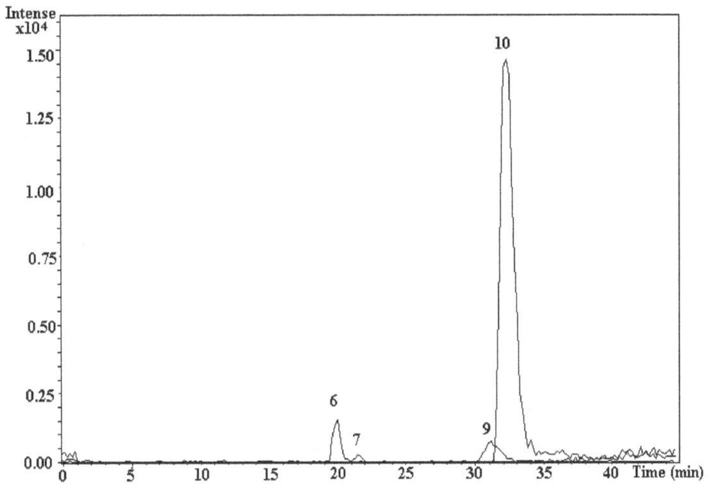

Figure 6. MRM chromatograms of putative four indoles in sugar cane juice by LC-MS/MS method. The LC-MS running conditions were described in Section 2.3.

3.6. Recovery Study of SPE Procedure

We investigated on the recovery of the SPE procedure by comparing the difference of the obtained analyte concentration in extracted spiked sample (sugar cane juice sample with spiked indole standards) and the extracted blank sample (sugar cane juice sample without spiked indole standards) with the concentration of the spiked analytes in the same volume of blank solution, i.e., ultrapure water.

As estimated in Table 4, the recovery yields of the 10 different indole compounds ranged from 61% to 92%, which was considered reasonably high for a pre-concentration and sample clean-up procedure. Based on the recovery yields and the results of HPLC measurements, the estimated original concentrations of indoles in sugar cane juice were also given in Table 4.

4. Conclusions

This study reported the development and validation of new and relatively simple HPLC and LC-MS/MS methods for the simultaneous analysis of various indole compounds with high detection sensitivity. The separation of the 10 standards, including 5 isomers (indole-2-carboxylic acid, indole-3-carboxylic acid, indole-4-carboxylic acid, indole-5-carboxylic acid, and indole-6-carboxylic acid), was completed in 35 min with well-resolved peaks. Running time was decreased and the separation of different indole compounds was optimized by increasing the strength of the organic solvent. The optimum separation conditions developed in this project will be useful for future studies, as typical separation of different indole compounds including isomers and IAA, being one of the most important auxins in a single analysis was rarely, if not never, reported in the literature.

Furthermore, a new SPE method for the pre-concentration and purification of indole compounds was developed by using C_{18} SPE columns. The pH of water solution used for the washing step and the composition of eluent were also optimized to achieve higher recovery rates by reducing the effects of the interfering substances in the final extraction.

The effectiveness of the new pre-concentration SPE method and analytical HPLC method was evaluated by screening for the putative indole compounds in sugar cane juice. The presence of indole compounds in sugar cane juice, first detected by HPLC, was further confirmed independently by LC-MS/MS experiments. The results from the HPLC and LC-MS/MS indicated the presence of IAA, 3-acetyindole, IPA, and indole-2-carboxylica acid in sugar cane juice samples. This is also the first report of indole compounds in sugar cane juice. The detection of different indole compounds is valuable in plant science and biomedical science, due to the different important biological functions of these indole compounds. The presence of IAA in sugar cane juice can partially explain why sugar cane juice can be used as a growth supplement in plant tissue culture. The presence of IPA in sugar cane juice can partly confirm the relationship between IAA and IPA. In addition, as IPA has medically useful properties such as bringing relief from anxiety and insomnia [13–16], sugar cane juice may be a good beverage for people who suffer from sleep related problems. The presence of 3-acetylindole and indole-2-carboxylic acid may provide more potential usage of sugar cane juice besides being the main ingredient for the sugar industry. It is also plausible that some other indole compounds may be present in sugar cane juice, which we were unable to detect due to the current detection limits of current HPLC and LC-MS/MS methods.

Nevertheless, due to the important bioactivities driven by indole compounds, more work should be carried out on indole compounds naturally present in plants. Moving forward, UPLC-TOF-MS analytical approach can also be evaluated in order to develop shorter analytical duration, higher separation resolution, and higher sensitivity for the indole compounds [39]. Furthermore, the outstanding performances provided by core-shell particles column on a traditional HPLC instruments are comparable to those obtained with a costly UPLC instrumentation, making this novel column a promising tool for separation of indole compounds [40]. Also, sugar cane juice can be analyzed for the other bioactive compounds, such as other phytohormones.

Author Contributions: Jean Wan Hong Yong conceived and designed the experiments; Liya Ge set up the experiment and prepared manuscript; Wei San Wong analyzed the data; Zhen Ma performed some experiments; Swee Ngin Tan reviewed the data and revised the manuscript.

Conflicts of Interest: The authors declare no conflict of interest. The founding sponsors had no role in the design of the study; in the collection, analyses, or interpretation of data; in the writing of the manuscript; or in the decision to publish the results.

References

1. Duarte-Almeida, J.M.; Novoa, A.V.; Linares, A.F.; Lajolo, F.M.; Genovese, M.I. Antioxidant activity of phenolics compounds from sugar cane (*Saccharum officinarum* L.) juice. *Plant Food Hum. Nutr.* **2006**, *61*, 187–192. [CrossRef] [PubMed]
2. McGhie, T.K. Analysis of sugar cane flavonoids by capillary zone electrophoresis. *J. Chromatogr.* **1993**, *634*, 107–112. [CrossRef]
3. Naik, G.R. *Sugarcane Biotechnology*; Agritech Publications: Shrub Oak, NY, USA, 2001.
4. Arditti, J.; Ernst, R. *Micropropagation of Orchids*; John Wiley & Sons, Inc.: New York, NY, USA, 1993; pp. 47–48.
5. Ang, S.L.P.; Yong, J.W.H. A protocol for in vitro germination and sustainable growth of two tropical mistletoes. *Plant Cell Tissue Organ Cult.* **2005**, *80*, 221–228. [CrossRef]
6. Available online: http://en.wikipedia.org/wiki/Indole (accessed on 1 December 2016).
7. Guerrero, J.R.; Garcia-Ruiz, P.; Sanchez-Bravo, J.; Acosta, J.M.; Arnao, M.B. Quantification of indole-3-acetic acid by LC with electrochemical detection in etiolated hypocotyls of Lupinus Albus. *J. Liq. Chromatogr. Relat. Technol.* **2001**, *24*, 3095–3104. [CrossRef]

8. Davies, P.J. *Plant Hormones: Biosynthesis, Signal Transduction, Action!* Kluwer Academic Publishers: Norwell, MA, USA, 2004.
9. Kende, H.; Zeevarrt, J.A.D. The five "classical" plant hormones. *Plant Cell* **1997**, *9*, 1197–1210. [CrossRef] [PubMed]
10. Manulis, S.; Shafrir, H.; Epstein, E.; Lichter, A.; Barash, I. Biosynthesis of indole-3-acetic acid via the indole-3-acetamide pathway in *Streptomyces* spp. *Microbiology* **1994**, *140*, 1045–1050. [CrossRef] [PubMed]
11. Yagi, K.; Matsumoto, T.; Chujo, T.; Nojiri, H.; Omori, T.; Minamisawa, K.; Nishiyama, M.; Yamane, H. Isolation and characterization of low-indole-3-acetic acid-producing mutants from *Bradyrhizobium elkanii*. *Biosci. Biotechnol. Biochem.* **2001**, *65*, 1265–1269. [CrossRef] [PubMed]
12. Pollmann, S.; Neu, D.; Weiler, E. Molecular cloning and characterization of an amidase from Arabidopsis thaliana capable of converting indole-3-acetamide into the plant growth hormone, indole-3-acetic acid. *Phytochemistry* **2003**, *62*, 293–300. [CrossRef]
13. Silvestri, R.; Mento, G.; Raffaele, M.; Luca, G.D.; Buttini, G.; Casella, C.; Tisano, A.; Domenico, P.D.; Rosa, A.D.; Perri, R.D. Indole-3-pyruvic acid as a possible hypnotic agent in insomniac subjects. *J. Int. Med. Res.* **1991**, *19*, 403–409. [CrossRef] [PubMed]
14. Lapin, I.P.; Politi, V. Anxiolytic effect of indole-3-pyruvic acid (IPA) in mice. *Pharmacol. Res.* **1993**, *28*, 129–134. [CrossRef] [PubMed]
15. Politi, V.; Deluca, G.; Gallai, V.; Puca, O.; Comin, M. Clinical experiences with the use of indole-3-pyruvic acid. *Adv. Exp. Med. Biol.* **1999**, *467*, 227–232. [PubMed]
16. Bartolini, B.; Corniello, C.; Sella, A.; Somma, F.; Politi, V. The enol tautomer of indole-3-pyruvic acid as a biological switch in stress responses. *Adv. Exp. Med. Biol.* **2003**, *527*, 601–608. [PubMed]
17. Kaminski, R.; Wolinska, E.; Gasior, M.; Kleinrok, Z.; Czuczwar, S.J. NMDA receptor glycine site antagonist (indole-2-carboxylic acid) enhances the anticonvulsive activity of antiepileptic drugs. *Neurosci. Res. Commun.* **1998**, *22*, 137–145. [CrossRef]
18. Stetinova, V.; Smetanova, L.; Grossmann, V.; Anzenbacher, P. In vitro and in vivo assessment of the antioxidant activity of melatonin and related indole derivatives. *Gen. Physiol. Biophys.* **2002**, *21*, 153–162. [PubMed]
19. Rajendra, N.G.; Aditi, S. Oxidation chemistry of indole-2-carboxylic acid: Mechanism and products formed in neutral aqueous solution. *Electrochim. Acta* **2004**, *50*, 2135–2143.
20. Rani, P.; Srivastava, V.K.; Kumar, A. Synthesis and antiinflammatory activity of heterocyclic indole derivatives. *Eur. J. Med. Chem.* **2004**, *39*, 449–452. [CrossRef] [PubMed]
21. Siatra-Papastaikoudi, T.; Tsotinis, A.; Raptopoulou, C.; Sambani, C.; Thomou, H. Synthesis of new alkylaminoalkyl thiosemicarbazones of 3-acetylindole and their effect on DNA synthesis and cell proliferation. *Eur. J. Med. Chem.* **1995**, *30*, 107–114. [CrossRef]
22. Moussata, J.; Wang, Z.; Wang, J. Development and validation of an HPLC method for the simultaneous quantification of indole-3-carbinol acetate, indole-3-carbinol, and 3,3′-diindolylmethane in mouse plasma, liver, and kidney tissues. *J. Chromatogr. B Anal. Technol. Biomed. Life Sci.* **2014**, *958*, 1–9.
23. Gomez, F.J.V.; Hernández, I.G.; Cerutti, S.; Silva, M.F. Solid phase extraction/cyclodextrin-modified micellar electrokinetic chromatography for the analysis of melatonin and related indole compounds in plants. *Microchem. J.* **2015**, *123*, 22–27. [CrossRef]
24. Grunwald, C.; Lockard, R.G. Analysis of indole acid derivatives by gas chromatography using liquid phase OV-101. *J. Chromatogr. A* **1970**, *52*, 491–493. [CrossRef]
25. Gallagher, C.A.; Hough, L.B.; Keefner, S.M.; Seyed-Mozaffari, A.; Archer, S.; Glick, S.D. Identification and quantification of the indole alkaloid ibogaine in biological samples by gas chromatography-mass spectrometry. *Biochem. Pharmacol.* **1995**, *49*, 73–79. [CrossRef]
26. Müller, A.; Düchting, P.; Weiler, E. A multiplex GC-MS/MS technique for the sensitive and quantitative single-run analysis of acidic phytohormones and related compounds, and its application to Arabidopsis thaliana. *Planta* **2002**, *216*, 44–56. [CrossRef] [PubMed]
27. Ludwig-Müller, J.; Georgiev, M.; Bley, T. Metabolite and hormonal status of hairy root cultures of Devil's claw (*Harpagophytum procumbens*) in flasks and in a bubble column bioreactor. *Process Biochem.* **2008**, *43*, 15–23. [CrossRef]
28. Tan, S.N.; Yong, J.W.H.; Ge, L. Analyses of phytohormones in coconut (*Cocos Nucifera* L.) water using capillary electrophoresis-tandem mass spectrometry. *Chromatography* **2014**, *1*, 211–226. [CrossRef]

29. Fei, X.Q.; Li, C.; Yu, X.D.; Chen, H.Y. Determination of heterocyclic amines by capillary electrophoresis with UV-DAD detection using on-line preconcentration. *J. Chromatogr. B* **2007**, *854*, 224–229. [CrossRef] [PubMed]
30. Graffeo, A.P.; Karger, B.L. Analysis for indole compounds in urine by high-performance liquid chromatography with fluorometric detection. *Clin. Chem.* **1976**, *22*, 184–187. [PubMed]
31. Hanna, M.; Alicja, K.; Ksenia, P. Determination of indole-3-acetic acid in the Gulf of Gdańsk by high-performance liquid chromatography of its 4-methyl-7-methoxycoumarin derivative. *J. Chromatogr. A* **1997**, *766*, 261–266.
32. Matsuda, F.; Miyazawa, H.; Wakasa, K.; Miyagawa, H. Quantification of indole-3-acetic acid and amino acid conjugates in rice by liquid chromatography-electrospray ionization-tandem mass spectrometry. *Biosci. Biotechnol. Biochem.* **2005**, *69*, 778–783. [CrossRef] [PubMed]
33. Verheyden, K.; Noppe, H.; Aluwé, M.; Millet, S.; Bussche, J.V.; Brabander, H.F.D. Development and validation of a method for simultaneous analysis of the boar taint compounds indole, skatole and androstenone in pig fat using liquid chromatography-multiple mass spectrometry. *J. Chromatogr. A* **2007**, *1174*, 132–137. [CrossRef] [PubMed]
34. Mattivi, F.; Vrhovsek, U.; Versini, G. Determination of indole-3-acetic acid, tryptophan and other indoles in must and wine by high-performance liquid chromatography with fluorescence detection. *J. Chromatogr. A* **1999**, *855*, 227–235. [CrossRef]
35. Gerasimenko, I.; Sheludko, Y.; Unger, M.; Stöckigt, J. Development of an efficient system for the separation of indole alkaloids by high performance liquid chromatography and its applications. *Phytochem. Anal.* **2001**, *12*, 96–103. [CrossRef] [PubMed]
36. Yusof, S.; Shian, L.S.; Osman, A. Changes in quality of sugar-cane juice upon delayed extraction and storage. *Food Chem.* **2000**, *68*, 395–401. [CrossRef]
37. Ma, Z.; Ge, L.; Lee, A.S.Y.; Yong, J.W.H.; Tan, S.N.; Ong, E.S. Simultaneous analysis of different classes of phytohormones in coconut (*Cocos nucifera* L.) water using high-performance liquid chromatography and liquid chromatography-tandem mass spectrometry after solid-phase extraction. *Anal. Chim. Acta* **2008**, *610*, 274–281. [CrossRef] [PubMed]
38. Frankenberger, W.T.; Arshad, M. (Eds.) *Phytohormones in Soils: Microbial Production and Function*; Marcel Dekker, Inc.: New York, NY, USA, 1995; pp. 17–136.
39. Sun, J.; Baker, A.; Chen, P. Profiling the indole alkaloids in yohimbe bark with ultra-performance liquid chromatography coupled with ion mobility quadrupole time-of-flight mass spectrometry. *Rapid Commun. Mass Spectrom.* **2011**, *25*, 2591–2602. [CrossRef] [PubMed]
40. Preti R. Core-shell columns in high-performance liquid chromatography: Food analysis applications. *Int. J. Anal. Chem.* **2016**, *2016*, 1–9. [CrossRef] [PubMed]

19

Liquid Chromatography Tandem Mass Spectrometry Analysis of Synthetic Coccidiostats in Eggs

Francesca Buiarelli [1,*], Patrizia Di Filippo [2], Carmela Riccardi [2], Donatella Pomata [2], Luigi Giannetti [3], Bruno Neri [3] and Daniela Rago [1]

1 Department of Chemistry, University "Sapienza", P.le Aldo Moro, 5-0185 Rome, Italy; chemical.lady@gmail.com
2 Inail DIT—Via Roberto Ferruzzi, 38-00143 Rome, Italy; p.difilippo@inail.it (P.D.F.); ca.riccardi@inail.it (C.R.); d.pomata@inail.it (D.P.)
3 Istituto Zooprofilattico e Sperimentale Regioni Lazio e Toscana—Via Appia Nuova, 1411-00178 Rome, Italy; luigi.giannetti@izslt.it (L.G.); bruno.neri@izslt.it (B.N.)
* Correspondence: francesca.buiarelli@uniroma1.

Academic Editor: Michael J. Dunphy

Abstract: Coccidiostats are synthetic drugs administered to animals, especially to poultry, to cure coccidiosis. In this paper, we present a selective liquid chromatography-tandem mass spectrometry (LC-MS/MS) method to analyze residues of five synthetic coccidiostats in eggs: clazuril, diclazuril, robenidine, nicarbazin, toltrazuril and its two metabolites. The extraction efficiency was evaluated by testing several solvents, pH, different volumes and time of extraction. The clean-up procedures were optimized using different solid phase extraction cartridges and different eluants. The chromatographic separation was achieved in reversed phase using a gradient of 0.1% formic acid in water and acetonitrile, whereas the MS detection was performed in negative electrospray ionization (ESI) for all the analytes, except for the robenidine. The developed method has been validated according to Commission Decision 2002/657/CE. The validation parameters, as linearity, precision, recovery, specificity, decision limit (CCα), detection capability (CCβ), and robustness have been determined. The proposed method resulted simple, fast, and suitable for screening and confirmation purposes.

Keywords: synthetic coccidiostats; eggs; mass spectrometry

---

1. Introduction

Coccidiosis is a parasitic disease caused by the development and multiplication of coccidia in the intestine cells. It is caused by the development and multiplication of coccidian protozoa belonging to the *Eimeria* (the most predominant) or *Cryptosporidium* species. This infection is not lethal in healthy animals, but it can cause weight loss, low growth, and intestinal lesions, and every year, it leads to severe losses in meat and egg production [1].

Coccidiostats are drugs that are administered against the coccidiosis both for prophylactic chemotherapy and for health care [2]. To this aim, two different kinds of drugs are employed: ionophores and synthetic drugs. They are administrated in feed and/or in zootechnical supplements and some of them are included in the European Union register of Feed Additives [3]. In this paper, we investigated the simultaneous detection and quantification of clazuril, diclazuril, robenidine, nicarbazine and toltrazuril (including its main metabolites) in eggs (Figure 1).

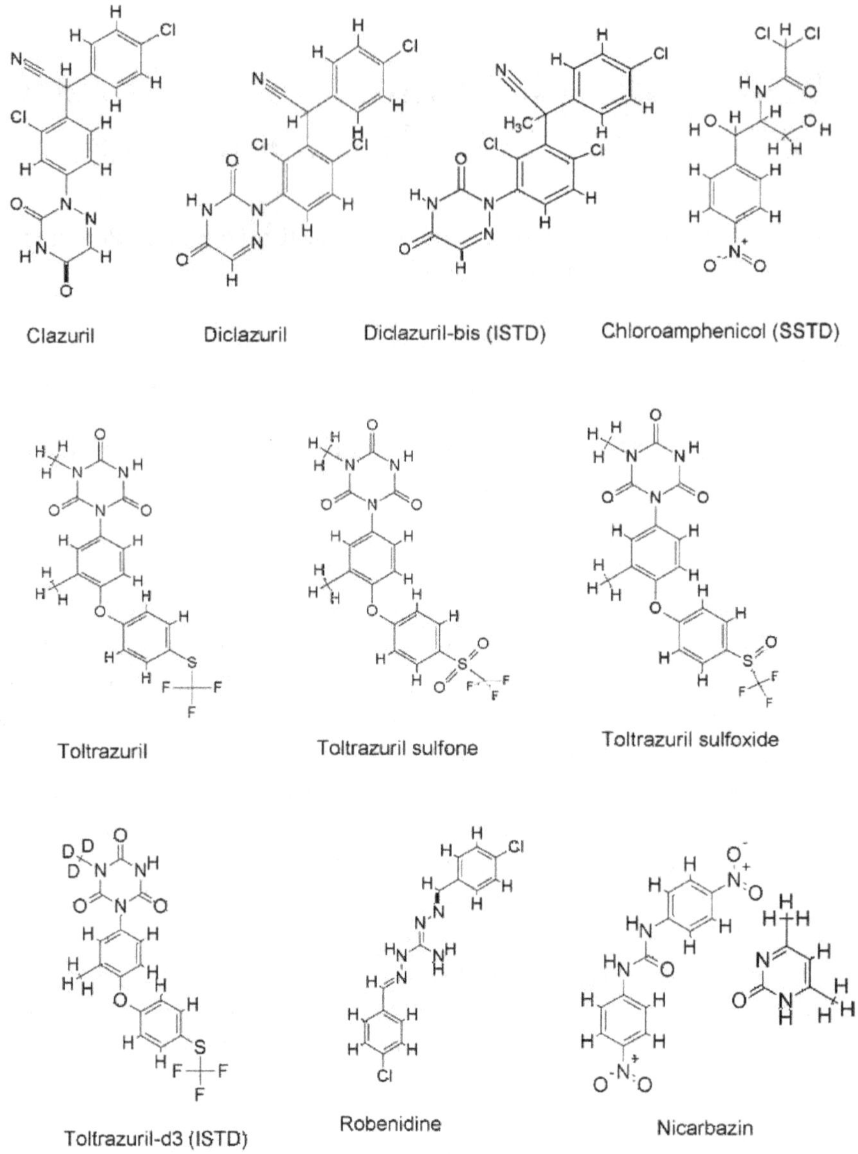

Figure 1. Molecular structures of the investigated coccidiostats and of the internal (ISTD) and external standards (SSTD).

Toltrazuril after administration to chicken by drinking water, is absorbed, metabolized and excreted as toltrazuril sulfone (main metabolite) and toltrazuril sulfoxide. Toxicology, metabolism and its characteristics are reported in the Summary Report on toltrazuril of the European agency for the evaluation of medicinal products [4,5]. Its use is regulated by Commission Regulation N 37/2010 [6].

Diclazuril is administered orally, and it is excreted by feces [7]. Diclazuril and toltrazuril act effectively against a large spectrum of coccidia. The first one is usually added to feed (about 1 ppm) for prevention purposes, whereas the other is added to water for disease care. Clazuril shows the same chemical-physical properties and acting way of diclazuril [8].

Nicarbazin is the equi-molar complex of 4,4-dinitrocarbanilide (DNC) and 2-hydroxy-4,6 dimetylpyrimidine (HDP) and it is administered by feed.

The DNC moiety is metabolized and excreted more slowly respect to HDP; consequently most of residue analyses for nicarbazin are based on determination methods for DNC moiety. Nicarbazin, being strongly electrostatic, can lead to cross contamination of feed production lines after milling of medicated feed [9].

Robenidine is administered by feed, like nicarbazin, and may give rise to phenomena of cross contamination [10]. In the past, it was discarded because of phenomenon of drug resistance, but it was then reintroduced for coccidian resistance to ionophores.

As a result of cross-contamination in feed intended for different species, the Commission decided to publish a Regulation 610/2012 [11], reviewing the allowable value of certain coccidiostats, previously set by the Regulation 124/2009 [12].

Several analytical methods have been developed for the analysis of one or more coccidiostats in different biological matrixes and with different techniques, as reported by Mortier et al. [13]. Most of the works use a chromatographic technique, especially high-performance liquid chromatography coupled with ultraviolet detector or mass spectrometric detector [14–28].

High resolution liquid chromatography-UV technique was used for the determination of nicarbazin and robenidine in eggs and in feeds [14–17]. Several authors used an HPLC-MS technique for single and multiresidual determination of these compounds in eggs, in muscle and in feedstuffs [18–25] reaching a lower Limit of Detection (LOD) than HPLV-UV. Sample preparation is a critical step in the analysis of coccidiostats [26], and sometimes highly specific and selective methods like immunoaffinity chromatography (IAC) can be used, [27], but this procedure, however, is very long, complex and expensive for routine analysis.

In 2012, a multiresidue method, including 20 coccidiostats, in eggs [28] was published. Five grams of egg were extracted by 20 mL of CH_3CN and after evaporation of the volume to dryness, the residue was reconstituted and injected to HPLC-MS/MS. Clazuril was not included in this study. Recently, an exhaustive review [29] summarized the analyses of coccidiostats in meat and other food.

At the moment, there are few papers about toltrazuril and its metabolites in eggs or in feedstuffs [30–32] and, to our knowledge, there are no scientific works about these five synthetic coccidiostats and the toltrazuril metabolites all together.

Due to the massive use of these drugs in poultry, in this study, we proposed a method for a simultaneous determination of the mentioned five synthetic coccidiostats in eggs together with toltrazuril metabolites, in order to be applied in research and routine control laboratories.

Starting from the literature about the topic and according to other authors [33], considering the ion suppression in the electrosray ionization to be one of the main problems (when analyzing drugs in complex matrices like eggs), we optimized a liquid–liquid extraction and solid phase extraction (SPE) purification of analytes before the analysis by liquid chromatography-tandem mass spectrometry (LC-MS/MS) in multiple reaction monitoring MRM mode. To this aim, several extraction solvents, extraction time, solvent volume, pH of extraction, clean-up cartridges and solvent for the elution were investigated for the highest recovery of the analytes from the eggs, and the results were evaluated to study the efficacy of the method. At the end, after the optimization of the mass spectrometric parameters and chromatographic conditions, the proposed method was validated according to Commission Decision 2002/657/CE [33,34].

2. Materials and Methods

2.1. Chemicals, Reagents and Solutions

Toltrazuril (TOL), Robenidine (ROB) hydrochloride, Chloroamphenicol (CAP) (External Standard ES), Diclazuril (DCLAZ) were from VETRANAL®, Fluka, Sigma-Aldrich (Milwaukee, WI, USA); Clazuril (CLAZ), Diclazuril (DCLAZbis) bis (Internal Standard ISTD) were from Janssen Animal Health (Beerse, Belgium); Nicarbazin (NIC) was from Sigma-Aldrich (Milan, Italy); Toltrazuril-d_3 (TOL-d_3) (ISTD), Toltrazuril sulfone, Toltrazuril sulfoxide were from Witega Laboratorien Berlin-Adlershof GmbH, Magnustrasse, (Berlin, Germany). Acetonitrile (AcN), methanol (MeOH), Ethyl-acetate (HPLC grade), acetone, ammonium acetate, ammonium formate (RPE), ammonium hydroxide, ammonium sulphate (RPE) were from Carlo Erba reagents (Milan, Italy). Dimethylformamide (DMF) was from R.P. Normapur AR, VWR (West Chester, PA, USA). Formic acid (98%–100%) and acetic acid were from Merck (Darmstadt, Germany). Water was HPLC grade (generated by Milli-Q biocel, Millipore, purification system (Merck S.p.a. Vimodrone (MI) Italy).

Individual stock standard solutions were prepared at a concentration of 1 mg/mL.

Solvents used were: DMF for clazuril, diclazuril e nicarbazin and methanol for robenidine, toltrazuril, toltrazuril sulfone and toltrazuril sulfoxide. The stock solutions were stored at −20°C.

Working solutions were prepared by diluting stock solutions in acetonitrile to obtain solutions from 100 to 10 μg/mL. Standard mixtures were daily prepared by mixing adequate volume of each working solution and diluting in water-acetonitrile (30/70 *v*/*v*) up to a final concentration between 0.001 and 1.0 μg/mL. To enhance the quantification and the robustness of the method two internal standards were used: toltrazuril-d_3 for toltrazuril, toltrazuril sulfone and toltrazuri lsulfoxide; and diclazuril-bis for diclazuril, clazuril and robenidine. Stock solution were prepared in DMF for diclazuril bis and in methanol for toltrazuril-d_3 and stored at −20 °C.

2.2. HPLC-MS/MS Apparatus and Conditions

The LC system was a PE LC-200 Micro Pump (binary pump, vacuum degasser, autosampler AS 200) (Perkin-Elmer Sciex Instruments, Milan, Italy) coupled to a Triple Quadrupole LC-MS/MS Mass Spectrometer API2000 (Applied Biosystem, Foster, CA, USA) The MS system was controlled by Analyst software (version, Applied Biosystem, Foster, CA, USA). A chromatographic separation was achieved on reversed phase system using a C_{18} column (5 μm, 150 mm × 2.1 mm), Gemini, Phenomenex (Torrance, CA, USA) protected by a guard column containing the same packing material. Gradient elution was performed, starting from 70% of 0.1% aqueous formic acid (A) and acetonitrile (B) up to 20% A and 80% B in 15 min. Flow rate was 0.2 mL/min. A syringe pump (Harvard apparatus, Holliston, MA, USA) was connected to the interface for tuning purposes and to add ammonium hydroxide (0.08%) post column, before the mass spectrometer. All the experiments were performed in ESI (electrospray ionization) in negative mode, except for the robenidine, acquired in positive ionization. For each compound, the protonated or deprotonated molecular ion, $[M + H]^+$ or $[M - H]^-$, was chosen as a precursor and was subsequently fragmented by nitrogen, also used as drying and nebulising gas. The source block and desolvatation temperature were set at 120 °C and 380 °C, respectively. Dwell time was around 100 ms for each analyte.

2.3. Sample Preparation

One gram of homogenized eggs, purchased from the local market, was weighed in a falcon tube. Furthermore, 50 μL of the internal standard solution of TOL-d_3 and DCLAZbis were added to all samples. At this step, if necessary, samples were spiked by a specific amount of analyte standard solution mixture. The samples were then vortexed and allowed to stand for 15 min. Samples were extracted by 5 mL of AcN, placed in a ultrasonic bath for 5 min, then placed in a horizontal shaker for 20 min and finally centrifuged for 5 min at 4000 rpm. The supernatant (5 mL) was transferred into a graduated tube, diluted up to 25 mL withMilli-Q water and passed through a polymeric cartridge SPE Varian (60 mg) (Varian, Palo Alto, CA, USA), pre-activated with 3 mL of methanol and washed with 5 mL water. Two washing steps, using 5 mL of water and subsequently 5% aqueous MeOH, were tested before the elution with 5 mL of MeOH. The eluate was evaporated to dryness, using nitrogen at 40 °C. In addition, 50 μL of CAP solution (100 ng/mL) in aqueous AcN (70%) was used to reconstitute the dried residue and aliquots of 5 μL were injected into the LC–ESI-MS/MS system on a C_{18} column.

3. Result and Discussion

3.1. Optimization of the Chromatographic and Mass Spectrometric Conditions

Firstly, all the chromatographic conditions were optimized. Since all of the investigated compounds have an absorption in the UV region (between 240 and 350 nm), preliminary experiments were performed by HPLC with a diode array detector to select the column and the most suitable chromatographic conditions for the purpose (data not shown). Three C_{18} columns were tested: a Discovery, Supelco, (Sigma Aldrich, Milan, Italy), a Zorbax (Agilent, 20063 Cernusco sul Naviglio, MI, Italy), anda Gemini (Phenomenex srl 40013—Castel Maggiore, Bo Italy) 4.6 × 150 mm, 5 μm at a flow rate of 1 mL·min^{-1}. The best selectivity was obtained by Gemini C_{18} column. Since the mobile phase composition has a significant effect on the peak shape and on the retention behavior of the analyte in the LC column, as well as on the MS response, modifications of the mobile phase (water-MeOH and water-AcN) in isocratic and gradient conditions with different additives were also tested.

The best performance, in terms of the mobile phase, was obtained by 0.1% formic acid-AcN in gradient elution with the addition of ammonium hydroxide (0.08%) at the exit of the column, prior to the mass spectrometer, in order to increase the ionization response.

As expected, the limits of quantifications (LOQs) were too high for the legal limits imposed for these substances; therefore, the method was further readapted to LC-MS/MS analysis and then validated according to the Commission Decision 2002/657/CE. The development of an MRM LC/MS/MS method requires experiments carried out by infusion at 10 μL·min^{-1} of standard solutions (2.5–10 ng·μL^{-1}), in order to determine suitable source parameters for the best sensitivity and signal to noise S/N ratio, as well as the molecule-related ions.

First of all, we selected the more effective ionization mode (ESI vs. Atmospheric pressure chemical ionization—APCI), enhancing the formation of protonated/deprotonated molecular ions of the target analytes. Experiments were carried out in negative and positive polarity using different mobile phase mixtures (MeOH and AcN) and additives such as: formic and acetic acid, ammonium formate and acetate, NH_3 and water. All the instrumental parameters, and potentials, such as ion source voltage (IS), cone potential (CP), and collision energy (CE), were optimized in order to maximize the quasi-molecular ion intensities on Q_1 and the MS/MS transitions in Q_3.The best results were obtained operating with ammonium formate 20 mM (Figure 2) in negative ion mode for all compounds, except for robenedine, more efficiently ionized in positive ion mode.

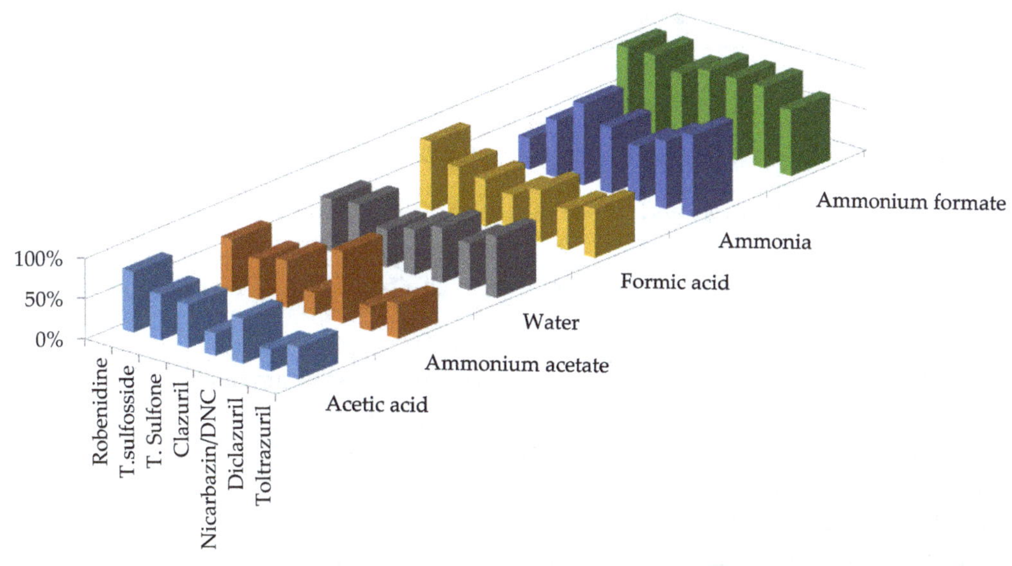

Figure 2. Effect of different additives in the mass spectrometric response of the investigated analytes.

For qualitative purposes, a minimum of 3.5 identification points are necessary. In LC-MS/MS, the transition of one precursor ion into two product ions corresponds to four identification points, and this criterion is accomplished except for diclazuril, toltrazuril and toltrazuril sulfone because they have only one fragment as a product ion. In this case, we used two precursor ions and one fragment from each precursor.

All the transitions and the optimized electrical parameters are reported in Table 1. For the development of the method to detect nicarbazin, we focused on the DNC moiety.

Finally, all of the analyses were performed in HPLC-MS/MS in MRM mode, according to Section 2.2 and acquired the most intense transitions from Table 1. To gain detection sensitivity, NH_3 was added post column to the eluate prior to the mass spectrometer. The analytes were eluted in less than 12 min.

Table 1. Summary of electric parameters, precursor and fragment ions chosen for the multiple reaction monitoring (MRM) analysis and chromatographic retention times.

Compound	Ionization	*m*/*z* Precursor Ion (Intensity %)	Capillary (kV)	Cone Voltage (V)	*m*/*z* Product Ion (Intensity %)	Collision Energy (eV)	Retention Time R_t (min)
Chloroamphenicol CAP (SSTD)	ES−	325.0 (10)	−4200	−35	194.0 (100) 152.0 (45)	−18 −20	5.7
Robenidine	ES+	334.1 (58)	+5300	50	138.0 (90) 155.0 (100)	35 28	6.7
Toltrazuril Sulfoxide	ES−	440.3 (60)	−4200	−40	42.1 (80) 371.0(100)	−25 −22	8.6
Toltrazuril Sulfone	ES−	456.2 (10) 456.2 (10)	−4200	−40	42.1 (100) 456.0 (100)	−30 −10	9.8
Clazuril	ES−	371.2 (9)	−4200	−40	299.9 (100) 265.0 (39)	−22 −22	9.8
Nicarbazin/DNC	ES−	301.0 (37)	−4200	−40	107.0(5) 137.0 (100)	−43 −20	10.1
Diclazuril	ES−	405.0 (6) 406.9 (6)	−4200	−40	333.8 (100) 335.8 (100)	−25 −25	10.5
Diclazuril-bis (ISTD)	ES−	419.0 (20)	−4200	−40	321.0 (80) 348.0(100)	−22 −20	10.8
Toltrazuril	ES−	424.0 (37) 424.0 (37)	−4200	−40	42.1 (100) 424.0 (100)	−38 −9	10.9
Toltrazuril-d3 (ISTD)	ES−	427.0 (30)	−4200	−40	42.1 (100)	−38	10.9

3.2. Optimization of the Clean-Up Procedure

The clean-up procedure was structured in the following steps:

- optimization of the liquid–liquid extraction (nature and volume of solvent, time and pH of extraction)
- optimization of the purification of the extracts (choice of the SPE cartridge, washing and elution conditions)

Preliminary experiments were done directly on the standard solution of analytes for the choice of the conditions to meet the goals of the best extraction and purification steps.

Successively, in order to optimize the clean-up procedure, 1 g of egg, spiked with each analyte at different level, was used. In order to find the best extraction conditions, we tested different kinds of solvents in different volumes: AcN, ethyl acetate, and a mixture of ethyl acetate:acetone (50:50, v/v), different pH, extraction times, and extraction in ultrasonic bath and in the horizontal shaker. In Table 2, the extraction efficiency for the tested organic solvents is shown. The best recovery was obtained with AcN, on average, 85% for all the compounds. Good results were obtained with ethyl acetate too, except for robenedine, where the recovery was only 10%. The mixture of ethyl acetate/acetone 50:50 v/v improved the extraction of toltrazuril and its metabolites (80%), but decreased the recovery of the other compounds to 30%. The main concern using the ethyl acetate/acetone mixture is due to a higher extraction of the lipid components as well, which can cause interferences and higher background.

Table 2. Extraction efficiency comparing different extraction solvents: ethyl acetate, acetonitrile, and ethyl acetate-acetone = 50:50.

Compound	Ethyl Acetate ($x_m \pm \sigma$)%	Acetonitrile ($x_m \pm \sigma$)%	Ethyl Acetate–Acetone = 50:50 ($x_m \pm \sigma$)%
Robenidine	10 ± 3.2	82 ± 1.0	33 ± 2.0
ToltrazurilSulfoxide	75 ± 2.2	72 ± 0.9	81 + 2.1
ToltrazurilSulfone	73 ± 1.5	70 ± 1.3	85 ± 1.5
Clazuril	95 ± 3.1	95 ± 0.8	45 ± 1.0
Nicarbazin/DNC	90 ± 2.8	98 ± 2.1	25 ± 4.2
Diclazuril	88 ± 1.0	95 ± 1.1	31 ± 0.5
Toltrazuril	75 ± 0.5	70 ± 1.5	82 ± 3.0

x_m average of three different tests. σ standard deviation.

The optimum conditions in terms of time/volume of extraction were, respectively, 5 min in ultrasonic bath with 5 mL of solvent. An additional 20 min in the horizontal shaker improved the extraction efficiency. The pH was another parameter under study. Acidifying at pH 5 or alkalizing at pH 9.5, the egg, before the extraction, did not improve the extraction efficiency (data not shown). In order to improve the sample clean-up, reducing the lipid components, a SPE procedure was mandatory. Different brands of polymeric SPE cartridges were tested: OASIS (Waters spa Milan, Italy), Strata X (Phenomenex srl 40013—Castel Maggiore, Bo, Italy) and Varian. The extracts were diluted five times with Milli-Q water before being loaded in the cartridge. Two washing steps (5 mL each), the first one with water and the second one with aqueous methanol (5%, 10%, and 15%) were done before the elution step. The higher concentration of MeOH should facilitate the clean-up. It was observed that 5% aqueous MeOH was the best compromise to avoid the loss of analytes (data not shown). Both MeOH and AcN were tested for the elution step, but MeOH gave the best results. Strata X and Varian (60 mg) cartridges were comparable, but the Varian was more reproducible and gave slightly higher recovery for the target analytes, on average 80% with a coefficient of variation CV, calculated over three replicates, below 15%. The eluate was dried under nitrogen, reconstituted and analyzed in HPLC/ESI-MS/MS in MRM mode, as described above. We experimentally calculated the loss due to the evaporation step between 1% and 5%.

The acquisition has been divided into three periods: The first one between 0 and 6.2 min in negative ionization where the CAP is acquired, the second one between 6.2 and 7.5 min in positive mode where the robenidine is acquired and finally the third one between 7.5 and 12.0 min, where clazuril, diclazuril, toltrazuril sulfone, toltrazuril sulfoxide and nicarbazin are acquired.

4. Validation Study

The validation method was carried out according to the Revision of Commission Decision93/256/EC [32]; therefore, instrumental linearity, specificity, recovery, precision, CCα and CCβ and robustness were studied.

4.1. *Linearity and Matrix Effect*

The matrix effect can greatly affect the reproducibility and accuracy of the method. The linearity was checked for three days by calibration curve both in solution and in matrix (adding analytes and ISTDs at the end of the clean-up) in the following range of concentrations: 1–10 $\mu g \cdot kg^{-1}$ for CLAZ and DCLAZ, (corresponding to 20–200 ng/mL in solution), and 5–20 $\mu g \cdot kg^{-1}$ for TOL and its metabolites (corresponding to 100–400 ng/mL in solution), 5–40 $\mu g \cdot kg^{-1}$ for ROB (corresponding to 100–800 ng/mL in solution) and 10–45 $\mu g \cdot kg^{-1}$ for NIC (corresponding to 200–900 ng/mL in solution). The linearity in the investigated ranges was very good, as demonstrated by the correlation factors $R^2 \geq 0.999$ (data not shown). To evaluate matrix effects (ME) the slope of matrix matched calibration curves and the slope of standard calibration curves were calculated. The slope ratio (R)

×100 is defined as the matrix effect (ME %). A value of 100% indicates that there is no matrix effect. There is signal enhancement if the value is >100% and signal suppression if the value is <100% [35]. Thanks to the proposed clean-up procedure, reducing the interfering compounds in the matrix that alter the ionization in the source of the mass spectrometer, the solution and matrix-matched calibration curve equations had an ME % within 10%. Therefore, for quantitative purposes, we used the solvent calibration curves.

4.2. Specificity

Specificity was tested on 20 representative blank samples, compared to spiked ones with analytes and ISTD. The acquisition, as discussed previously, was divided into three periods according to the retention time of the analytes. In the target regions, no interfering peaks were observed.

4.3.Recovery

Once the best extraction and clean-up conditions were chosen, the recovery of the whole procedure was determined forsix spiked blank samples for each concentration level for three days. The levels were: 1, 2, 3, 4, 6 $\mu g \cdot kg^{-1}$ for diclazuril and clazuril and 5, 7.5, 10 and 15 $\mu g \cdot kg^{-1}$ for TOL and its metabolites. For ROB, the levels were 5, 12.5, 25, 37.5 $\mu g \cdot kg^{-1}$, whereas, for NIC, the levels were 100, 150, 300, 450 $\mu g \cdot kg^{-1}$ (applying the dilution factor of 10 to be inside the calibration curve range). The recovery was calculated by differences among nominal values of spiked blank samples and experimental values. The results were obtained by a solvent curve, since the matrix effect was not considerable, and the average recovery was about 80%, considering 62% as a minimum for the robenidine and 95% as a maximum for the toltrazuril, as shown in Table 3.

Table 3. Results of recovery and precision of the method.

Analyte	Validation Level	Experimental Concentration	Recovery	Repeatability			Reproducibility
	($\mu g \cdot kg^{-1}$)	($\mu g \cdot kg^{-1}$) ± σ (n = 18)	(%)	(CV%) (n = 6) 1	2	3	(CV%) (n = 18)
Robenedine	5.0	3.1 ± 0.4	62	13.7	14.7	11.9	13.9
	12.5	8.5 ± 0.8	68	11.5	10.1	9.0	11
	25.0	18.5 ± 1.2	74	7.1	11	7.0	9
	37.5	28.5 ± 2.5	76	8.0	10	13	12.0
Toltrazuril Sulfoxide	5.0	4.0 ± 0.5	80	13.0	7.5	7.5	13.6
	7.5	6.4 ± 0.6	85	5.7	4.6	10.5	10.2
	10.0	7.5 ± 0.6	75	9.4	8.8	8.4	9.2
	15.0	11.4 ± 1.3	76	8.2	13.2	12.5	11.5
Toltrazuril Sulfone	5.0	4.4 ± 0.5	88	9.6	4.2	11.5	11.6
	7.5	6.6 ± 0.6	88	9.1	5.4	11.4	9.2
	10.0	8.1 ± 0.9	81	12.2	7.8	8.4	10.5
	15.0	12.7 ± 1.1	85	5.0	10.1	11.4	8.6
Clazuril	1.0	0.8 ± 0.1	80	12.5	11.9	12.1	12.0
	2.0	1.7 ± 0.2	85	12.4	7.9	11.0	11.0
	3.0	2.4 ± 0.2	80	6.1	8.4	8.5	8.9
	4.0	3.2 ± 0.3	80	3.2	10.3	10.3	10.2
	6.0	5.4 ± 0.3	90	2.9	4.8	8.0	8.7
Nicarbazin	100.0	75.0 + 9.0	75	11.0	9.0	13.0	12.0
	150.0	110.0 + 11.0	73	9.0	10.0	12.0	11.0
	300.0	234.0 + 10.0	78	5.0	8.0	7.0	8.0
	450.0	315.0 + 28	70	9.0	8.0	12.0	9.0
	100.0	75.0 + 9.0	75	11.0	9.0	13.0	12.0

Table 3. *Cont.*

Analyte	Validation Level ($\mu g \cdot kg^{-1}$)	Experimental Concentration ($\mu g \cdot kg^{-1}$) $\pm \sigma$ (n = 18)	Recovery (%)	Repeatability (CV%) (n = 6) 1	2	3	Reproducibility (CV%) (n = 18)
Diclazuril	1.0	0.8 ± 0.1	80	11.0	13.0	12.1	12.1
	2.00	1.8 ± 0.3	90	7.0	5.6	8.8	13.7
	3.0	2.6± 0.2	87	7.2	5.5	7.0	10.2
	4.0	3.2 ± 0.3	80	9.2	9.1	8.9	9.5
	6.0	5.1 ± 0.5	85	9.9	12.2	8.4	11.5
Toltrazuril	5.0	4.4 ± 0.3	88	9.3	7.7	8.8	14.1
	7.5	7.1 ± 0.5	95	5.5	4.1	9.5	6.4
	10.0	8.7 ± 0.9	87	11.7	12.4	10.1	10.9
	15.0	12.7 ± 1.2	85	8.1	9.2	11.3	9.8

4.3. Precision: Repeatability and Reproducibility Intra-Laboratory

Precision has been evaluated as repeatability and reproducibility intra-laboratory. Repeatability is measured as relative standard deviation (CV%) of repeated measures of six aliquots for each concentration level for three consecutive days, and the reproducibility as CV% of repeated measures of all three days (18 samples for each concentration level).The results are shown in Table 3; they are good for all analytes with CV% below 20% and they respect the recommendations of the 657/2002 Directive.

4.4. CCα and CCβ and Robustness (Minor Changes)

The decision limit (CCα) and the detection capability (CCβ) has been estimated according to Lynn Vanhaecke et al. [36] and Commission Decision 2002/657/EC [34].

The CCα is between 2.2 $\mu g \cdot kg^{-1}$ for diclazuril and 320 $\mu g \cdot kg^{-1}$ for nicarbazine, whereas the CCβ is between 2.2 $\mu g \cdot kg^{-1}$ for diclazuril and 350 $\mu g \cdot kg^{-1}$ for nicarbazine. The results are reported in Table 4. Robustness has been evaluated by Youden test on eight spiked blank samples at the CCβ concentration level. The parameters shown in Table 5 were chosen to evaluate how small changes can affect the proposed method, so one variable was chosen in the extractive step (extraction volume) and the other five variables in the purification step, including the stability of analytes at the boiling point of the solvent. The effect of each factor was calculated by determining the difference between the value of the variable at the highest and at the lowest level. Differences were not significant; in other words, the investigated parameters have no effect on the characteristics of the method, and, therefore, the method can be defined robust.

Table 4. CCα and CCβ values for the investigated analytes.

Analyte	CCα ($\mu g \cdot kg^{-1}$)	CCβ ($\mu g \cdot kg^{-1}$)
Robenidine	28.0	30.0
Toltrazuril sulfoxide	5.1	6.1
Toltrazuril sulfone	5.8	6.7
Clazuril	2.2	2.6
Nicarbazin/DNC	320.0	350.0
Diclazuril	2..2	2.6
Toltrazuril	6.0	6.9

CCα Decision limit.

Table 5. Youden experiment table for the robustness of the method.

Variable Number	Experiment Number							
	1	2	3	4	5	6	7	8
Lot of SPE	lotA	lotA	lotA	lotA	lotA	lotA	lotA	lotA
Extraction Volume Acetonitrile (mL)	5.5	5.5	4.5	4.5	5.5	5.5	4.5	4.5
Dilution Volume with Water (mL)	30	20	30	20	30	20	30	20
Washing Volume for SPE with Water (mL)	5.5	5.5	4.5	4.5	4.5	4.5	5.5	5.5
Washing Volume for SPE with MeOH 5% (mL)	5.5	4.5	5.5	4.5	4.5	5.5	4.5	5.5
Elution Volume for SPE (mL)	5.5	4.5	4.5	5.5	5.5	4.5	4.5	5.5
Drying temperature (°C)	44	36	36	44	36	44	44	36

SPE solid phase extraction cartridge.

5. Conclusions

Due to themassive use of coccidiostats in poultry farming, for the prevention and treatment of coccidiosis, we developed a sensitive, simple, rapid and robust LC-MS/MS method, to detect, in eggs, simultaneously different synthetic coccidiostats for a total of seven analytes (five compounds and two metabolites), usually not analysed all together. Possible critical factors were examined and several clean-up strategies were tested to find the parameters and the conditions to meet the goals of the best extraction and purification step for screening and confirmation purposes. Experimental data showed that, thanks to a proper sample preparation, the matrix effect was drastically decreased, reducing the endogenous substances liable to interfere with the assay. This circumstance allowed us to use the solvent calibration curve for quantitative purposes.

The method has been validated in conformity with the main lines of the UE requirements for detecting residues of veterinary drugs in animal products and can be used to detect residues, in eggs, of the five coccidiostats at the level of μg/kg.

Acknowledgments: All sources of funding of the study should be disclosed. Please clearly indicate grants that you have received in support of your research work. Clearly state if you received funds for covering the costs to publish in open access.

Author Contributions: Francesca Buiarelli, Bruno Neri and Luigi Giannetti conceived and designed the experiments and provided materials and reagents. Daniela Rago analysed samples and data. Donatella Pomata, Patrizia Di Filippo and Carmela Riccardi performed some experiments and calculations. Francesca Buiarelli and Daniela Rago wrote the article. Francesca Buiarelli, Bruno Neri and Luigi Giannetti reviewed and revised the article. All authors have read and approved the final manuscript.

Conflicts of Interest: The authors declare no conflict of interest.

References

1. Booth, N.H.; Mc Donald, L.E. *Farmacologia e Terapeutica Veterinaria*, 1st ed.; EMSI: Roma, Italy, 1991.
2. Gerhold, R.W., Jr. Overview of coccidiosis in poultry. In *Merck Manual—Veterinary Manual*, 10th ed.; Merck & Co., Inc.: Kenilworth, NJ, USA, 2015.
3. Regulation (EC) No. 1831/2003. European Union Register of Feed Additives. Edition 254. Appendixes 3e, 4 23.03.2017 Annex 1 List of Additives. Available online: http://ec.europa.eu/food/safety/animal-feed/feed-additives/index_en.htm (accessed on 28 March 2017).
4. The European Agency for the evaluation of Medicinal Products; Veteriny Medicine Evaluation Unit. Committee for Medicinal Products for Veterinary Use Toltrazuril; Summary Report (1). Available online: http://www.eudra.org/emea.html (accessed on 1 February 2017).
5. European Medicine Agency, Veterinary Medicines and Inspections; Committee for Medicinal Products for Veterinary Use Toltrazuril. Extension to Cattle and Extrapolation to All Mammalian Food-Producing Species and Poultry; Summary Report (5). Available online: http://www.emea.eu.int (accessed on 30 December 2016).
6. Regulation, C. (EU) No. 37/2010 of 22 December 2009 on Pharmacologically Active Substances and Their Classification Regarding Maximum Residue Limits in Foodstuffs of Animal Origin. Available online: http://

ec.europa.eu/health//sites/health/files/files/eudralex/vol-5/reg_2010_37/reg_2010_37_en.pdf (accessed on 30 December 2016).
7. European medicines agency, Veterinary Medicines Evaluation Unit Committee for Medicinal Products for Veterinary Use Diclazuril Summary Report. Available online: http://www.emea.eu.int (accessed on 30 December 2016).
8. Committee for Medicinal Products for Veterinary Use Clazuril Summary Report. Available online: http://www.eudra.org/emea.html (accessed on 30 December 2016).
9. European Food safety Authority Question No. EFSA-Q-2005-220K Adopted on 9 April 2008. Available online: http://www.efsa.europa.eu/sites/default/files/scientific_output/files/main_documents/690.pdf (accessed on 30 December 2016).
10. European Food safety Authority Question No. EFSA-Q-2005-220g Adopted 19 February 2008. Available online: http://onlinelibrary.wiley.com/doi/10.2903/j.efsa.2008.655/pdf (accessed on 28 December 2016).
11. Commission Regulation (EU) No. 610/2012 amending Regulation (EC) No. 124/2009 of 10 February 2009 Setting Maximum Levels for the Presence of Coccidiostats or Histomonostats in Food Resulting from the Unavoidable Carry-over of These Substances in Non-Target Feed. Available online: https://www.fsai.ie/uploadedFiles/Reg610_2012.pdf (accessed on 28 December 2016).
12. Commission Regulation (EC) No. 124/2009 Setting Maximum Levels for the Presence of Coccidiostats or Histomonostats in Food Resulting from the Unavoidable Carry-over of These Substances in Non-Target Feed. Available online: http://eur-lex.europa.eu/LexUriServ/LexUriServ.do?uri=OJ:L:2009:040:0007:0011:en:PDF (accessed on 27 December 2016).
13. Mortier, L.; Daeseleire, E.; Delahaut, P. Simultaneous detection of five coccidiostats in eggs by liquid chromatography-tandem mass spectrometry. *Anal. Chim. Acta* **2003**, *483*, 27–37. [CrossRef]
14. Stahl, R.S.; Ver Cauteren, K.; Buettgenbach, T.L.; Johnston, J.J. Determination of 4,4′-Dintrocarbanilide (DNC), a Component of Nicarbazin, in Canada Goose (Brantacanadensis) Eggs hells Using High Performance Liquid Chromatography. *J. Agric. Food Chem.* **2003**, *51*, 1130–1135. [CrossRef] [PubMed]
15. Wilga, J.; Wasik, A.K.; Namiesnik, J. Comparison of extraction techniques of robenidine from poultry feed Samples. *Talanta* **2007**, *73*, 812–819. [CrossRef] [PubMed]
16. Dowling, G.; Keeffe, M.O.; Smyth, M.R. Determination of robenidine in eggs by liquid chromatography with UV spectrophotometric detection. *Anal. Chim. Acta* **2005**, *539*, 31–34. [CrossRef]
17. Draisci, R.; Lucentini, L.; Boria, P.; Lucarelli, C. Micro high-performance liquid chromatography for the determination of nicarbazin in chicken tissues, eggs, poultry feed and litter. *J. Chromatogr. A* **1995**, *697*, 407–414. [CrossRef]
18. Dubois, M.; Pierret, G.; Delhaut, P. Efficient and sensitive detection of nine coccidiostats in egg by liquid chromatography—Electrospray tandem mass spectrometry. *J. Chromatogr. B* **2004**, *813*, 181–189. [CrossRef] [PubMed]
19. Olejnik, M.; Szprengier-Juszkiewicz, T.; Jedziniak, P.; Sledzińska, E.; Szymanek-Bany, I.; Korycińska, B.; Pietruk, K.; Zmudzki, J. Residue control of coccidiostats in food of animal origin in Poland during 2007–2010. *Food Addit. Contam. Part B* **2011**, *4*, 259–267. [CrossRef] [PubMed]
20. Mortier, L.; Daeseleire, E.; Van Peteghem, C. Liquid chromatographic tandem mass spectrometry determination of five coccidiostats in poultry eggs and feed. *J. Chromatogr. B* **2005**, *820*, 261–270. [CrossRef] [PubMed]
21. Clarke, L.; Moloney, M.; O'Mahony, I.; O'Kennedy, R.; Danaher, M. Determination of 20 coccidiostats in milk, duck muscle, and non avian muscle tissue. *Food Addit. Contam. Part A* **2013**, *30*, 958–969. [CrossRef] [PubMed]
22. Galarini, R.; Fioroni, L.; Moretti, S.; Pettinacci, L.; Dusi, G. Development and validation of a multi-residue liquid chromatography–tandem mass spectrometry confirmatory method for eleven coccidiostats in eggs. *Anal. Chim. Acta* **2010**, *700*, 167–176. [CrossRef] [PubMed]
23. Broekaert, N.; Van Peteghem, C.; Daeseleire, E.; Sticker, D.C.; Van Poucke, C. Development and validation of an UPLC-MS/MS method for the determination of ionophoric and synthetic coccidiostats in vegetables. *Anal. Bioanal. Chem.* **2011**, *401*, 3335–3344. [CrossRef] [PubMed]
24. Olejnik, M.; Szprengier-juszkiewicz, T.; Jedziniak, P. Confirmatory method for determination of coccidiostats in eggs. *Bull. Vet. Inst. Pulawy* **2010**, *54*, 327–333.

25. Chico, J.; Rúbies, A.; Centrich, F.; Companyó, R.; Prat, M.D.; Granados, M. Use of gel permeation chromatography for clean-up in the analysis of coccidiostats in eggs by liquid chromatography–tandem mass Spectrometry. *Anal. Bioanal. Chem.* **2013**, *405*, 4777–4786. [CrossRef] [PubMed]
26. Ha, J.; Song, G.; Ai, L.F.; Li, J.C. Determination of six polyether antibiotics residues in foods of animal origin by solid phase extraction combined with liquid chromatography-tandem mass spectrometry. *J. Chromatogr. B* **2016**, *1017*, 187–194. [CrossRef] [PubMed]
27. Connolly, L.; Fodey, T.L.; Crooks, S.R.; Delahaut, P.; Elliott, C.T.J. The production and characterisation of dinitrocarbanilide antibodies raised using antigen mimics. *Immunol. Methods* **2002**, *264*, 45–51. [CrossRef]
28. Moloney, M.; Clarke, L.; O'Mahony, J.; Gadaj, A.; O'Kennedy, R.; Danaher, M. Determination of 20 coccidiostats in egg and avian muscle tissue using ultra high performance liquid chromatography tandem mass spectrometry. *J. Chromatogr. A* **2012**, *1253*, 94–104. [CrossRef] [PubMed]
29. Clarke, L.; Fodey, T.L.; Crooks, S.R.H.; Moloney, M.; O'Mahony, J.; Delahaut, P.; Gadaj, A.; O'Kennedy, R.; Danaher, M. A review of coccidiostats and the analysis of their residues in meat and other food. *Meat Sci.* **2014**, *97*, 358–374. [CrossRef] [PubMed]
30. Martínez-Villalba, A.; Moyano, E.; Martins, C.P.; Galceran, M.T. Fast liquid chromatography/tandem mass spectrometry (highly selective selected reaction monitoring) for the determination of toltrazuril and its metabolites in food. *Anal. Bioanal. Chem.* **2010**, *397*, 2893–2901. [CrossRef] [PubMed]
31. Ai, L.; Sun, H.; Wang, F.; Chen, R.; Guo, C. Determination of diclazuril, toltrazuril and its two metabolites in poultry tissues and eggsby gel permeation chromatography–liquid chromatography-tandem masss pectrometry. *J. Chromatogr. B* **2011**, *879*, 1757–1763. [CrossRef] [PubMed]
32. Mulder, P.P.J.; Balzer-Rutgers, P.; Brinke, E.M.; Bolck, Y.J.C.; Berendsen, B.J.A.; Gerçek, H.; Schat, B.; van Rhijn, J.A. Deposition and depletion of the coccidiostats toltrazuril and halofuginone in eggs. *Anal. Chim. Acta* **2005**, *529*, 331–337. [CrossRef]
33. Dubreil-Chéneau, E.; Bessiral, M.; Roudaut, B.; Verdon, E.; Sanders, P. Validation of multi-residue liquid chromatography-tandem mass spectrometry confirmatory method for 10 anticoccidials in eggs according to Commission Decision 2002/657/EC. *J. Chromatogr. A* **2009**, *1216*, 8149–8157. [CrossRef] [PubMed]
34. Commission Decision 2002/657/EC of 12 August 2002. Available online: http://eur-lex.europa.eu/legal-content/EN/TXT/PDF/?uri=CELEX:32002D0657&from=it (accessed on 20 December 2016).
35. Buiarelli, F.; Di Filippo, P.; Riccardi, C.; Pomata, D.; Giannetti, L.; Neri, B. Analytical method for the determination of mycotoxins in indoor/outdoor airborne particulate matter by HPLC-MS-MS. *Int. J. Environ. Anal. Chem.* **2015**, *95*, 713–729. [CrossRef]
36. Vanhaecke, L.; Gowik, P.; Le Bizec, B.; Van Ginkel, L.; Bichon, E.; Blokland, M.; De Brabander, H.F. European Analytical Criteria: Past, Present, and Future. *J. AOAC Int.* **2011**, *94*, 360–372.

Permissions

All chapters in this book were first published in SEPARATIONS, by MDPI; hereby published with permission under the Creative Commons Attribution License or equivalent. Every chapter published in this book has been scrutinized by our experts. Their significance has been extensively debated. The topics covered herein carry significant findings which will fuel the growth of the discipline. They may even be implemented as practical applications or may be referred to as a beginning point for another development.

The contributors of this book come from diverse backgrounds, making this book a truly international effort. This book will bring forth new frontiers with its revolutionizing research information and detailed analysis of the nascent developments around the world.

We would like to thank all the contributing authors for lending their expertise to make the book truly unique. They have played a crucial role in the development of this book. Without their invaluable contributions this book wouldn't have been possible. They have made vital efforts to compile up to date information on the varied aspects of this subject to make this book a valuable addition to the collection of many professionals and students.

This book was conceptualized with the vision of imparting up-to-date information and advanced data in this field. To ensure the same, a matchless editorial board was set up. Every individual on the board went through rigorous rounds of assessment to prove their worth. After which they invested a large part of their time researching and compiling the most relevant data for our readers.

The editorial board has been involved in producing this book since its inception. They have spent rigorous hours researching and exploring the diverse topics which have resulted in the successful publishing of this book. They have passed on their knowledge of decades through this book. To expedite this challenging task, the publisher supported the team at every step. A small team of assistant editors was also appointed to further simplify the editing procedure and attain best results for the readers.

Apart from the editorial board, the designing team has also invested a significant amount of their time in understanding the subject and creating the most relevant covers. They scrutinized every image to scout for the most suitable representation of the subject and create an appropriate cover for the book.

The publishing team has been an ardent support to the editorial, designing and production team. Their endless efforts to recruit the best for this project, has resulted in the accomplishment of this book. They are a veteran in the field of academics and their pool of knowledge is as vast as their experience in printing. Their expertise and guidance has proved useful at every step. Their uncompromising quality standards have made this book an exceptional effort. Their encouragement from time to time has been an inspiration for everyone.

The publisher and the editorial board hope that this book will prove to be a valuable piece of knowledge for researchers, students, practitioners and scholars across the globe.

List of Contributors

Cody J. Peer and William D. Figg
Clinical Pharmacology Program, Office of the Clinical Director, National Cancer Institute, 9000 Rockville Pike, Building 10, Room 5A01, Bethesda, MD 20892, USA

Lukas Ronner and Louis Rodgers
Clinical Pharmacology Program, Office of the Clinical Director, National Cancer Institute, 9000 Rockville Pike, Building 10, Room 5A01, Bethesda, MD 20892, USA
Neuro-Oncology Branch, National Cancer Institute, Bethesda, MD 20892, USA

Katherine E. Warren
Neuro-Oncology Branch, National Cancer Institute, Bethesda, MD 20892, USA
Pediatric-Oncology Branch, National Cancer Institute, Bethesda, MD 20892, USA

Cynthia M. Lester McCully
Pediatric-Oncology Branch, National Cancer Institute, Bethesda, MD 20892, USA

Mireia Farrés-Cebrián and Raquel Seró
Department of Chemical Engineering and Analytical Chemistry, University of Barcelona, Martí i Franquès 1-11, E08028 Barcelona, Spain

Javier Saurina
Department of Chemical Engineering and Analytical Chemistry, University of Barcelona, Martí i Franquès 1-11, E08028 Barcelona, Spain
Research Institute in Food Nutrition and Food Safety, University of Barcelona, Recinte Torribera, Av. Prat de la Riba 171, Edifici de Recerca (Gaudí), E08921 Santa Coloma de Gramenet, Barcelona, Spain

Oscar Núñez
Department of Chemical Engineering and Analytical Chemistry, University of Barcelona, Martí i Franquès 1-11, E08028 Barcelona, Spain
Research Institute in Food Nutrition and Food Safety, University of Barcelona, Recinte Torribera, Av. Prat de la Riba 171, Edifici de Recerca (Gaudí), E08921 Santa Coloma de Gramenet, Barcelona, Spain
Serra Húnter Program, Generalitat de Catalunya, Rambla de Catalunya 19-21, E08007 Barcelona, Spain

Peter H. Tang
Departments of Pediatrics, Pathology and Laboratory Medicine, Cincinnati Children's Hospital Medical Center, University of Cincinnati College of Medicine, 3333 Burnet Avenue, Cincinnati, OH 45229-3039, USA

Jing Tan and Chunyan Zhang
Abbott Nutrition Research and Development, 20 Biopolis Way #09-01/02, Centros Building, Singapore 138668, Singapore

Jason Gek Leong Neo
Department of Chemistry, Faculty of Science, National University of Singapore, Singapore 117543, Singapore

Tania Setiawati
Division of Chemistry and Biological Chemistry, School of Physical and Mathematical Sciences, Nanyang Technological University, Singapore 637371, Singapore

Shivender Singh Saini, Avasarala Lakshmi Jagannadha Rao and Ashok Kumar Malik
Department of Chemistry, Punjabi University, Patiala 147002, India

Abuzar Kabir and Kenneth G. Furton
International Forensic Research Institute, Department of Chemistry and Biochemistry, Florida International University, Miami, FL 33193, USA

Robert A. Downs, John R. Dean and Justin J. Perry
Department of Applied Sciences, Northumbria University, Newcastle upon Tyne NE1 8ST, UK

Adrian Downer
AkzoNobel, Stoneygate Lane, Felling Gateshead, Tyne & Wear NE10 0JY, UK

Taito Kobayashi, Kei Kumakura, Wataru Kobayashi and Hiroki Matsuoka
Department of Health and Nutrition, Takasaki University of Health and Welfare, 37-1 Nakaorui-machi, Takasaki-shi, Gunma 370-0033, Japan

Asaka Takahashi
Higashinihon College of Nutrition and Pharmaceutical, 1098-1 Koyahara-machi, Maebashi-shi, Gunma 379-2184, Japan

Anh Van Nguyen, Victor Deineka and Lumila Deineka
Institute of Engineering Technologies and Natural Sciences of Belgorod National Research University, Pobeda str. 85, Belgorod 308015, Russia

Anh Vu Thi Ngoc
Pharmaceutical Department of of Buon Ma Thuot University, 298 Ha Huy Tap, Buon Ma Thuot, Daclak 630000, Vietnam

Mansi Shah
Department of Obstetrics & Gynecology, University of Texas Medical Branch, 301 University Boulevard, Galveston, TX 77555-1062, USA

Erik Rytting
Department of Obstetrics & Gynecology, University of Texas Medical Branch, 301 University Boulevard, Galveston, TX 77555-1062, USA
Department of Pharmacology and Toxicology, University of Texas Medical Branch, Galveston, TX 77555-1062, USA

Luke Bourner and Sanaalarab Al-Enazy
Department of Pharmacology and Toxicology, University of Texas Medical Branch, Galveston, TX 77555-1062, USA

Shariq Ali
Department of Pharmacology and Toxicology, University of Texas Medical Branch, Galveston, TX 77555-1062, USA
School of Medicine, University of Texas Medical Branch, Galveston, TX 77555-1062, USA

Menatallah M. Youssef
Department of Pharmaceutical Analytical Chemistry, Ain-Shams University, Cairo 1156, Egypt

Morgan Fisler
Department of Obstetrics & Gynecology, Medical University of South Carolina, Charleston, SC 29425-2221, USA

Angelos T. Rigopoulos and Maria Touraki
Laboratory of General Biology, Division of Genetics, Development and Molecular Biology, Department of Biology, School of Sciences, Aristotle University of Thessaloniki (A.U.TH.), 54 124 Thessaloniki, Greece

Victoria F. Samanidou
Laboratory of Analytical Chemistry, Department of Chemistry, School of Sciences, Aristotle University of Thessaloniki (A.U.TH.), 54 124 Thessaloniki, Greece

Abul K. Mallik
Department of Applied Chemistry and Chemical Engineering, University of Dhaka, Dhaka 1000, Bangladesh

Hiroki Noguchi, Yige Han and Yutaka Kuwahara
Department of Applied Chemistry and Biochemistry, Kumamoto University, 2-39-1 Kurokami, Chuo-ku, Kumamoto 860-8555, Japan

Makoto Takafuji and Hirotaka Ihara
Department of Applied Chemistry and Biochemistry, Kumamoto University, 2-39-1 Kurokami, Chuo-ku, Kumamoto 860-8555, Japan
Kumamoto Institute for Photo-Electro Organics (Phoenics), 3-11-38 Higashimachi, Higashi-ku, Kumamoto 862-0901, Japan

Nadia Spano, Paola Meloni, Ilenia Idda, Alberto Mariani, Maria Itria Pilo and Gavino Sanna
Dipartimento di Chimica e Farmacia, Università degli Studi di Sassari, via Vienna 2, 07100 Sassari, Italy

Valeria Marina Nurchi and Joanna Izabela Lachowicz
Dipartimento di Scienze Chimiche e Geologiche, Cittadella Universitaria di Monserrato, 09042 Monserrato, Italy

Ernesto Rivera
Instituto de Investigaciones en Materiales UNAM, Circuito Exterior Ciudad Universitaria, CP 04510 Mexico DF, Mexico

Ancelmo Orona-Espino
Facultad de Ciencias Biológicas, Universidad Juárez del Estado de Durango, 35010 Gómez Palacio, Mexico

Dimitrios Bitas and Victoria Samanidou
Laboratory of Analytical Chemistry, Department of Chemistry, Aristotle University of Thessaloniki, 54124 Thessaloniki, Greece

Abuzar Kabir
International Forensic Research Institute, Department of Chemistry and Biochemistry, Florida International University, 11200 SW 8th St, Miami, FL 33199, USA

Marcello Locatelli
Department of Pharmacy, University "G. d'Annunzio" of Chieti-Pescara, 66100 Chieti, Italy

Hiroki Noguchi and Tianhang Liu
Department of Applied Chemistry and Biochemistry, Kumamoto University; 2-39-1 Kurokami, Chuo-ku, Kumamoto 860-8555, Japan

Makoto Takafuji
Department of Applied Chemistry and Biochemistry, Kumamoto University; 2-39-1 Kurokami, Chuo-ku, Kumamoto 860-8555, Japan
Kumamoto Institute for Photo-electro Organics (PHOENICS), Kumamoto 862-0901, Japan

Hirotaka Ihara
Kumamoto Institute for Photo-electro Organics (PHOENICS), Kumamoto 862-0901, Japan
Department of New Frontier Sciences, Kumamoto University, 2-39-1 Kurokami, Chuo-ku, Kumamoto 860-8555, Japan

Troy Purvis
Humco Compounding, 313 East 12th Street, Suite 230, Austin, TX 78701, USA
Humco, 7400 Alumax Road, Texarkana, TX 75501, USA

Rafaela de Paiva Gomes, Ananda Lima Sanson, Fabiana Aparecida Lobo, Robson José de Cássia Franco Afonso and Mauricio Xavier Coutrim
Department of Chemistry, Institute of Exact and Biological Sciences, Federal University of Ouro Preto, Campus Morro do Cruzeiro, s/n, Bauxita, 35400-000 Ouro Preto, MG, Brazil

Matthias Schöbinger, Oskar-James Klein and Goran Mitulović
Clinical Institute of Laboratory Medicine, Medical University of Vienna, Währinger Gürtel 18-20, A-1090 Wien, Austria

Jean Wan Hong Yong
ARC Centre for Mine Site Restoration, School of Biological Sciences, University of Western Australia, 35 Stirling Highway, Crawley, Perth 6009,Western Australia, Australia

Liya Ge
Residues and Resource Reclamation Centre, Nanyang Environment and Water Research Institute, Nanyang Technological University, 1 Cleantech Loop, Singapore 637141, Singapore

Francesca Buiarelli
Department of Chemistry, University "Sapienza", P.le Aldo Moro, 5-0185 Rome, Italy

Patrizia Di Filippo
Inail DIT—Via Roberto Ferruzzi, 38-00143 Rome, Italy

Natalia Manousi
Institute of Chemistry, Karl-Franzens-Universität Graz, 8010 Graz, Austria
Laboratory of Analytical Chemistry, Department of Chemistry, Aristotle University of Thessaloniki, 54124 Thessaloniki, Greece

Georg Raber
Institute of Chemistry, Karl-Franzens-Universität Graz, 8010 Graz, Austria

Ioannis Papadoyannis
Laboratory of Analytical Chemistry, Department of Chemistry, Aristotle University of Thessaloniki, 54124 Thessaloniki, Greece

Alain Berthod
Institut des Sciences Analytiques, Université de Lyon 1, CNRS, 5 rue de la Doua, Villeurbanne 69100, France

Chrysostomi Zisi, Athina Maria Mangipa, Eleftheria Boutou and Adriani Pappa-Louisi
Department of Chemistry, Aristotle University of Thessaloniki, 54124 Thessaloniki, Greece

Index

CPSIA information can be obtained
at www.ICGtesting.com
Printed in the USA
BVHW011717020122
625129BV00003B/63

9 781639 892693